AF688865

Photovoltaic Systems Engineering

Photovoltaic Systems Engineering

Edited by Fraser Cox

SYRAWOOD
PUBLISHING HOUSE
New York

Published by Syrawood Publishing House,
750 Third Avenue, 9th Floor,
New York, NY 10017, USA
www.syrawoodpublishinghouse.com

Photovoltaic Systems Engineering
Edited by Fraser Cox

© 2022 Syrawood Publishing House

International Standard Book Number: 978-1-64740-123-8 (Hardback)

This book contains information obtained from authentic and highly regarded sources. Copyright for all individual chapters remain with the respective authors as indicated. All chapters are published with permission under the Creative Commons Attribution License or equivalent. A wide variety of references are listed. Permission and sources are indicated; for detailed attributions, please refer to the permissions page and list of contributors. Reasonable efforts have been made to publish reliable data and information, but the authors, editors and publisher cannot assume any responsibility for the validity of all materials or the consequences of their use.

Trademark Notice: Registered trademark of products or corporate names are used only for explanation and identification without intent to infringe.

Cataloging-in-Publication Data

Photovoltaic systems engineering / edited by Fraser Cox.
p. cm.
Includes bibliographical references and index.
ISBN 978-1-64740-123-8
1. Photovoltaic power systems. 2. Dwellings--Power supply.
3. Building-integrated photovoltaic systems. I. Cox, Fraser.
TK1087 .P46 2022
621.312 44--dc23

TABLE OF CONTENTS

PREFACE

A photovoltaic system is the power system that supplies usable solar power through photovoltaics. It involves the conversion of the sun's radiation into usable electricity. Photovoltaic systems can be categorized into grid-connected, rack mounted, distributed, ground-mounted, building-integrated photovoltaic systems, etc. It comprises of various components such as solar array, cabling, mounting, tracker, etc. A photovoltaic system can range from building-integrated systems having a capacity of several tens of kilowatts to large utility-scale power stations having a capacity of hundreds of megawatts. This book aims to shed light on some of the unexplored aspects of the photovoltaic systems and the recent researches in this field. The various studies that are constantly contributing towards advancing technologies and the evolution of such systems are examined in detail. Researchers and students in this field will be assisted by this book.

This book has been the outcome of endless efforts put in by authors and researchers on various issues and topics within the field. The book is a comprehensive collection of significant researches that are addressed in a variety of chapters. It will surely enhance the knowledge of the field among readers across the globe.

It gives us an immense pleasure to thank our researchers and authors for their efforts to submit their piece of writing before the deadlines. Finally in the end, I would like to thank my family and colleagues who have been a great source of inspiration and support.

Editor

1

Design and Simulation of a Solar Chimney PV/T Power Plant in Northwest China

Qingjun Liu,[1,2] Fei Cao,[1,3] Yanhua Liu,[1] Tianyu Zhu,[1] and Deyou Liu[2]

[1]*College of Mechanical and Electrical Engineering, Hohai University, Changzhou 213022, China*
[2]*College of Water Conservancy and Hydropower Engineering, Hohai University, Nanjing 210098, China*
[3]*Sunshore Solar Energy Company Limited, Nantong 226300, China*

Correspondence should be addressed to Fei Cao; yq.cao@hotmail.com

Academic Editor: Angel A. Bayod-Rújula

A solar chimney PV/T power plant (SCPVTPP) is proposed. Mathematical models are established for the PV/T solar collector, the chimney, and the power conversion unit, respectively. Performances of the designed SCPVTPP are then simulated. The SCPVTPPs with different PV module areas are finally discussed. It is found that the PV cells hold the highest temperature in the solar collector. Temperature rise of the PV module has significant influences to its power generation. Without cooling, the PV power capacity has an average decrease of 28.71%. The contradictory influences of temperature rise and airflow cooling lead to an 11.81% decrease of the average power capacity. By adding the power generated by PVT, the total PV-related power contribution increases by 4.72%. With the increase of the solar collector ratio, the temperature rise and the wind velocity both first decrease then increase, the SCPP power productivity decreases linearly, and the PV power productivity increases linearly, whereas the PVT power productivity first increases linearly then increases superlinearly. There is a reversed solar collector ratio, exceeding which the PV generates most power. In this study, solar thermal power takes the major role when the solar PV area ratio is smaller than 0.055.

1. Introduction

The solar chimney power plant (SCPP) consists of three essential parts: a solar collector, a chimney, and a power conversion unit. The schematic of a SCPP is shown in Figure 1(a). Sunlight transfers through the transparent collector cover and heats the ground below. The ground temperature increases and heats the air above it through heat convection. The air temperature increases, leading to the decrease of air density. The density difference then is generated between the ambient air and the air in the solar collector. With the chimney, the air flows towards to the center of the solar collector under buoyancy effect. The air flows through the chimney and runs out of the SCPP at the top of the chimney.

Theoretical, experimental, and case studies of the SCPPs all around the world have concluded that the SCPP is with low power efficiency [1–3], huge solar collector area [4–6], and high chimney [6–9]. Some case studies of SCPPs are summarized in Table 1. Our previous studies have concluded that the reason of SCPP's low efficiency is a compound result of the air characteristics, the solar radiation, and the chimney height [10, 11]. Air is the only practical working fluid of the SCPP because of the huge air mass flow rate inside the SCPP [11]. For a given location, the solar radiation is fixed. Therefore, it is concluded that the SCPP power capacity can be increased through enlarging the solar collector area or the chimney height [4, 8]. However, there must be a limit for both these solutions from the techno-economical and safety points of view. A solution that can increase the power capacity but would not increase the chimney height and collector area, or even decrease both of them, is of high significance from the engineering points of view. We then noticed that the

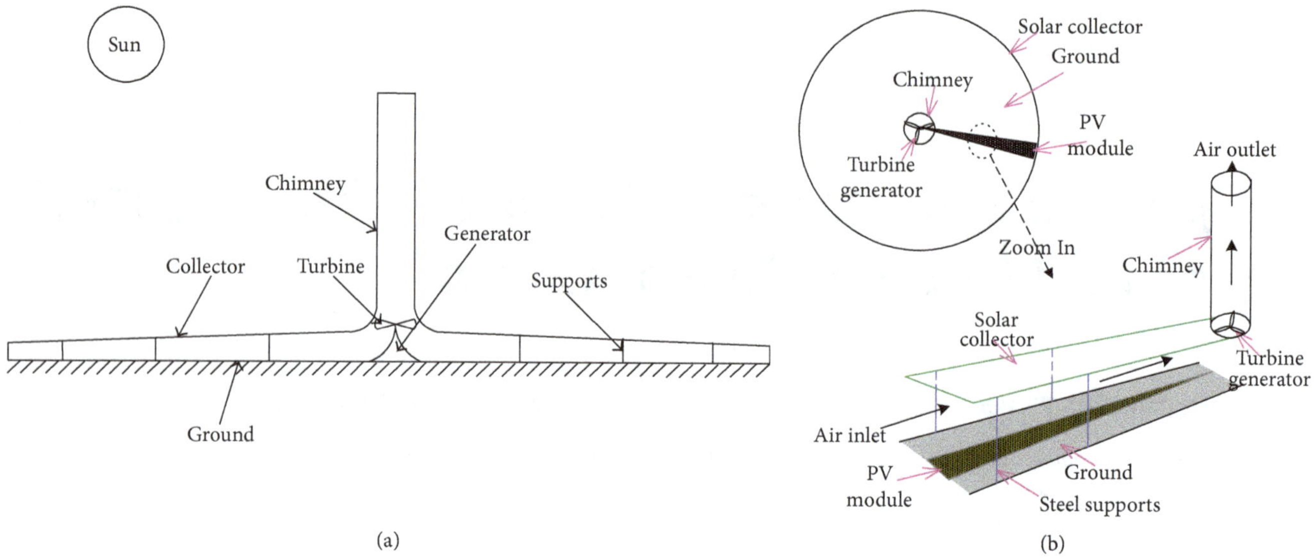

FIGURE 1: (a) Schematic of a conventional solar chimney power plant (SCPP). (b) Schematic of a SCPVTPP.

TABLE 1: Parameters of some case studies of SCPPs.

Year	Location	Weather condition Solar radiation	Collector radius	SCPP parameters Chimney height	Power capacity	Power efficiency
1983	Manzanares, Spain [20, 21]	1000 W/m^2	122 m	194.6 m	50 kW	0.11%
2003	Yinchuan, China [22]	600 W/m^2	250 m	200 m	110–190 kW	0.09%
2010	Adrar, Algeria [23]	800 W/m^2	250 m	200 m	140–200 kW	0.09%
2010	Qinghai-Tibet Plateau [24][1]	807 W/m^2	2825 m	1000 m	92.4 MW	0.89%
2012	7 cities in Iran [25][2]	640 W/m^2	122 m	194.6 m	75.9 kW	0.08%

[1]The authors of [24] did not supply the detailed solar radiation and power capacity in their literature. We calculated the results according to the results supplied in their paper and the solar duration supplied by the China Meteorological Information Center (CMIC). [2]The solar radiation and power capacity are calculated according to the data supplied by the authors.

huge solar collector, which is empty for the airflow, can be used to place the solar PV modules. On one side, the solar collector area can be used, not influencing the airflow. On the other side, the PV module can be cooled by the airflow in the solar collector by forced heat convection. Correspondingly, a new solar chimney PV/T power plant (SCPVTPP) is proposed in Figure 1(b) to fulfill this target. In the system, some solar PV modules are set on the ground of the solar collector. Due to the considerations of safety and construction cost, there are no PV modules in the vertical chimney. The solar radiation transfers through the transparent glass cover and then reaches the PV modules. The insulating layer and the wire connection are set between the PV modules and the ground. Solar radiation transfers through the glass cover and reaches the PV modules. Electricity is generated directly from the PV modules. Meanwhile, some solar radiation is converted into the thermal energy and heats the PV models. The high-temperature PV modules then heat the air above it through heat convection. The other area of the solar collector without the PV modules absorbs the solar radiation directly, which leads to the temperature increase of the ground. The ground then heats the air above it through heat convection. The high-temperature air above the ground and the PV modules then flows into the chimney through buoyancy effect. The solar PV/T technology or device was proposed for the purpose of cooling the PV module to increase its power efficiency [12] and recovering the waste heat to increase the system's total efficiency [13]. It has been widely utilized in the solar water heating (SWH) [14], heating ventilation air conditioning (HVAC) [15], solar-powered buildings [16], and so on. However, the large-scale PV/T system has not been investigated. Considering this, the main tasks in this study include (1) to build a mathematical model for the SCPVTPP, (2) to analyze the performance of the SCPVTPP, and (3) to study the SCPVTPP with different PV areas.

2. Mathematical Model

A mathematical model is built for the SCPVTPP. The mathematical model can be divided into three parts, namely, the PVT solar collector model, the chimney model, and the power conversion unit (PCU) model. Practically, the SCPVTPP has a large solar collection area, high chimney height, high PV surface temperature, and high

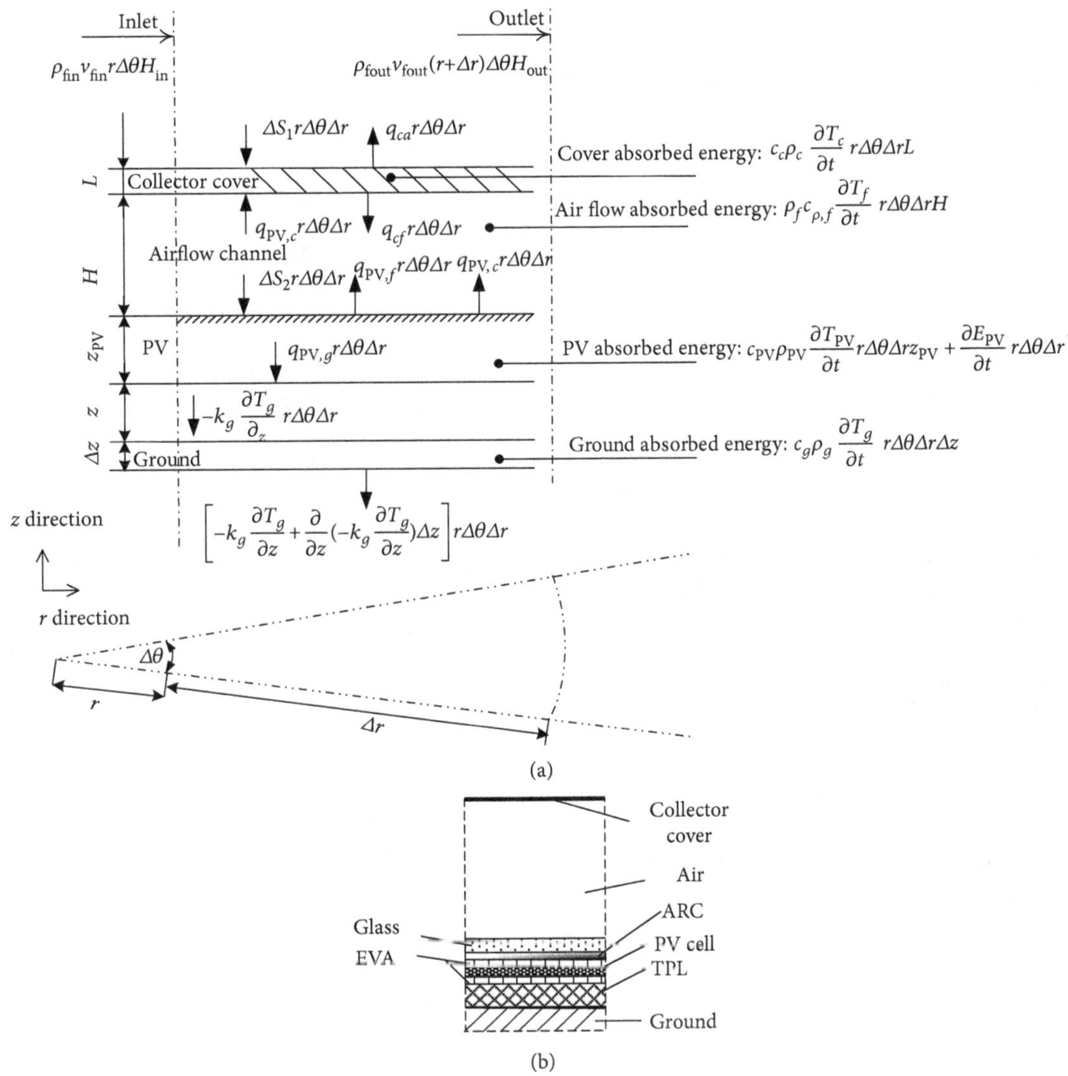

FIGURE 2: Energy balance of a representative elemental volume in the solar collector and PV: (a) full view and (b) enlarged view of the PVT section.

airflow velocity, compared with the collector height, the collector cover thickness, and the temperature rise in the solar collector are much smaller. Correspondingly, the next assumptions can be made: (a) temperature rises linearly along the airflow direction in the solar collector; (b) ignore the velocity and temperature gradient on the vertical direction in the solar collector; (c) ignore the velocity and temperature gradient on the cross section of the chimney; (d) ignore the temperature difference between the collector upper and back surface; (e) airflow is under adiabatic condition in the chimney; and (f) PV lower surface and ground up surface are connected together.

2.1. PVT Solar Collector Model. Energy balance of a representative elemental volume in the solar collector and PV is established (Figure 2). In the model, there are three energy balance equations presented in (1), (2), and (3), respectively.

Continuity equation:

$$\frac{\partial \rho}{\partial t} + \frac{1}{r}\frac{\partial}{\partial r}(\rho r v_r) = 0. \tag{1}$$

Momentum equation:

$$\frac{\partial}{\partial t}(\rho v_r) + \rho v_r \frac{\partial v_r}{\partial r} - -\frac{\partial p}{\partial r} + \mu\left[\frac{\partial}{\partial r}\left(\frac{1}{r}\frac{\partial}{\partial r}(r v_r)\right)\right] \tag{2}$$

Energy equation:

$$\frac{\partial}{\partial t}(\rho c_p T) + \frac{\partial}{\partial r}(c_p \rho v_r T) = \frac{\partial S_\Phi}{\partial t}. \tag{3}$$

Considering the energy balance of a representative elemental volume, as shown in Figure 2(a), the energy source $\partial S_\Phi/\partial t$ is

$$\frac{\partial S_\Phi}{\partial t} = q_{cf} r\Delta\theta\Delta r + q_{\mathrm{PV},f} r\Delta\theta\Delta r + \rho_f c_{p,f}\frac{\partial T_f}{\partial t} r\Delta\theta\Delta r H, \tag{4}$$

where the first and the second terms on the right side of (4) are the heat exchange between the collector cover and airflow and between the PV and the airflow, which can be expressed as

$$q_{cf} r\Delta\theta\Delta r = \frac{\partial S_1}{\partial t} r\Delta\theta\Delta r + q_{\mathrm{PV},c} r\Delta\theta\Delta r - q_{ca} r\Delta\theta\Delta r - c_c\rho_c \frac{\partial T_c}{\partial t} r\Delta\theta\Delta r L,$$
$$q_{\mathrm{PV},f} r\Delta\theta\Delta r = \frac{\partial S_2}{\partial t} r\Delta\theta\Delta r + k_{\mathrm{PV},1} \frac{\partial T_{\mathrm{PV}}}{\partial z}\Big|_{z=0} r\Delta\theta\Delta r - q_{\mathrm{PV},c} r\Delta\theta\Delta r, \quad (5)$$

where $q_{\mathrm{PV},c} r\Delta\theta\Delta r$ is the radiation heat exchange between the PV and the collector cover, $q_{ca} r\Delta\theta\Delta r$ is the convection and radiation heat exchange between the collector cover and the ambient, $k_{\mathrm{PV},1}(\partial T_{\mathrm{PV}}/\partial z)|_{z=0} r\Delta\theta\Delta r$ is the heat contribution from the PV, S_1 is the solar radiation absorbed by the collector cover, and S_2 is the solar radiation absorbed by the PV module or the ground. The PV modules are laid on the ground directly as shown in Figure 2(b). There are six layers in the PV modules, namely, the glass, the anti-reflective coating (ARC), the EVA (ethylene-vinyl acetate), the PV cell, and the Tedlar Polymer Layer (TPL).

Taking a depth of Δz of the ith layer inside the PV layer into consideration, we obtain

$$-k_{\mathrm{PV},i} \frac{\partial T_{\mathrm{PV}}}{\partial z_{\mathrm{PV}}} r\Delta\theta\Delta r\Delta z_{\mathrm{PV}} = \left[-k_{\mathrm{PV},i} \frac{\partial T_{\mathrm{PV}}}{\partial z_{\mathrm{PV}}} + \frac{\partial}{\partial z_{\mathrm{PV}}}\left(-k_{\mathrm{PV},i} \frac{\partial T_{\mathrm{PV}}}{\partial z_{\mathrm{PV}}}\right)\Delta z_{\mathrm{PV}}\right]\Big|_{z=z+\Delta z_{\mathrm{PV}}} r\Delta\theta\Delta r\Delta z_{\mathrm{PV}} + \frac{\partial E_{\mathrm{PV}}}{\partial t} r\Delta\theta\Delta r\Delta z_{\mathrm{PV}} + c_{\mathrm{PV},i}\rho_{\mathrm{PV},i} \frac{\partial T_{\mathrm{PV}}}{\partial t} r\Delta\theta\Delta r\Delta z_{\mathrm{PV}}, \quad (6)$$

where $[-k_{\mathrm{PV},i}(\partial T_{\mathrm{PV}}/\partial z_{\mathrm{PV}}) + (\partial/\partial z_{\mathrm{PV}})(-k_{\mathrm{PV},i}(\partial T_{\mathrm{PV}}/\partial z_{\mathrm{PV}}))\Delta z_{\mathrm{PV}}]|_{z=z+\Delta z_{\mathrm{PV}}} r\Delta\theta\Delta r\Delta z_{\mathrm{PV}}$ is the heat contribution from upper surface z to the lower surface $z + \Delta z_{\mathrm{PV}}$ and $E_{\mathrm{PV}} = 0$ for the layers excluding the PV cell. And at the depth of $z = -z_{\mathrm{PV}}$, we obtain

$$-k_{\mathrm{PV},6} \frac{\partial T_{\mathrm{PV}}}{\partial z_{\mathrm{PV}}} r\Delta\theta\Delta r\Delta z_g\Big|_{z=-z_{\mathrm{PV}}} = \left[-k_g \frac{\partial T_g}{\partial z_g} + \frac{\partial}{\partial z_g}\left(-k_g \frac{\partial T_g}{\partial z_g}\right)\Delta z_g\right]\Big|_{z=z+\Delta z_g} r\Delta\theta\Delta r\Delta z_g + c_g\rho_g \frac{\partial T_g}{\partial t} r\Delta\theta\Delta r\Delta z_g. \quad (7)$$

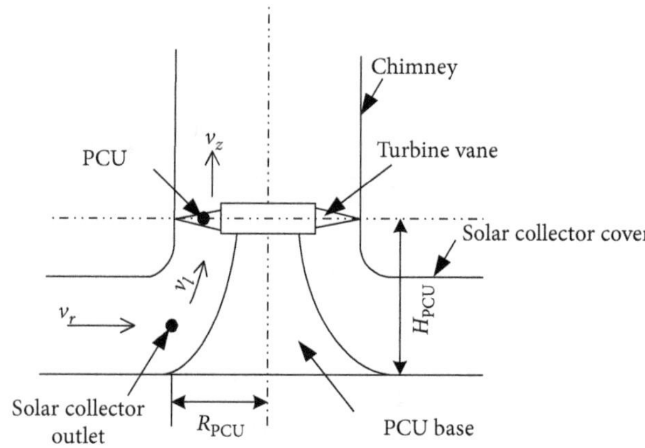

Figure 3: Cross-section of the PCU.

And taking a depth of Δz inside the ground into consideration, we obtain

$$-k_g \frac{\partial T_g}{\partial z} r\Delta\theta\Delta r\Delta z = \left[-k_g \frac{\partial T_g}{\partial z} + \frac{\partial}{\partial z}\left(-k_g \frac{\partial T_g}{\partial z}\right)\Delta z\right] r\Delta\theta\Delta r\Delta z + c_g\rho_g \frac{\partial T_g}{\partial t} r\Delta\theta\Delta r\Delta z. \quad (8)$$

$\partial E_{\mathrm{PV}}/\partial t$ is the energy output of the PV and can be calculated according to [17]:

$$E_{\mathrm{PV}} = S_2\tau_c\eta_{\mathrm{ref}}(1 - \mathrm{Br}(T_{\mathrm{PV}} - T_{\mathrm{ref}})), \quad (9)$$

where τ_c is the transmittance of the PV cover, η_{ref} is the efficiency of PV under standard conditions, Br is the temperature coefficient, and T_{ref} is the standard testing temperature of PV.

2.2. Chimney. Continuity equation:

$$\frac{\partial\rho}{\partial t} + \frac{\partial}{\partial z}(\rho v_z) = 0. \quad (10)$$

Momentum equation:

$$\frac{\partial}{\partial t}(\rho v_z) + \rho v_z \frac{\partial v_z}{\partial z} = -\frac{\partial p}{\partial z} - \rho g_z + \mu \frac{\partial^2 v_z}{\partial z^2}. \quad (11)$$

Energy equation:

$$\frac{\partial}{\partial t}(c_p\rho T) + \frac{\partial}{\partial z}(c_p\rho v_z T) = 0. \quad (12)$$

For a vertical adiabatic chimney, the pressure difference created in the chimney is

$$\Delta P_{\mathrm{chi}} = (\rho_a - \rho_o) g H_{\mathrm{chi}}. \quad (13)$$

And the pressure difference between the inlet and outlet of the solar collector is calculated as

$$\Delta P = \int_{\mathrm{inlet}}^{\mathrm{outlet}} g(\rho_a - \rho(z))dz, \quad (14)$$

where z denotes the height. Comparing with the chimney

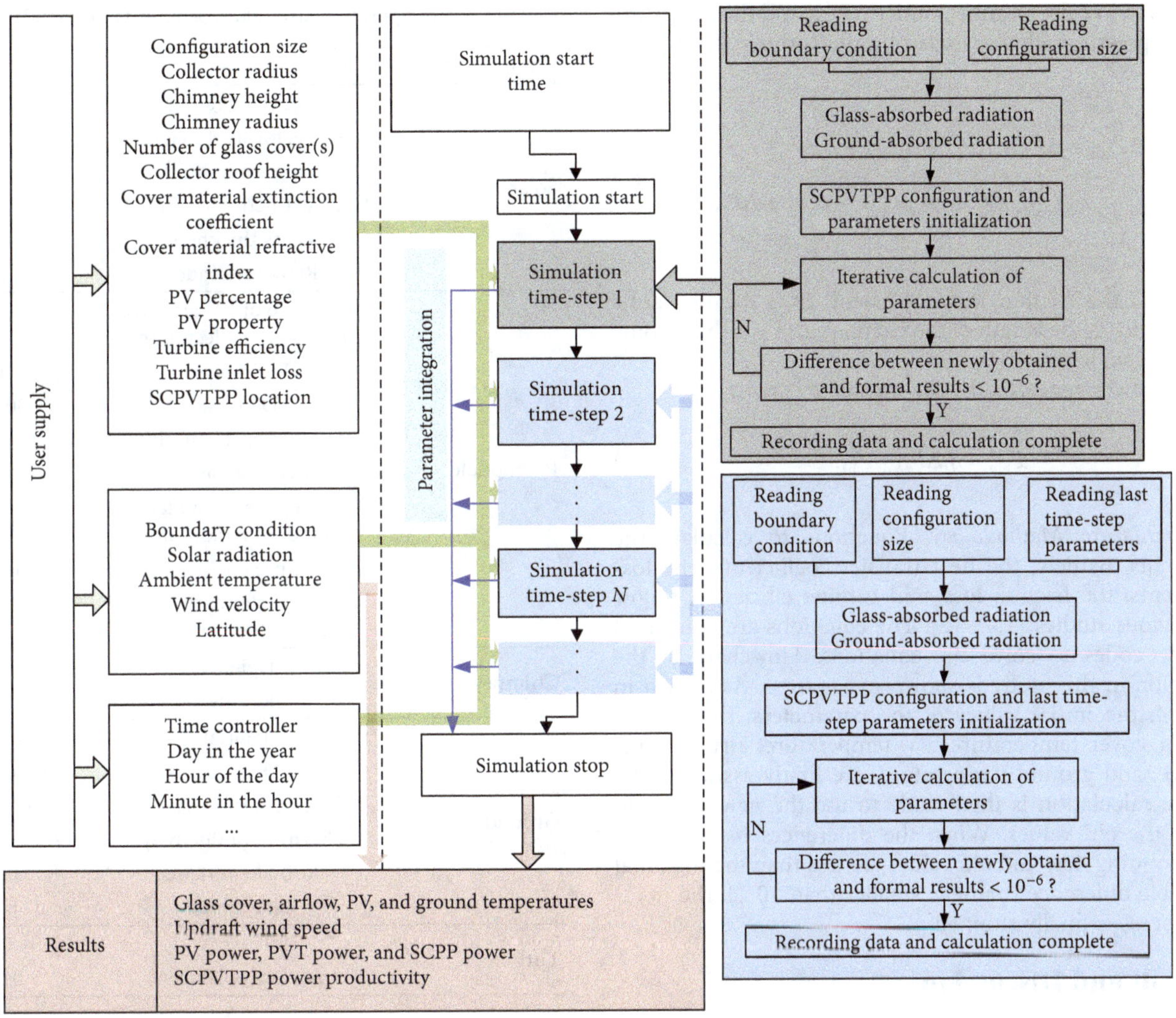

FIGURE 4: Flowchart of the simulation process.

height, the solar collector height is much smaller. Thus, $H_{col} \approx 0$.

According to Boussinesq assumption and (13) and (14), the momentum equation of the chimney can be simplified as

$$-\frac{\partial p}{\partial z} - \rho g_z + \mu \frac{\partial^2 v_z}{\partial z^2} = 0. \tag{15}$$

2.3. PCU. As the connection section of the PCU is irregular, a randomized coordinate along the power generator surface is built for this area (Figure 3). And we can obtain the following:

Continuity equation:

$$\frac{\partial \rho}{\partial t} + \frac{\partial}{\partial l}(\rho v_l) = 0. \tag{16}$$

Momentum equation:

$$\frac{\partial}{\partial t}(\rho v_l) + \rho v_l \frac{\partial v_l}{\partial l} = -\frac{\partial p}{\partial l} - \rho g_l + \mu \frac{\partial^2 v_l}{\partial l^2}. \tag{17}$$

Energy equation:

$$\frac{\partial}{\partial t}(c_p \rho T) + \frac{\partial}{\partial l}(c_p \rho v_l T) = \frac{\partial}{\partial l}\left(k \frac{\partial T}{\partial l}\right). \tag{18}$$

As $H_{PCU} << H_{chi}$ and $R_{PCU} << R_{col}$, we can obtain that

$$\begin{aligned} v_r|_{r=R_{PCU}} &= v_l|_{l=0}, \\ v_z|_{z=H_{PCU}} &= v_l|_{l=L_{PCU}}, \\ H_{PCU} &\approx 0, \end{aligned} \tag{19}$$

where H_{chi} is the chimney height, R_{col} is the collector radius, and L_{PCU} is the length of the PCU channel.

Consequently, the momentum equation can be simplified as

$$-\frac{\partial p}{\partial l} + \mu \frac{\partial^2 v_l}{\partial l^2} = 0. \tag{20}$$

The generated pressure is consumed by four parts, that is, the friction losses in the collector and the chimney ΔP_f, the kinetic energy losses at the turbine inlet ΔP_{in}, the kinetic energy losses at the chimney outlet ΔP_{out}, and the rest is the effective pressure which is used by the turbine to generate electricity ΔP_t.

$$\begin{aligned}\Delta P_{chi} &= \Delta P_f + \Delta P_{in} + \Delta P_{out} + \Delta P_t \\ &= f\frac{L_{th}}{2D}\rho v_l^2 + \gamma\frac{1}{2}\rho v_z^2 + \frac{1}{2}\rho_a v_z^2 + \Delta P_t,\end{aligned} \tag{21}$$

where f is the friction loss coefficient, D is the hydraulic diameter, L_{th} is the length of the channel, and γ is the turbine inlet loss coefficient.

The power generated by the turbine P_{ele} is

$$P_{ele} = \eta_t \Delta P_t v_o A_{chi}. \tag{22}$$

2.4. Simulation Methodology. Equations to calculate the coefficients, namely, the heat transfer coefficients, the loss coefficients, the friction loss, and turbine efficiency, follow the previous studies [1, 4, 18]. The equations are converted into the codes to solve the equations. Flowchart of the simulation in this study is shown in Figure 4. As shown in Figure 4, the initial values of the parameters, that is, the collector cover temperature, PV temperature, airflow temperature, and ground temperature, are firstly assumed. An iterative calculation is then made to use the new values to replace the old values. When the differences between any corresponding new and old values are less than the maximal acceptable difference (namely, smaller than 10^{-6}), the iteration process is finally stopped.

3. Result and Discussion

3.1. Configuration Sizes of the SCPVTPP. The configuration sizes of the SCPVTPP are shown in Table 2. The SCPVTPP is designed according to 5 and MW SCPP to [19]. And the SCPVTPP is assumed to be operated in Lanzhou. Lanzhou (103.50°E, 36.03°N) is located in the geographical center of Northwest China. And it has typical Northwestern Chinese climate, that is, strong solar radiation, rare rainfall that diminishes from east to west, dry and cold winter, hot summer, and broad daily temperature width. Its annual global solar radiation is more than 5020 MJ/m^2 and sunshine duration is over 2600 h per year. Its annual mean temperature is 9.8°C. The properties of six layers in the PV module are shown in Table 3.

Table 2: Configuration sizes and coefficients of the SCPVTPP.

Component		Value
Solar collector	Collector radius	625 m
	Collector inlet height	3 m
	Collector cover emittance	0.87
	Glass extinction coefficient	32 m^{-1}
	Glass thickness	5 mm
	Refractive index	1.526
PV module	Percentage of PV area over whole solar collector area	30%
	Emittance	0.9
	Density	2330 kg/m^3
	Specific heat capacity	677 J/(kg·K)
	PV cover transmittance	0.912
	PV efficiency under standard conditions	0.115
	Temperature coefficient (Br)	0.0045 K^{-1}
	PV standard testing temperature	298 K
Chimney	Height	550 m
	Radius	22.5 m
Ground	Material	Granite
	Density	2640 kg/m^3
	Specific heat capacity	820 J/(kg·K)
	Thermal conductivity	1.73 W/(m·K)
	Normal emittance	0.92
	Reflectance	0.25
Turbine	Efficiency	0.8
	Inlet loss coefficient	0.056

3.2. Temperature Increase and Wind Velocity of the SCPVTPP. As the SCPP is dominated by the buoyancy effect, temperature rise in the solar collector is of high significance. Moreover, two contradictory phenomena occur concerning the temperature rise in the PV/T solar collector. On one side, the PV modules are sensitive to the temperature rise. The PV power generation would decrease with the rise of temperature. On the other hand, the updraft wind would enhance with the temperature rising and the generated wind would reduce the temperature of the PV modules. Considering this, the meteorological data, temperatures, and wind velocity inside the SCPVTPP are calculated and the results are shown in Figure 5.

Solar radiation and ambient temperature of Lanzhou is shown in Figure 5(a). It is found from the figure that the solar radiation first increases from January to June, with the peak of 689.73 W/m^2. Then the solar radiation decreases till December. The tendency of the ambient temperature is similar with the solar radiation, whereas the highest ambient temperature appears at July. Temperatures of the glass cover, airflow, PV module, and ground and the temperature increase inside the collector at different months in the year are shown in Figure 5(b). It is found from the figure that the temperatures of the glass cover, airflow, PV module, and ground have the same tendency with the ambient temperature, with the peak values at July. Moreover, the PV cell holds the highest temperature among the four temperature groups throughout the year, followed by the ground temperature, the glass cover temperature, and the airflow temperature. Temperature differences between each two groups of temperatures increase from January to July and then decrease till December.

3.3. Power Generation. There are three parts contributing to the power generation in the SCPVTPP, namely, the PV

Table 3: Photovoltaic layer properties.

Layer	Glass	ARC	PV cell	EVA	TPL
Thickness δ (m)	0.003	100×10^{-9}	225×10^{-6}	500×10^{-6}	0.0001
Thermal conductivity k (W/m·K)	1.8	32	148	0.35	0.2

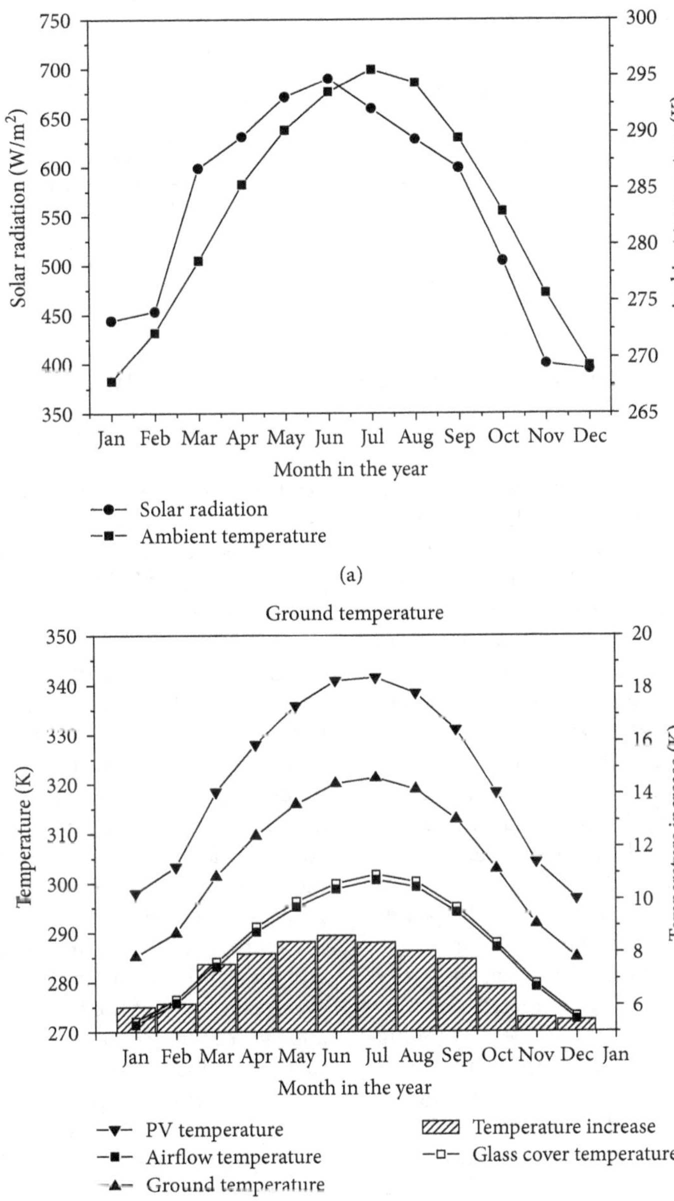

Figure 5: (a) Solar radiation and ambient temperature of the SCPVTPP. (b) Temperatures of the glass cover, airflow, PV module, and ground and temperature increase inside the chimney during different months in the year.

power capacity, the SCPP power capacity, and the PVT power capacity, which, respectively, refers to the power generation directly from the PV modules, the power contribution by the airflow from the solar collector without PV modules, and the power contribution by the airflow heated by the PV modules. Power generations from the PV, the PVT, and the SCPP are shown in Figure 6. It is found from the figure that PV generates much higher power than PVT and SCPP. SCPP generates more power productivity than PVT. The power generates from the solar thermal effect, namely, the power generation from PVT and SCPP is much smaller than that from the PV. This is reasonable as the energy conversion efficiency of SCPP is at 1% level [1–3], whereas the energy conversion efficiency of PV is near

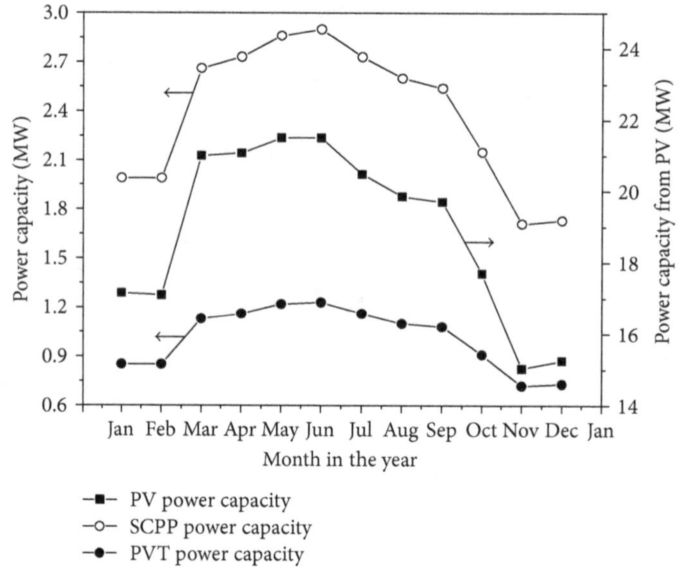

Figure 6: Power capacity of the PV, SCPP, and PVT.

16% [17]. Power productivities from the PV, PVT, and SCPP have similar tendencies with the solar radiation, except the start and the end of the year for the PV power productivity. It is found in Figure 5(a) that the solar radiation increases from January to February. However, power productivities decrease from January to February in Figure 6. The reason is that the power productivity is a compound result of the solar radiation and the ambient temperature. It is believed that less power is generated as the PV cell temperature rises according to (9). And the tendency of the ambient temperature rise is much larger than the increase of solar radiation from January to February (Figure 5(a)). The two effects lead to the decrease of power productivity of PV from January to February in Figure 6. This also explains the tendency of PV power productivity from November to December in Figure 6.

As mentioned above, the temperature takes contradictory effect on the PV power generation. The temperature influences on the PV power generation is thus analysed. By taking the PV power generation under ambient temperature as a reference, the PV model power, the PVT power contribution, and the total PV-related power contribution (PV + PVT) in the present study and the PV power generation without airflow cooling are then compared in Table 4. In can be found from Table 4 that the temperature rise of the PV module has significant influences to its power generation. Without cooling, the PV power capacity has an average decrease of 28.71%. The contradictory influences of temperature rise and airflow cooling lead to an 11.81% decrease of the average power capacity, reflecting the temperature rise which takes the major role in the SCPVTPP. The PVT power contribution is much smaller than the others. But adding the power generated by the PVT effect, the total PV-related power contribution would increase by 4.72%. And there are some conditions, that is, in January, February, November, and December, the total PV-related power contribution is larger than the reference P_{PV_Ta}. The reason is that the PV models' temperatures are low in these months, due to also low ambient temperatures. And the system takes full advantage of PV modules and generated updraft winds in these months.

3.4. SCPVTPP with Different PV Areas. The performances of the SCPVTPP is discussed in Figures 5(b) and 6 for the case that PV area takes 30% of the whole solar collector. However, there are some cases that the PV area can take more or less percentage to the whole solar collector. The SCPVTPP performances and power productivities under different solar collector area ratios are then calculated with the same methodology as the case of 30%. The results are shown in Figures 7(a) and 7(b). It is found in Figure 7(a) that with the increase of the solar collector ratio, the temperature rise and the wind velocity both first decrease then increase. However, the valleys of the temperature rise and the wind velocity appear at different solar collector ratios. The lowest temperature rise is located at the solar collector ratio of 0.4, whereas the lowest wind velocity is located at the solar collector ratio of 0.6. Moreover, tendencies of the temperature rise and the wind velocity differ from each other. The temperature rise first decreases gradually and then increases gradually. But the wind velocity first decreases slowly and then increases quickly. It is found from Figure 7(b) that with the increase of the solar collector ratio, the SCPP power productivity decreases linearly and the PV power productivity increases linearly, whereas the PVT power productivity first increases linearly then increases superlinearly. The total power productivity increases with the solar collector ratio rises. There is a reversed solar collector ratio, exceeding the PV which generates most of the power. With detailed calculation, it is found that the solar collector ratio is 0.055, which means that solar thermal power takes the major role of power generation when the solar PV area ratio is smaller than 0.055.

TABLE 4: Comparison of PV model power, PVT power contribution, and total PV-related power contribution (PV + PVT) in the present study and the PV power generation without airflow cooling by taking PV power generation under ambient temperature as a reference.

Month		1	2	3	4	5	6	7	8	9	10	11	12	Average
PV power generation under ambient temperature, P_{PV_Ta}	Value/MW	17.15	17.51	23.13	24.36	25.91	26.64	25.46	24.25	23.14	19.48	15.46	15.17	21.47
	Percentage to P_{PV_Ta}/%	100.00	100.00	100.00	100.00	100.00	100.00	100.00	100.00	100.00	100.00	100.00	100.00	100.00
PV model power in the present study	Value/MW	17.15	17.09	21.00	21.07	21.50	21.50	20.48	19.85	19.70	17.70	15.03	15.17	18.94
	Percentage to P_{PV_Ta}/%	100.00	97.60	90.79	86.49	82.98	80.71	80.44	81.86	85.13	90.86	97.22	100.00	88.19
PVT power contribution in the present study	Value/MW	0.85	0.85	1.13	1.16	1.22	1.23	1.16	1.10	1.08	0.91	0.72	0.73	1.01
	Percentage to P_{PV_Ta}/%	4.96	4.85	4.89	4.76	4.71	4.62	4.56	4.54	4.67	4.67	4.66	4.81	4.72
Total PV-related power contribution (PV + PVT)	Value/MW	18.00	17.94	22.13	22.23	22.72	22.73	21.64	20.95	20.78	18.61	15.75	15.90	19.95
	Percentage to P_{PV_Ta}/%	104.96	102.46	95.68	91.26	87.69	85.32	85.00	86.39	89.80	95.53	101.88	104.81	92.91
PV power generation without airflow cooling	Value/MW	12.27	12.77	15.56	16.67	17.68	18.29	18.17	17.62	16.69	14.57	12.00	11.39	15.31
	Percentage to P_{PV_Ta}/%	71.55	72.93	67.27	68.43	68.24	68.66	71.37	72.66	72.13	74.79	77.62	75.08	71.29

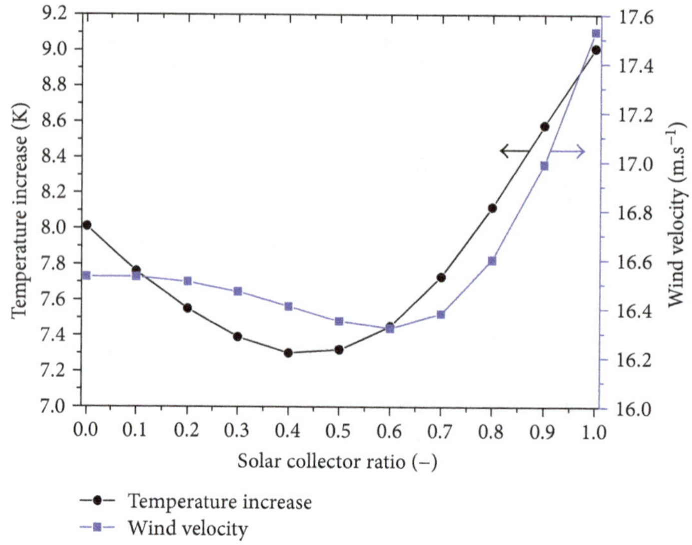

(a)

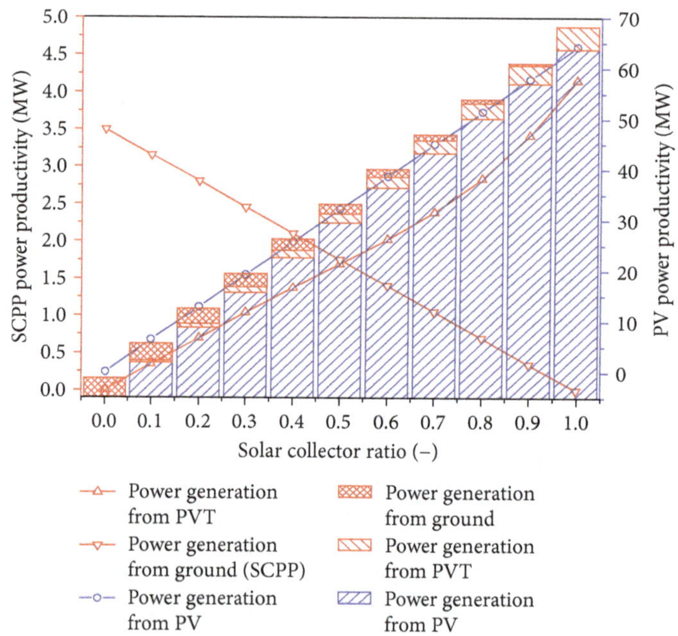

(b)

FIGURE 7: (a) Temperature increase and wind velocity in the SCPVTPP under different solar collector ratios. (b) Power generation from PVT, PV, and ground and the total power generation under different solar collector ratios.

4. Conclusions

The SCPP has a large solar collector area, which can be used to lay the PV modules. A SCPVTPP is proposed and studied in this study. A mathematical model is established for the proposed system. The SCPVTPP performances and the power productivities under different solar collector ratios are then discussed. It can be concluded from this study that

(1) The PV cells hold the highest temperature in the solar collector throughout the year, followed by the ground temperature, the glass cover temperature, and the airflow temperature. The temperature rise of the PV module has significant influences to its power generation. Without cooling, the PV power capacity has an average decrease of 28.71%. The contradictory influences of temperature rise and airflow cooling lead to an 11.81% decrease of the average power capacity. By adding the power generated by PVT, the total PV-related power contribution increases by 4.72%.

(2) With the increase of the solar collector ratio, the temperature rise and the wind velocity both first decrease then increase. The lowest temperature rise is located at the solar collector ratio of 0.4, whereas

the lowest wind velocity is located at the solar collector ratio of 0.6.

(3) With the increase of the solar collector ratio, the SCPP power productivity decreases linearly and the PV power productivity increases linearly, whereas the PVT power productivity first increases linearly then increases superlinearly.

(4) There is a reversed solar collector ratio, exceeding the PV which generates most of the power. In this study, solar thermal power takes the major role when the solar PV area ratio is smaller than 0.055.

Nomenclature

A: Area (m^2)
c_p: Specific heat ($J \cdot m^{-2} \cdot K^{-1}$)
E_{PV}: PV power (W)
f: Friction loss coefficient (–)
g: Acceleration of gravity ($m \cdot s^{-2}$)
h: Heat transfer coefficient ($W \cdot m^{-2} \cdot K^{-1}$)
H: Height (m); solar radiation ($W \cdot m^{-2}$)
K: Extinction coefficient (m^{-1})
L: Length (m); channel length (m)
P: Pressure (Pa); electrical power generation (W)
Q: Energy flux ($J \cdot hr^{-1}$)
r: Direction (–)
R: Radius (m)
S_1: Solar radiation absorb by glass (W/m^2)
S_2: Solar radiation absorb by PV (W/m^2)
T: Temperature (K)
U: Compound heat coefficient ($W \cdot m^{-2} \cdot K^{-1}$)
v: Air velocity ($m \cdot s^{-1}$)
ΔP_{tot}: Total pressure difference (Pa).

Greek Symbols

Δ: Difference (–)
ρ: Air density ($kg \cdot m^{-3}$); reflectance (–)
θ: Angle (°); direction (–)
γ: Turbine inlet loss coefficient (–)
μ: Dynamic viscosity ($kg \cdot m^{-1} \cdot s^{-1}$)
η_t: Turbine efficiency.

Subscripts

a: Ambient
c: Collector cover
col: Collector
chi: Chimney
f: Airflow
gro: Ground
in: Turbine inlet
o: Outlet of the solar collector
out: Outlet of the chimney
PV: Photovoltaic.

Conflicts of Interest

The authors declare that they have no conflicts of interest.

Acknowledgments

This research was supported by the National Natural Science Foundation of China (Grant no. 51506043), the Natural Science Foundation of Jiangsu Province (Grant no. BK20141153), the "Dayu Scholar" Foundation of Hohai University, and the China Scholarship Council (Grant no. 201706715058).

References

[1] F. Cao, L. Zhao, and L. Guo, "Simulation of a sloped solar chimney power plant in Lanzhou," *Energy Conversion and Management*, vol. 52, no. 6, pp. 2360–2366, 2011.

[2] T. Z. Ming, Y. Zheng, C. Liu, W. Liu, and Y. Pan, "Simple analysis on thermal performance of solar chimney power generation systems," *Journal of the Energy Institute*, vol. 83, no. 1, pp. 6–11, 2010.

[3] X. Zhou, F. Wang, and R. M. Ochieng, "A review of solar chimney power technology," *Renewable and Sustainable Energy Reviews*, vol. 14, no. 8, pp. 2315–2338, 2010.

[4] S. Larbi, A. Bouhdjar, and T. Chergui, "Performance analysis of a solar chimney power plant in the southwestern region of Algeria," *Renewable and Sustainable Energy Reviews*, vol. 14, no. 1, pp. 470–477, 2010.

[5] R. Sangi, "Performance evaluation of solar chimney power plants in Iran," *Renewable and Sustainable Energy Reviews*, vol. 16, no. 1, pp. 704–710, 2012.

[6] C. D. Papageorgiou and P. P. Katopodis, "A modular solar collector for desert floating solar chimney technology," in *Energy, Environment, Ecosystems, Development and Landscape Architecture*, pp. 126–132, WSEAS Press, Athens, Greece, 2009.

[7] X. Zhou, J. Yang, B. Xiao, G. Hou, and Y. Wu, "Numerical investigation of a compressible flow through a solar chimney," *Heat Transfer Engineering*, vol. 30, no. 8, pp. 670–676, 2009.

[8] X. Zhou, M. A. D. S. Bernardes, and R. M. Ochieng, "Influence of atmospheric cross flow on solar updraft tower inflow," *Energy*, vol. 42, no. 1, pp. 393–400, 2012.

[9] S. Nizetic, N. Ninic, and B. Klarin, "Analysis and feasibility of implementing solar chimney power plants in the Mediterranean region," *Energy*, vol. 33, no. 11, pp. 1680–1690, 2008.

[10] F. Cao, H. Li, L. Zhao, T. Bao, and L. Guo, "Design and simulation of the solar chimney power plants with TRNSYS," *Solar Energy*, vol. 98, pp. 23–33, 2013.

[11] F. Cao, L. Zhao, H. Li, and L. Guo, "Performance analysis of conventional and sloped solar chimney power plants in China," *Applied Thermal Engineering*, vol. 50, no. 1, pp. 582–592, 2013.

[12] J. A. Duffie and W. A. Beckman, *Solar Engineering of Thermal Processes*, John Wiley & Sons, New York, NY, USA, 2006.

[13] E. C. J. Kern and M. C. Russell, "Combined photovoltaic and thermal hybrid collector systems," in *IEEE Photovoltaic Specialists Conference*, vol. 1, pp. 1153–1157, Washington, DC, USA, 1978.

[14] L. W. Florschuetz, "Extension of the Hottel-Whillier model to the analysis of combined photovoltaic/thermal flat plate collectors," *Solar Energy*, vol. 22, no. 4, pp. 361–366, 1979.

[15] A. A. Hegazy, "Comparative study of the performances of four photovoltaic/thermal solar air collectors," *Energy Conversion and Management*, vol. 41, no. 8, pp. 861–881, 2000.

[16] H. Yang, J. Burnett, and J. Ji, "Simple approach to cooling load component calculation through PV walls," *Energy and Buildings*, vol. 31, no. 3, pp. 285–290, 2000.

[17] G. Chao, *Numerical and Experimental Study of Tri-Functional Photovoltaic/Thermal Collector [Ph.D. Thesis]*, University of Science and Technology of China, China, 2015.

[18] X. Zhou and Y. Xu, "Solar updraft tower power generation," *Solar Energy*, vol. 128, pp. 95–125, 2016.

[19] J. Schlaich, R. Bergermann, W. Schiel, and G. Weinrebe, "Design of commercial solar updraft tower systems—utilization of solar induced convective flows for power generation," *Journal of Solar Energy Engineering*, vol. 127, no. 1, pp. 117–124, 2005.

[20] W. Haaf, "Solar chimneys. Part II: preliminary test results from the Manzanares pilot plant," *International Journal of Solar Energy*, vol. 2, no. 2, pp. 141–161, 1984.

[21] J. Schlaich, *The Solar Chimney: Electricity from the Sun*, Axel Menges, Stuttgart, Germany, 1995.

[22] Y. J. Dai, H. B. Huang, and R. Z. Wang, "Case study of solar chimney power plants in Northwestern regions of China," *Renewable Energy*, vol. 28, no. 8, pp. 1295–1304, 2003.

[23] M. O. Hamdan, "Analysis of a solar chimney power plant in the Arabian gulf region," *Renewable Energy*, vol. 36, no. 10, pp. 2593–2598, 2011.

[24] X. Zhou, F. Wang, J. Fan, and R. M. Ochieng, "Performance of solar chimney power plant in Qinghai-Tibet plateau," *Renewable and Sustainable Energy Reviews*, vol. 14, no. 8, pp. 2249–2255, 2010.

[25] A. Asnaghi and S. M. Ladjevardi, "Solar chimney power plant performance in Iran," *Renewable and Sustainable Energy Reviews*, vol. 16, no. 5, pp. 3383–3390, 2012.

2

Fabrication and Characterization of $CH_3NH_3PbI_3$ Perovskite Solar Cells Added with Polysilanes

Takeo Oku[1] **Junya Nomura,**[1] **Atsushi Suzuki,**[1] **Hiroki Tanaka,**[1] **Sakiko Fukunishi,**[2] **Satoshi Minami,**[2] **and Shinichiro Tsukada**[2]

[1]*Department of Materials Science, The University of Shiga Prefecture, Hikone, Shiga 522-8533, Japan*
[2]*Frontier Materials Laboratories, Osaka Gas Chemicals Co., Ltd., Osaka 554-0051, Japan*

Correspondence should be addressed to Takeo Oku; oku@mat.usp.ac.jp

Academic Editor: Věra Cimrová

Effects of polysilane additions on $CH_3NH_3PbI_3$ perovskite solar cells were investigated. Photovoltaic cells were fabricated by a spin-coating method using perovskite precursor solutions with polymethyl phenylsilane, polyphenylsilane, or decaphenyl cyclopentasilane (DPPS), and the microstructures were examined by X-ray diffraction and optical microscopy. Open-circuit voltages were increased by introducing these polysilanes, and short-circuit current density was increased by the DPPS addition, which resulted in the improvement of the photoconversion efficiencies to 10.46%. The incident photon-to-current conversion efficiencies were also increased in the range of 400~750 nm. Microstructure analysis indicated the formation of a dense interfacial structure by grain growth and increase of surface coverage of the perovskite layer with DPPS, and the formation of PbI_2 was suppressed, leading to the improvement of photovoltaic properties.

1. Introduction

Inorganic-organic solar cells have been developed and studied as next-generation systems. Since organic thin-film solar cells have the advantages of low cost, flexibility, and being lightweight, the photovoltaic and optical properties of the organic solar cells have been studied [1–7]. Polysilane is a p-type semiconductor and has been used as an electrical conductive material and photovoltaic systems [7–14]. Polysilanes are known as σ-conjugate polymers, and their hole mobility is 10^{-4} cm^2 V^{-1} s^{-1} [7]. Although the polysilanes could be applied to p-type semiconductors on organic thin-film solar cells, few studies on polysilane-based solar cells have been reported [8, 14, 15].

Recently, inorganic-organic $CH_3NH_3PbI_3$-based perovskite solar cells have been widely studied because of its easy fabrication process and the high conversion efficiencies compared to the common organic solar cells [16–19]. After a conversion efficiency reached 15% in 2013 [20], higher efficiencies have been achieved for various processes and device structures, and the conversion efficiency increased above 20% [21–25]. The photovoltaic properties of the perovskite solar cells are fairly dependent on compositions and the crystal structures of the perovskite compounds. Halogen atom doping, such as bromine (Br) or chlorine (Cl), at the iodine (I) sites of the perovskite crystals have been investigated. The doped Cl ions could lengthen the diffusion length of excitons, which resulted in the improvement in the conversion efficiency [26–30]. In addition, studies on metal atom doping, such as tin (Sn) [31, 32], germanium [33, 34], antimony (Sb) [35–40], copper (Cu) [41–44], arsenic (As) [45], or thallium (Tl) [34], at the lead (Pb) sites have been performed. The optical absorption ranges of the perovskite solar cells were expanded by Sn or Tl doping [31, 32, 34], and the photoconversion efficiencies were improved by Sb-, Cu-, or As doping. Further doping of cesium [23], rubidium [24], or formamidinium ($NH{=}CHNH_3$) [21, 22] also improved the conversion efficiencies of the perovskite solar cells. Detailed studies on the elemental doping at the I, Pb, and/or CH_3NH_3 sites are fascinating for effects on the photovoltaic properties and microstructures, and tolerance factors are also valuable to estimate stabilities of the perovskite crystals [46].

Carrier transport layers are also important for the improvement of the perovskite solar cells. Commercial 2,2′,7,7′-tetrakis[N,N-di-pmethoxyphenylamino]-9,9′-spirobifluorene (spiro-OMeTAD) has been normally used as the hole transport layers. However, spiro-OMeTAD is an expensive organic compound and not so much stable at elevated temperatures, which is one of the crucial issues for the development of the perovskite-type solar cells. In our previous works, several polysilanes have been applied for perovskite solar cells [47]. The conversion efficiencies of $CH_3NH_3PbI_3$-based solar cells using polysilane-doped hole transport layers were improved compared with those of a conventional $CH_3NH_3PbI_3$-based solar cells using spiro-OMeTAD.

The purpose of the present work is to investigate photovoltaic properties of $CH_3NH_3PbI_3$-based solar cells introduced with various polysilanes in the perovskite precursor solutions. Effects of polysilane addition on the photovoltaic properties were investigated by current density-voltage curves and incident photon-to-current conversion efficiencies. Microstructures of the thin films were investigated by X-ray diffraction (XRD) and optical microscopy.

2. Materials and Methods

Figure 1 shows a schematic illustration of fabricated device structure of the present photovoltaic devices. The detailed fabrication process is described in our previous reports [47–49], except for the process of polysilane doping into $CH_3NH_3PbI_3$ solutions. Fluorine-doped tin oxide- (FTO-) coated glass substrates were cleaned in an ultrasonic bath containing acetone and methanol. After rinsing with ultrapure water and drying under nitrogen gas, the substrates were treated with an ultraviolet-ozone cleaner (Asumi Giken Ltd., ASM 401N) for 15 min. For a compact TiO_2 layer, 0.15 and 0.30 M TiO_x precursor solutions were prepared from titanium diisopropoxide bis(acetylacetonate) (Sigma-Aldrich, 0.055 and 0.110 mL) as a solute and 1-butanol (Nacalai Tesque, 1 mL) as a solvent. The precursor solutions were stirred for 12 h at room temperature. The 0.15 M TiO_x precursor layer was spin-coated onto the cleaned FTO surface by a spin coater (Mikasa, MS-A100) at 3000 rpm for 30 s and was heated onto a hot plate (As One, ND-1) at 125°C for 5 min. Then, the 0.30 M TiO_x precursor layer was spin-coated on the 0.15 M TiO_x layer at 3000 rpm for 30 s. This process was repeated twice. After the spin-coating, the 0.30 M TiO_x layer was heated onto the hot plate at 125°C for 5 min. The TiO_x on the FTO was annealed in an electric furnace (As One, SMF-1) at 500°C for 30 min to form a compact TiO_2 layer. After cooling to room temperature, a mesoporous TiO_2 layer was spin-coated on the compact TiO_2 layer at 5000 rpm for 30 s, where the TiO_2 paste was prepared by diluting TiO_2 powder (Aerosil, P-25, 100 mg) in ultrapure water and adding poly(ethylene glycol) (molecular weight: 20000, Nacalai Tesque, 10 mg), acetylacetone (Sigma-Aldrich, 10 mL), and surfactant (Sigma-Aldrich, Triton X-100, 5 mL) for 30 min. After heating at 125°C for 5 min, the samples were annealed in the electric furnace at 500°C for 30 min. For the preparation of $CH_3NH_3PbI_3$ layer, CH_3NH_3I (Showa Chemical, 98.8 mg) and PbI_2 (Sigma-Aldrich, 289.3 mg) powders were dissolved in γ-butyrolactone (Wako Pure Chemical Industries, 0.275 mL), and *N,N*-dimethylformamide (Sigma-Aldrich, 0.225 mL) was prepared at 60°C. The molar ratio of the solutes is 1 : 1. Three kinds of polysilanes, poly(methyl phenylsilane) (PMPS, Osaka Gas Chemicals, OGSOL SI-10-10, 3 mg), poly(phenylsilane) (PPS, Osaka Gas Chemicals, OGSOL SI-20-10, 3 mg), and decaphenylcyclopentasilane (DPPS, Osaka Gas Chemicals, OGSOL SI-30-10, 3 mg), were used as additives to the perovskite solution, and structures of these polysilanes are shown in Figure 1. In the present work, the perovskite layers are described as $CH_3NH_3PbI_3$, +PMPS, +PPS, and +DPPS. The $CH_3NH_3PbI_3$ solution was spin-coated on the mesoporous TiO_2 layer at 2000 rpm for 60 s. The samples were annealed onto the hot plate at 100°C for 15 min to form a $CH_3NH_3PbI_3$ layer. After cooling to room temperature, hole transport layers were spin-coated. For the hole transport layer, a solution of spiro-OMeTAD (Sigma-Aldrich, 36.1 mg) in chlorobenzene (Wako Pure Chemical Industries, 0.5 mL) was mixed with a solution of lithium bis(trifluoromethylsulfonyl) imide (Li-TFSI, Tokyo Chemical Industry, 260 mg) in acetonitrile (Nacalai Tesque, 0.5 mL) for 12 h. The former solution with 4-tert-butylpyridine (Aldrich, 14.4 μL) was mixed with the Li-TFSI solution (8.8 μL) for 30 min at 70°C. All procedures for preparation of the thin films were carried out in ordinary air. Finally, gold (Au) electrodes were evaporated as top electrodes using a metal mask for the patterning.

The *J-V* characteristics of the photovoltaic devices were measured using a solar simulator composed of light source (San-ei Electric, XES-301S), where solar irradiance of 100 mW cm^{-2} was calibrated before measurements. The measurements were performed by source measure unit (Keysight, B2901A Precision SMU). The scan rate and sampling time were ~0.08 V s^{-1} and 1 ms, respectively. The photovoltaic devices were irradiated through the backside of FTO-coated glass substrate. The effective area of the devices was 0.090 cm^2. IPCE spectra of the devices were also collected using total quantum efficiency solutions for solar cells (Enli Technology, QE-R). The microstructures of the $CH_3NH_3PbI_3$ films were investigated using an X-ray diffractometer (Bruker, D2 PHASER) and an optical microscope (Nikon Eclipse E600). All measurements were performed at room temperature in ambient air.

3. Results and Discussion

Figure 2 shows the *J-V* characteristics of the TiO_2/$CH_3NH_3PbI_3$(polysilane)/spiro-OMeTAD photovoltaic cells under illumination, which indicate the effects of the polysilane addition. Forward and reverse scans are indicated in the figure as indicated by arrows, and the measured photovoltaic parameters (reverse scan) of the present $CH_3NH_3PbI_3$(polysilane) cells are summarized in Table 1. A small hysteresis of reverse and forward scans was observed for the cells. Ordinary

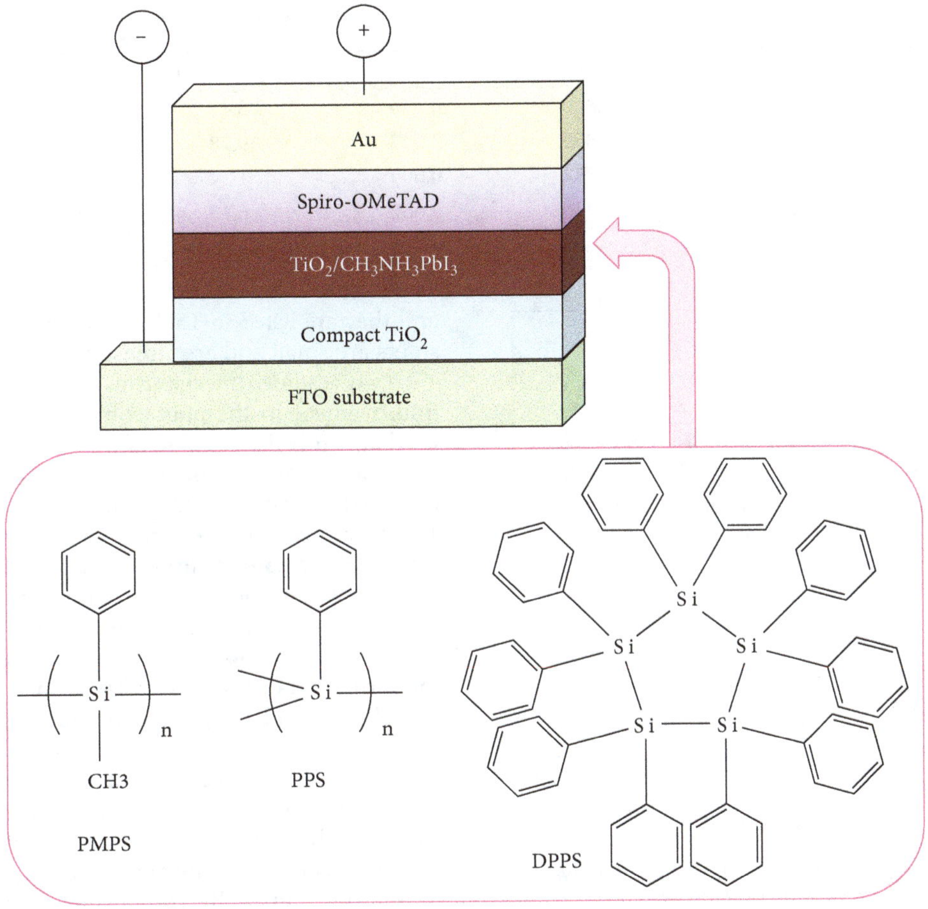

FIGURE 1: Schematic illustration of the present photovoltaic cells and structure models of the present polysilanes.

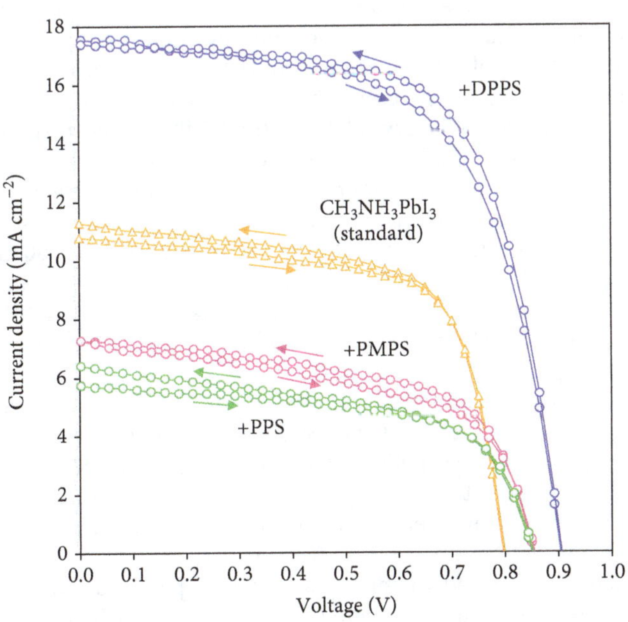

FIGURE 2: *J-V* characteristics of $CH_3NH_3PbI_3$ solar cells with polysilanes.

$CH_3NH_3PbI_3$ cells provided a power conversion efficiency (η) of 5.82%, as listed in Table 1. By adding the present three polysilanes, open-circuit voltages (V_{OC}) were increased, as observed in Figure 2. When the PMPS or PPS was added to the perovskite, short-circuit current density (J_{SC}) was decreased. On the other hand, by adding the DPPS, the J_{SC} value and shunt resistance (R_{sh}) were fairly increased, and the series resistance (R_s) was decreased compared with the PMPS and PPS, which resulted in the improvement of the conversion efficiency to 10.46%. The fill factor (FF) is almost the same as that of the ordinary cell.

Figure 3 shows IPCE spectra of the $CH_3NH_3PbI_3$(polysilane) devices. The present cells show photoconversion efficiencies between 320 and 800 nm, which corresponds to an energy gap of 1.55 eV for the $CH_3NH_3PbI_3$ [49]. The IPCE was improved in the range of 400~750 nm by adding DPPS to the perovskite layer, leading to the increase of J_{SC} values, as shown in Figure 2.

Optical microscopic images of ordinary $CH_3NH_3PbI_3$ cell are shown in Figure 4(a), and microparticles with sizes of 5~20 μm are observed. When the polysilanes were added to the perovskite phase, microparticles with homogeneous size distribution were formed, as shown in Figures 4(b)–4(d). In particular, by the DPPS addition, the crystal growth of the perovskite phase was observed, and the distances

Table 1: Measured photovoltaic parameters of $CH_3NH_3PbI_3$ solar cells with polysilanes.

Perovskite layer	J_{sc} (mA cm^{-2})	V_{oc} (V)	FF	R_s (Ω cm^2)	R_{sh} (Ω cm^2)	η (%)
$CH_3NH_3PbI_3$	10.8	0.800	0.674	5.47	700	5.82
+PMPS	7.28	0.855	0.547	9.78	479	3.40
+PPS	5.75	0.852	0.609	11.4	905	2.99
+DPPS	17.4	0.906	0.662	6.13	2190	10.46

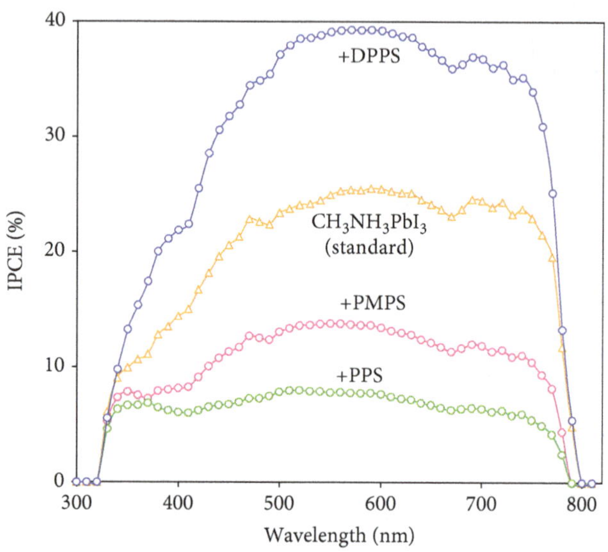

Figure 3: IPCE spectra of $CH_3NH_3PbI_3$ solar cells with polysilanes.

between the particles were decreased, leading formation of more dense microstructures, as observed in Figure 4(d).

Figure 5 shows XRD patterns of the present polysilane-added $CH_3NH_3PbI_3$ solar cells. When the $CH_3NH_3PbI_3$ has a tetragonal structure, an XRD peak corresponding to cubic 200 should be divided into tetragonal 004 and 220, as previously reported in [48]. The calculated XRD pattern also indicates splitting of diffraction reflections of 200 to 004/220 [48], and therefore the diffraction reflections can be indexed by a cubic crystal system ($Pm\bar{3}m$) for the present perovskite thin films. The diffraction peaks indicated by FTO and TiO_2 are from the FTO substrate and TiO_2 mesoporous layer, respectively. For the standard $CH_3NH_3PbI_3$ cell, a diffraction reflection of PbI_2 is observed, as indicated by an arrow. When the polysilanes were added to the perovskite phase, formation of the PbI_2 was suppressed. In addition, peak intensities of 100 of the perovskite phase increased with adding the polysilanes, which indicated crystal growth of the perovskite phase. It is considered that the crystal growth and PbI_2 suppression might be related with affinity between the γ-butyrolactone, *N,N*-dimethylformamide, CH_3NH_3I, PbI_2, and polysilanes. X-ray diffraction angles of the perovskite compounds were calibrated based on the 111 reflection of Au [50] of the cell electrodes. The XRD reflections due to the perovskite compounds were measured and indexed to determine the lattice distances. Then, the lattice constants of the perovskite crystals were determined from the lattice distances and indices by using a least squares method (CellCalc program [51]), and they are listed in Table 2. The differences of lattice constants are small and are almost constant for these cells.

Two assumed mechanisms could be considered for the improvement in the photoconversion efficiencies. The first mechanism is the increase and homogenization of the grain sizes of the perovskite crystals by adding polysilanes as observed optical microscopy images in Figure 4. This microstructure promotes the formation of dense interfacial structures and increase of surface coverage. When the grain sizes increase, grain boundary areas that cause carrier scattering decrease, which leads to the increase of the current density [52]. The present homogeneous structure could also suppress the formation of PbI_2, as observed in Figure 5. For the cells with PMPS and PPS, the perovskite domains are isolated, and the cell performances are severely affected by the coverage. Then, the pin holes provide short circuits between n-type semiconductor TiO_2 and p-type semiconductor spiro-MeOTAD, which resulted in the lower photovoltaic performances.

The second mechanism is the electronic structure change by the polysilanes. By adding polysilanes, the open-circuit voltages increased, as shown in Figure 2. From IPCE spectra in Figure 3, the bandgap energy of the perovskite crystals with PMPS and PPS increased from 1.55 eV to 1.57 eV, which leads to the increase of the V_{OC} values. This also corresponds to the decrease of lattice constants of cubic perovskite by addition of polysilanes. By adding DPPS, on the other hand, the bandgap energy seems to be almost the same as that of the standard perovskite crystal. Since the DPPS has more phenyl groups at the side of Si atoms compared with PMPS and PPS, carrier transport after the carrier separation might be facilitated by π-electrons on the phenyl groups, which affected the increase of IPCE in the range of 400~750 nm and the J_{SC} values. The existence of the phenyl groups might also reduce the electrical resistance of the DPPS-added $CH_3NH_3PbI_3$, which would result in the increase of the V_{OC} values.

Polymer-templated nucleation and growth had been reported as a method for crystal engineering of perovskites [53]. Poly(methyl methacrylate) (PMMA) provided heterogeneous nucleation which would be orders of magnitude faster than the homogeneous nucleation due to the lowering of the nucleation free energy barrier [54]. The PMMA also seems to slow down the growth rate of perovskite crystals by forming an intermediate adduct with PbI_2, which would provide the randomly formed nuclei, and would also lead to smooth thin films with few defects and large oriented perovskite grains [54–56]. The main improvement by the addition of the DPPS is an increase of the film coverage due to enlargement of the perovskite domains as observed in

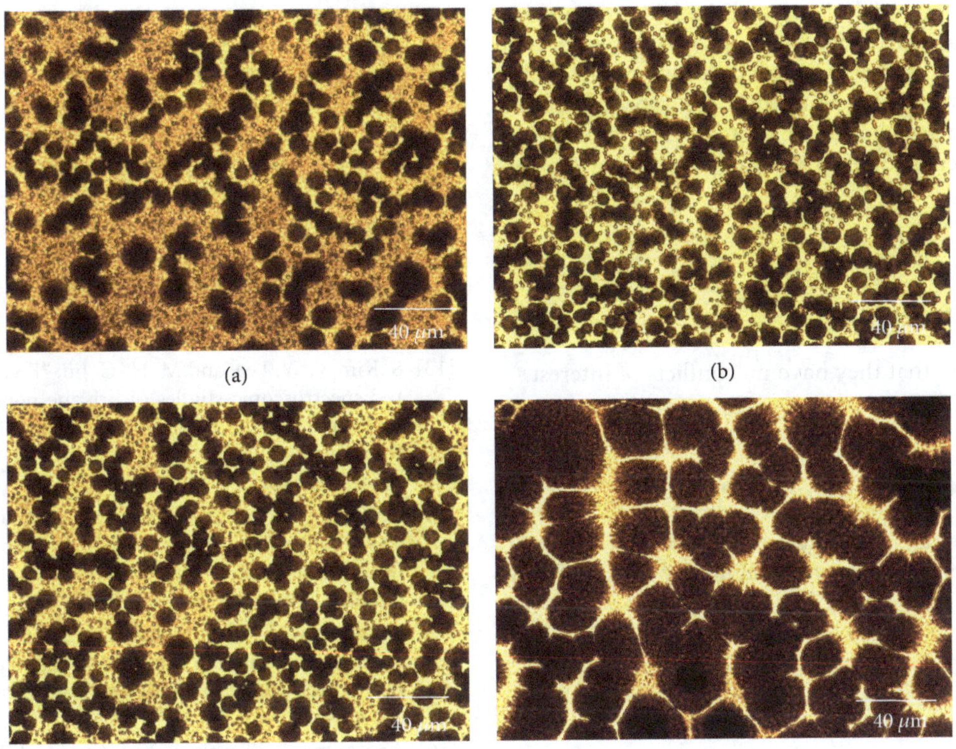

FIGURE 4: Optical microscopy images of (a) $CH_3NH_3PbI_3$ solar cells with (b) PMPS, (c) PPS, and (d) DPPS.

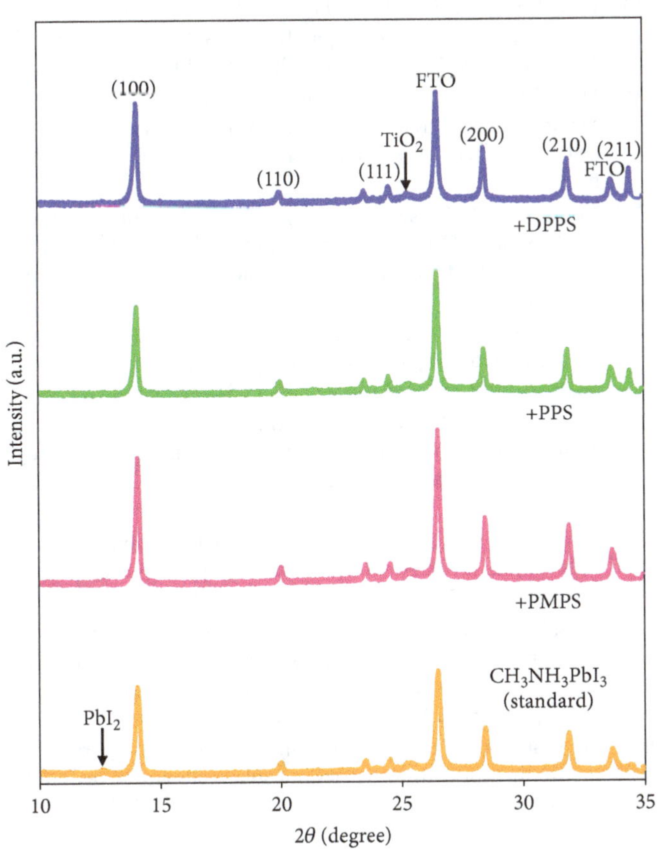

FIGURE 5: XRD patterns of $CH_3NH_3PbI_3$ solar cells with polysilanes.

TABLE 2: Measured lattice constants of cubic $CH_3NH_3PbI_3$ perovskite structures with polysilanes.

Perovskite layer	Lattice constant a (Å)
$CH_3NH_3PbI_3$	6.279
+PMPS	6.261
+PPS	6.276
+DPPS	6.276

the optical micrograph of Figure 4(d). Since the DPPS has more phenyl groups at the side of Si atoms compared with PMPS and PPS, molecular interactions created by the phenyl groups would affect the nucleation and crystal growth. Introduction of quantum dots such as PbS [57] or SnS [58] could also control the energy band structure. Further studies are needed for the growth of the perovskite phase and the control of the energy bands.

4. Conclusions

$CH_3NH_3PbI_3$ perovskite solar cells added with polysilanes were fabricated and characterized. Effects of PMPS, PPS, or DPPD addition to perovskite precursor solutions on the photovoltaic properties and the microstructures were investigated. The V_{OC} values were increased by applying these PMPS, PPS, or DPPD, and the IPCE was also increased by the DPPS addition in the range of 400~750 nm, leading to the improvement of the J_{SC} values. Formation of PbI_2 was also suppressed by the polysilane addition. Microstructure observation of the perovskite layer with DPPS indicated increase of surface coverage to form a dense interfacial structure by grain growth, which resulted in the improvement of conversion efficiencies above 10%. Lots of phenyl group in the DPPS might contribute to the carrier transport in the perovskite layers.

Conflicts of Interest

The authors declare that they have no conflicts of interest.

Acknowledgments

This work was partly supported by the Satellite Cluster Program of the Japan Science and Technology Agency (JST).

References

[1] N. S. Sariciftci, L. Smilowitz, A. J. Heeger, and F. Wudl, "Photoinduced electron transfer from a conducting polymer to buckminsterfullerene," *Science*, vol. 258, no. 5087, pp. 1474–1476, 1992.

[2] M. Granström, K. Petritsch, A. C. Arias, A. Lux, M. R. Andersson, and R. H. Friend, "Laminated fabrication of polymeric photovoltaic diodes," *Nature*, vol. 395, no. 6699, pp. 257–260, 1998.

[3] W. Ma, C. Yang, X. Gong, K. Lee, and A. J. Heeger, "Thermally stable, efficient polymer solar cells with nanoscale control of the interpenetrating network morphology," *Advanced Functional Materials*, vol. 15, no. 10, pp. 1617–1622, 2005.

[4] T. Oku, A. Takeda, A. Nagata et al., "Microstructures and photovoltaic properties of C_{60} based solar cells with copper oxides, $CuInS_2$, phthalocyanines, porphyrin, PVK, nanodiamond, germanium and exciton diffusion blocking layers," *Materials Technology*, vol. 28, no. 1-2, pp. 21–39, 2013.

[5] T. Oku, T. Noma, A. Suzuki, K. Kikuchi, and S. Kikuchi, "Fabrication and characterization of fullerene/porphyrin bulk heterojunction solar cells," *Journal of Physics and Chemistry of Solids*, vol. 71, no. 4, pp. 551–555, 2010.

[6] T. Oku, S. Hori, A. Suzuki, T. Akiyama, and Y. Yamasaki, "Fabrication and characterization of PCBM:P3HT:silicon phthalocyanine bulk heterojunction solar cells with inverted structures," *Japanese Journal of Applied Physics*, vol. 53, no. 5S1, article 05FJ08, 2014.

[7] S. M. Silence, J. C. Scott, F. Hache et al., "Poly(silane)-based high-mobility photorefractive polymers," *Journal of the Optical Society of America B*, vol. 10, no. 12, pp. 2306–2312, 1993.

[8] A. Rybak, J. Jung, W. Ciesielski, and J. Ulanski, "Photovoltaic effect in novel polysilane with phenothiazine rings and its blends with fullerene," *Material Science Poland*, vol. 24, no. 2, pp. 527–534, 2006.

[9] C.-Y. Liu, Z. C. Holman, and U. R. Kortshagen, "Hybrid solar cells from P3HT and silicon nanocrystals," *Nano Letters*, vol. 9, no. 1, pp. 449–452, 2009.

[10] M. Kakimoto, H. Kashihara, T. Kashiwagi, T. Takiguchi, J. Ohshita, and M. Ishikawa, "Visible light photoconduction of poly(disilanyleneoligothienylene)s and doping effect of C_{60}," *Macromolecules*, vol. 30, no. 25, pp. 7816–7820, 1997.

[11] Y. Haga and Y. Harada, "Photovoltaic characteristics of phthalocyanine-polysilane composite films," *Japanese Journal of Applied Physics*, vol. 40, 2A, Part 1, pp. 855–861, 2001.

[12] A. Watanabe and O. Ito, "Photoinduced electron transfer between C_{60} and polysilane studied by laser flash photolysis in the near-IR region," *The Journal of Physical Chemistry*, vol. 98, no. 31, pp. 7736–7740, 1994.

[13] S. Kim, C.-Y. Lee, and M. H.-C. Jin, "Fourier-transform infrared spectroscopic studies of pristine polysilanes as precursor molecules for the solution deposition of amorphous silicon thin-films," *Solar Energy Materials and Solar Cells*, vol. 100, pp. 61–64, 2012.

[14] T. Oku, J. Nakagawa, M. Iwase et al., "Microstructures and photovoltaic properties of polysilane-based solar cells," *Japanese Journal of Applied Physics*, vol. 52, no. 4S, article 04CR07, 2013.

[15] J. Nakagawa, T. Oku, A. Suzuki et al., "Effects of PBr3 addition to polysilane thin films on structures and photovoltaic properties," *Green and Sustainable Chemistry*, vol. 7, no. 1, pp. 20–34, 2017.

[16] A. Kojima, K. Teshima, Y. Shirai, and T. Miyasaka, "Organometal halide perovskites as visible-light sensitizers for photovoltaic cells," *Journal of the American Chemical Society*, vol. 131, no. 17, pp. 6050-6051, 2009.

[17] J.-H. Im, C.-R. Lee, J.-W. Lee, S.-W. Park, and N.-G. Park, "6.5% efficient perovskite quantum-dot-sensitized solar cell," *Nanoscale*, vol. 3, no. 10, pp. 4088–4093, 2011.

[18] H.-S. Kim, C.-R. Lee, J.-H. Im et al., "Lead iodide perovskite sensitized all-solid-state submicron thin film mesoscopic solar cell with efficiency exceeding 9%," *Scientific Reports*, vol. 2, no. 1, 2012.

[19] M. M. Lee, J. Teuscher, T. Miyasaka, T. N. Murakami, and H. J. Snaith, "Efficient hybrid solar cells based on meso-superstructured organometal halide perovskites," *Science*, vol. 338, no. 6107, pp. 643–647, 2012.

[20] J. Burschka, N. Pellet, S.-J. Moon et al., "Sequential deposition as a route to high-performance perovskite-sensitized solar cells," *Nature*, vol. 499, no. 7458, pp. 316–319, 2013.

[21] W. S. Yang, J. H. Noh, N. J. Jeon et al., "Solar cells. High-performance photovoltaic perovskite layers fabricated through intramolecular exchange," *Science*, vol. 348, no. 6240, pp. 1234–1237, 2015.

[22] D. Bi, W. Tress, M. I. Dar et al., "Efficient luminescent solar cells based on tailored mixed-cation perovskites," *Science Advances*, vol. 2, no. 1, article e1501170, 2016.

[23] M. Saliba, T. Matsui, J. Y. Seo et al., "Cesium-containing triple cation perovskite solar cells: improved stability, reproducibility and high efficiency," *Energy and Environmental Science*, vol. 9, no. 6, pp. 1989–1997, 2016.

[24] M. Saliba, T. Matsui, K. Domanski et al., "Incorporation of rubidium cations into perovskite solar cells improves photovoltaic performance," *Science*, vol. 354, no. 6309, pp. 206–209, 2016.

[25] Z. Tang, T. Bessho, F. Awai et al., "Hysteresis-free perovskite solar cells made of potassium-doped organometal halide perovskite," *Scientific Reports*, vol. 7, no. 1, p. 12183, 2017.

[26] S. D. Stranks, G. E. Eperon, G. Grancini et al., "Electron-hole diffusion lengths exceeding 1 micrometer in an organometal trihalide perovskite absorber," *Science*, vol. 342, no. 6156, pp. 341–344, 2013.

[27] D. Shi, V. Adinolfi, R. Comin et al., "Low trap-state density and long carrier diffusion in organolead trihalide perovskite single crystals," *Science*, vol. 347, no. 6221, pp. 519–522, 2015.

[28] Q. Dong, Y. Fang, Y. Shao et al., "Electron-hole diffusion lengths > 175 μm in solution-grown $CH_3NH_3PbI_3$single crystals," *Science*, vol. 347, no. 6225, pp. 967–970, 2015.

[29] T. Oku, Y. Ohishi, A. Suzuki, and Y. Miyazawa, "Effects of NH_4Cl addition to perovskite $CH_3NH_3PbI_3$ photovoltaic devices," *Journal of the Ceramic Society of Japan*, vol. 125, no. 4, pp. 303–307, 2017.

[30] T. Oku and Y. Ohishi, "Effects of annealing on $CH_3NH_3PbI_3$(cl) perovskite photovoltaic devices," *Journal of the Ceramic Society of Japan*, vol. 126, no. 1, pp. 56–60, 2018.

[31] F. Hao, C. C. Stoumpos, D. H. Cao, R. P. H. Chang, and M. G. Kanatzidis, "Lead-free solid-state organic–inorganic halide perovskite solar cells," *Nature Photonics*, vol. 8, no. 6, pp. 489–494, 2014.

[32] W. Liao, D. Zhao, Y. Yu et al., "Fabrication of efficient low-bandgap perovskite solar cells by combining formamidinium tin iodide with methylammonium lead iodide," *Journal of the American Chemical Society*, vol. 138, no. 38, pp. 12360–12363, 2016.

[33] T. Krishnamoorthy, H. Ding, C. Yan et al., "Lead-free germanium iodide perovskite materials for photovoltaic applications," *Journal of Materials Chemistry A*, vol. 3, no. 47, pp. 23829–23832, 2015.

[34] Y. Ohishi, T. Oku, and A. Suzuki, "Fabrication and characterization of perovskite-based $CH_3NH_3Pb_{1-x}Ge_xI_3$, $CH_3NH_3Pb_{1-x}Tl_xI_3$ and $CH_3NH_3Pb_{1-x}In_xI_3$ photovoltaic devices," *AIP Conference Proceedings*, vol. 1709, no. 1, article 020020, 2016.

[35] T. Oku, Y. Ohishi, and A. Suzuki, "Effects of antimony addition to perovskite-type $CH_3NH_3PbI_3$ photovoltaic devices," *Chemistry Letters*, vol. 45, no. 2, pp. 134–136, 2016.

[36] T. Oku, Y. Ohishi, A. Suzuki, and Y. Miyazawa, "Effects of Cl addition to Sb-doped perovskite-type $CH_3NH_3PbI_3$ photovoltaic devices," *Metals*, vol. 6, no. 7, p. 147, 2016.

[37] J. Zhang, M.-H. Shang, P. Wang et al., "n-type doping and energy states tuning in $CH_3NH_3Pb_{1-x}Sb_{2x/3}I_3$ perovskite solar cells," *ACS Energy Letters*, vol. 1, no. 3, pp. 535–541, 2016.

[38] T. Oku, Y. Ohishi, and A. Suzuki, "Effects of $SbBr_3$ addition to $CH_3NH_3PbI_3$ solar cells," *AIP Conference Proceedings*, vol. 1807, no. 1, article 020007, 2017.

[39] Y. Ando, T. Oku, and Y. Ohishi, "Rietveld refinement of crystal structure of perovskite $CH_3NH_3Pb(Sb)I_3$ solar cells," *Japanese Journal of Applied Physics*, vol. 57, no. 2S2, article 02CE02, 2018.

[40] T. Oku, M. Zushi, K. Suzuki, Y. Ohishi, T. Matsumoto, and A. Suzuki, "Nanostructured solar cells," in *Fabrication and characterization of element-doped perovskite solar cells*, N. Das, Ed., pp. 217–243, InTech, Croatia, 2017.

[41] Y. Shirahata and T. Oku, "Effects of copper addition on photovoltaic properties of perovskite $CH_3NH_3PbI_{3-x}Cl_x$ solar cells," *Physica Status Solidi A*, vol. 214, no. 10, 2017.

[42] Y. Shirahata and T. Oku, "Microstructures, optical and photovoltaic properties of $CH_3NH_3PbI_{3(1-x)}Cl_x$ perovskite films with CuSCN additive," *Materials Research Express*, vol. 5, no. 5, 2018.

[43] T. Oku, Y. Ohishi, and H. Tanaka, "Effects of CuBr addition to $CH_3NH_3PbI_3$(Cl) perovskite photovoltaic devices," *AIP Conference Proceedings*, vol. 1929, no. 1, article 020010, 2018.

[44] H. Tanaka, Y. Ohishi, and T. Oku, "Effects of Cu addition to perovskite $CH_3NH_3PbI_{3-x}Cl_x$ photovoltaic devices with hot airflow during spin-coating," *Japanese Journal of Applied Physics*, vol. 57, no. 8S3, article 08RE10, 2018.

[45] T. Hamatani and T. Oku, "Effects of halide addition to arsenic-doped perovskite photovoltaic devices," *AIP Conference Proceedings*, vol. 1929, no. 1, article 020018, 2018.

[46] H. Tanaka, T. Oku, and N. Ueoka, "Structural stabilities of organic–inorganic perovskite crystals," *Japanese Journal of Applied Physics*, vol. 57, no. 8S3, article 08RE12, 2018.

[47] Y. Shirahata, Y. Yamomoto, A. Suzuki, T. Oku, S. Fukunishi, and K. Kohno, "Effects of polysilane-doped spiro-OMeTAD hole transport layers on photovoltaic properties," *Physica Status Solidi A*, vol. 214, no. 3, 2017.

[48] T. Oku, M. Zushi, Y. Imanishi, A. Suzuki, and K. Suzuki, "Microstructures and photovoltaic properties of perovskite-type $CH_3NH_3PbI_3$ compounds," *Applied Physics Express*, vol. 7, no. 12, p. 121601, 2014.

[49] T. Oku, *Solar Cells and Energy Materials*, Walter de Gruyter, Berlin, 2017.

[50] M. Ellner, K. Kolatschek, and B. Predel, "On the partial atomic volume and the partial molar enthalpy of aluminium in some phases with Cu and Cu_3Au structures," *Journal of the Less Common Metals*, vol. 170, no. 1, pp. 171–184, 1991.

[51] H. Miura, "CellCalc: a unit cell parameter refinement program on Windows computer," *Nihon Kessho Gakkaishi*, vol. 45, no. 2, pp. 145–147, 2003.

[52] W. Nie, H. Tsai, R. Asadpour et al., "Solar cells. High-efficiency solution-processed perovskite solar cells with millimeter-scale grains," *Science*, vol. 347, no. 6221, pp. 522–525, 2015.

[53] D. Bi, C. Yi, J. Luo et al., "Polymer-templated nucleation and crystal growth of perovskite films for solar cells with efficiency greater than 21%," *Nature Energy*, vol. 1, no. 10, 2016.

[54] A. Cacciuto, S. Auer, and D. Frenkel, "Onset of heterogeneous crystal nucleation in colloidal suspensions," *Nature*, vol. 428, no. 6981, pp. 404–406, 2004.

[55] T. Oku, Y. Ohishi, and N. Ueoka, "Highly (100)-oriented $CH_3NH_3PbI_3$(Cl) perovskite solar cells prepared with NH_4Cl using an air blow method," *RSC Advances*, vol. 8, no. 19, pp. 10389–10395, 2018.

[56] N. Ueoka, T. Oku, H. Tanaka et al., "Effects of PbI_2 addition and TiO_2 electron transport layers for perovskite solar cells," *Japanese Journal of Applied Physics*, vol. 57, no. 8S3, p. 08RE05, 2018.

[57] Y. Yang and W. Wang, "Effects of incorporating PbS quantum dots in perovskite solar cells based on $CH_3NH_3PbI_3$," *Journal of Power Sources*, vol. 293, pp. 577–584, 2015.

[58] J. Han, X. Yin, H. Nan et al., "Enhancing the performance of perovskite solar cells by hybridizing SnS quantum dots with $CH_3NH_3PbI_3$," *Small*, vol. 13, no. 32, article 1700953, 2017.

3

Fabrication of Affordable and Sustainable Solar Cells Using NiO/TiO_2 *P-N* Heterojunction

Kingsley O. Ukoba, Freddie L. Inambao, and Andrew C. Eloka-Eboka

Mechanical Engineering, University of KwaZulu-Natal, Durban, South Africa

Correspondence should be addressed to Kingsley O. Ukoba; ukobaking@yahoo.com

Academic Editor: Ahmad Umar

The need for affordable, clean, efficient, and sustainable solar cells informed this study. Metal oxide TiO_2/NiO heterojunction solar cells were fabricated using the spray pyrolysis technique. The optoelectronic properties of the heterojunction were determined. The fabricated solar cells exhibit a short-circuit current of 16.8 mA, open-circuit voltage of 350 mV, fill factor of 0.39, and conversion efficiency of 2.30% under 100 mW/cm^2 illumination. This study will help advance the course for the development of low-cost, environmentally friendly, and sustainable solar cell materials from metal oxides.

1. Introduction

The need for affordable and sustainable electricity in developing nations has been an issue of concern to all stakeholders. Renewable energy has been identified as a viable solution to ending global electricity problems as its availability exceeds world energy demand [1]. Solar energy is a good source of renewable energy [2]. The hourly solar influx on the surface of the earth surpasses annual human energy needs [3]. Photovoltaic energy has received increasing interest caused by a decrease in module prices in countries like China [4]. Interest in these devices is due to improved reliability, efficiency, and costs in generating electricity [5]. Solar cells of high efficiency have been achieved with inorganic materials [6], but they require expensive materials of high purity and a technique that is energy intensive. There is, therefore, a need to explore ways of manufacturing solar cells that can scale-up to large volumes at low cost.

Metal oxide solar cells offer a good replacement for conventional silicon solar cells. This is because metal oxides are low-cost materials, have flexible optical properties, can be deposited using low-cost techniques, and are simple to scale-up to large volume production. They also display quantum confinement effects in two dimensions [7]. Nanostructured metal oxides are used in a wide range of device applications because of their broad composition and band structures [8–14]. The widely used oxides are ZnO [15], CuO [16], In_2O_3 [17], and TiO_2 [18], to mention but a few. They are widely applied in optoelectronic devices such as humidity sensors [19], photodiodes [20], solar cells [21], and photocatalysts [22].

NiO is a *P*-type semiconductor with a wide bandgap between 3.5 eV and 4.0 eV [23]. The excellent properties of NiO make it a promising material for solar cells [24]. Similarly, TiO_2 is a desirable material for harvesting solar energy because of its optoelectronic properties, high resistance to photocorrosion, affordability, stability in a wide range of pH, and nonpoisonous nature [25]. Various techniques are available for depositing metal oxides [26–31]. Low cost of equipment, ease of control of deposited film structure, and the ability to coat large areas in thin layers with uniform thickness [32–35] influenced the choice of the technique used in this study.

Heterojunctions are known to be the most competitive method of solar cell fabrication on account of being the simplest [36]. A *P-N* junction is created when *P*-type (NiO) and *N*-type (TiO_2) semiconductor materials are placed in contact with one another. A solar cell is basically a *P-N* junction with

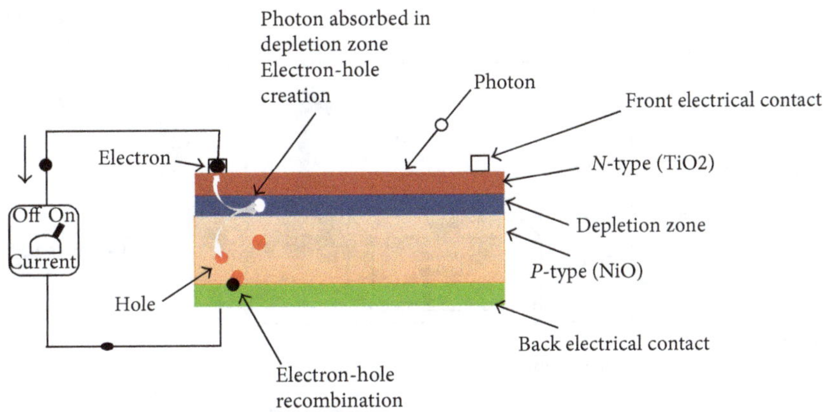

FIGURE 1: Solar cell generation of electricity using a *P-N* junction.

a large surface area. Figure 1 depicts generation of electricity by a solar cell using a *P-N* junction.

The overall aim of this research is the provision of affordable and sustainable solar panels for developing and low-income countries. This was achieved by fabricating nanostructured TiO_2/NiO heterojunction thin-film solar cells using the spray pyrolysis technique.

2. Methodology

2.1. Deposition. The chemicals used are of analytical reagent grade and were used without further purification. Distilled and deionized pure water were used during the course of the experiment.

The solar cell was fabricated using a modified spray pyrolysis technique (SPT) as reported by Ukoba et al. [37] and represented pictorially in Figure 2. Prior to sample preparation, the indium tin oxide- (ITO-) coated glass and soda lime glass used as substrate were clean ultrasonically as reported by Adeoye Abiodun and Salau [38]. The precursor for the window layer titanium oxide (TiO_2) nanostructure thin film was prepared by mixing 3 ml of titanium ethoxide with 30 ml of distilled water and ethanol mixture, and three droplets of acetic acid. This was stirred for one hour before spraying on cleaned indium tin oxide- (ITO-) coated glass substrates and soda lime glass substrates maintained at about 350°C. Also, deposition parameters such as substrate temperature, carrier gas flow rate, and pressure were optimized to obtain quality films.

The nanostructured nickel oxide (NiO) absorber layer was deposited on the prepared ITO/TiO_2 layers and empty soda lime glass using SPT, as shown in Figure 3. The precursor for NiO was obtained by preparing 0.05 M nickel acetate tetrahydrate in double distilled water.

The precursors were thoroughly stirred for several minutes prior to spraying onto preheated substrates maintained at about 350°C. Other deposition parameters were maintained to obtain good quality thin films. The optimized parameters used in the deposition of the NiO films are tabulated in Table 1. To complete the TiO_2/NiO heterojunction solar cell illustrated in Figure 4, gold (Au)

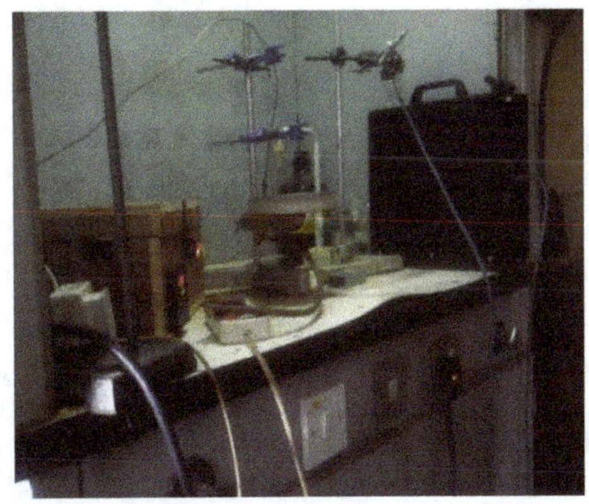

FIGURE 2: Pictorial representation of the experimental set-up of the spray pyrolysis technique.

metal contact was deposited as a back contact using DC magnetron sputtering.

2.2. Testing. The TiO_2 and NiO prepared on soda lime glass were used to study the elemental, morphological, and structural characteristics of TiO_2 and NiO using energy dispersive X-ray spectrometer (EDS or EDX: "AZtec Oxford Detector"), a ZEISS Ultra Plus field emission gun scanning electron microscope (FEGSEM), and Bruker AXS D8 Advance X-ray diffractometer (XRD) with Cu-Kα radiation, respectively. The *J-V* characteristics of the fabricated TiO_2/NiO heterojunction cell in dark and under illumination were done using the Keithley SourceMeter 2400, coupled with a two-point probe. Newport solar simulator of intensity (100 mW/cm^2) was used as the source of illumination.

3. Results and Discussion

3.1. Morphological Studies. Figures 5(a) and 5(b) show the scanning electron micrograph of the NiO thin film at lower and higher magnification, respectively. The micrograph reveals scattered distribution of the NiO particles across the surface of the film. The film has even distribution, is adherent to the film surface, and has no cracks. This represents a better surface morphology compared to that of NiO films reported by Sriram and Thayumanavan

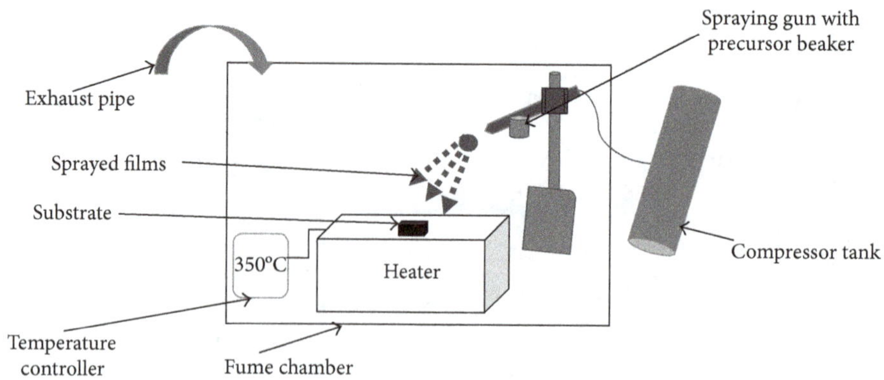

FIGURE 3: Experimental set-up of spray pyrolysis technique.

TABLE 1: Optimum deposition parameters of SPT NiO film.

Deposition parameter	Value
Substrate to nozzle height	20 cm
Rate of spray	1 ml/min
Spray time	1 min
Sprays interval	30 sec
Carrier gas	1 bar of filled compressed air

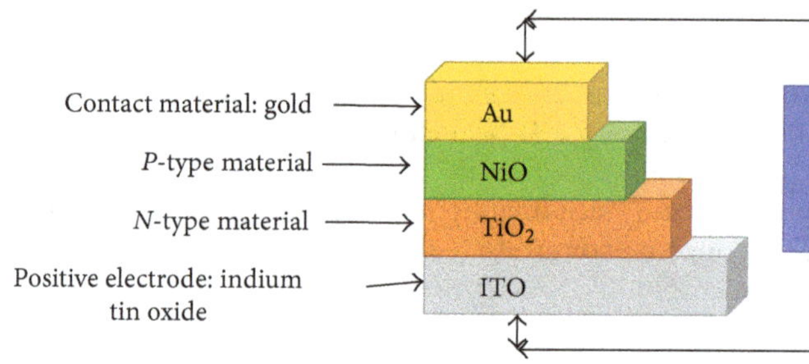

FIGURE 4: Schematic of the fabricated NiO/TiO_2 heterojunction solar cells.

[39]. The SEM shows the potential of NiO as an absorber layer in solar cell fabrication. Figures 5(c) and 5(d) show the scanning electron micrograph of the heterojunction of NiO/TiO_2. This micrograph was obtained by the SEM at the junction or point of interaction between the TiO_2 and NiO. It shows a polycrystalline structure. The micrograph shows the *P*-type NiO and *N*-type TiO_2 of the thin film with their polycrystalline structures. It shows complete penetration at the heterojunction.

3.2. Elemental Composition. Figure 6 shows the elemental composition of the NiO/TiO_2 heterojunction solar cell deposited on the ITO-coated glass substrate. Figure 6 shows the presence of Ti, O, and Ni for the TiO_2 and NiO, respectively, and the indium (In) representing the ITO-coated glass substrate. This confirms the presence of the metal oxides in the heterojunction.

3.3. Structural Analysis. Figure 7 shows the X-ray diffraction patterns of the fabricated $ITO/TiO_2/NiO$ heterojunction solar cell. The peaks corresponding to NiO and TiO_2 were determined with JCPDS patterns. The XRD spectrum indicates strong NiO peaks with (1 1 1), (2 0 0), and (2 2 0) preferential orientation. The patterns of the NiO thin film have peak diffractions at ($2\theta = 37°$, 43°, and 64°) for the (1 1 1), (2 0 0), and (2 2 0) plane. The XRD analysis confirms Bunsenite, which corresponds to the JCPDS card: 04-0835 for nickel oxide [40] confirming it as a good absorber layer of solar cells [41]. The TiO_2 spectrum also shows strong spectrum and polycrystalline structures typical of *N*-type in heterojunction solar cells. The structure of the heterojunction indicates that the film is polycrystalline and chemically pure.

3.4. Current Density-Voltage (J-V) Characterization. The *J-V* characteristic curve of the prepared TiO_2/NiO heterojunction thin-film solar cell under illumination and in the dark is depicted in Figure 8. The *J-V* characteristic at room temperature in the dark shows that the forward current of the cells increases slowly with increasing voltage. The solar cell has rectification properties since the dark *J-V* plots were similar to the Shockley diode characteristics, which can be expressed by the standard diode equation

$$J = J_O \left[\exp\left(\frac{qV}{AkT} \right) - 1 \right], \tag{1}$$

where q is the electronic charge, A is the diode quality factor (ideality factor), k is Boltzmann's constant, T is the absolute temperature, and J_o is the reverse saturation current.

The solar cell parameters evaluated from the *J-V* curve are presented in Table 2. The fabricated solar cell exhibits the short-circuit current (J_{sc}) of 16.8 mA, the open-circuit voltage (V_{oc}) of 350 mV, the fill factor (FF) of 0.39, and the conversion efficiency (η) of 2.30%. This is a marked improvement on the values of 0.33 V and 0.29 recorded by Georgieva and Tanusevski [42] for the open-circuit voltage and fill factor, respectively. It also showed improvement in the fill factor of 0.28 reported by Noda et al. [43].

4. Solar Cell Parameters

The primary parameters that describe the performance of a photovoltaic device are discussed below.

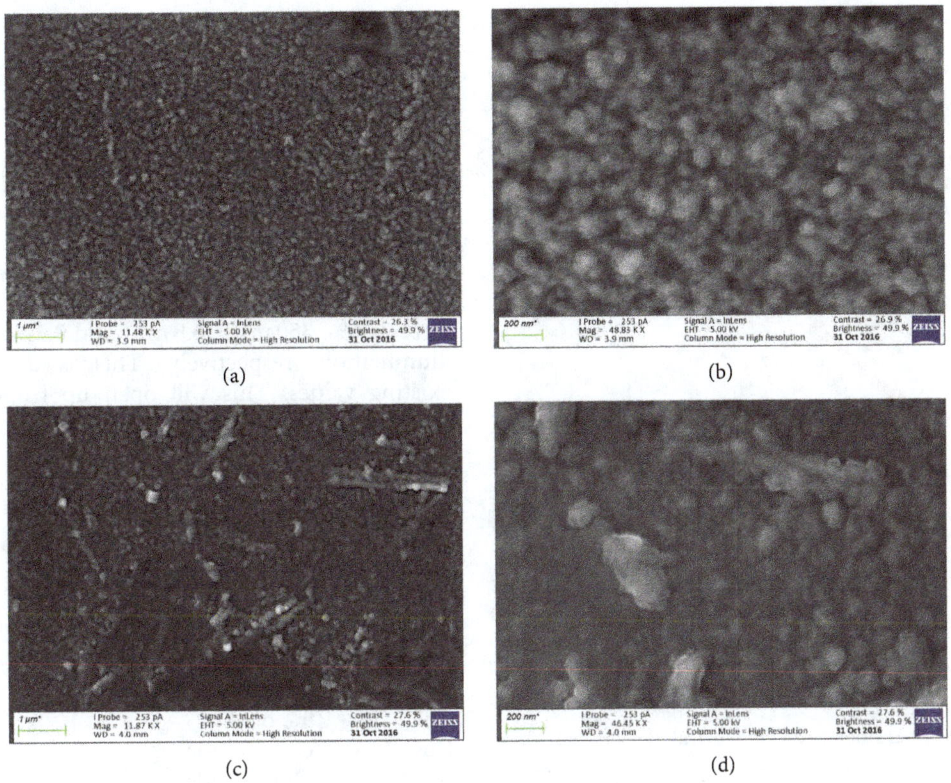

FIGURE 5: SEM of (a) NiO thin film at lower magnification, (b) NiO thin film at higher magnification, (c) fabricated NiO/TiO_2 heterojunction solar cell at lower magnification, and (d) fabricated NiO/TiO_2 heterojunction solar cell at higher magnification.

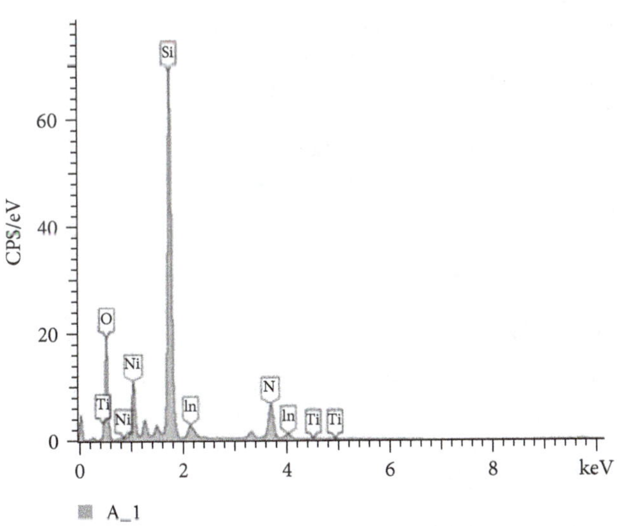

FIGURE 6: EDX of fabricated $ITO/TiO_2/NiO$ heterojunction solar cell.

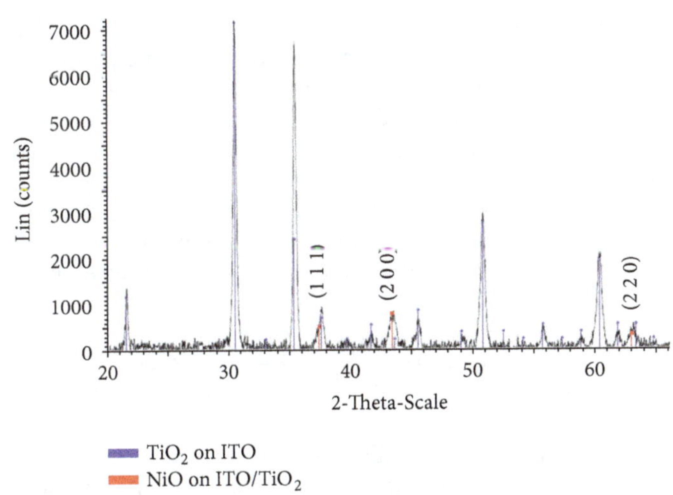

FIGURE 7: XRD of fabricated $ITO/TiO_2/NiO$ heterojunction solar cell.

4.1. Open-Circuit Voltage (V_{oc}). Open-circuit voltage is the applied voltage relative to an open circuit where no current flows through the device (i.e., the voltage across the device at zero current). V_{oc} is obtained at the point of intersection of the *I-V* curve under illumination at the voltage axis. Under open-circuit conditions, the structure has to bias itself to some voltage V_{oc} in order to counter the light-beam-induced current. The open-circuit voltage V_{oc} arises as a result of the built-in electric field present in the materials system and can be expressed as

$$V_{OC} = \frac{AkT}{q} \ln\left(\frac{J_L}{J_o} + 1\right). \quad (2)$$

This quantity is left unaffected by series resistance losses in the cell but is sensitive to shunt losses.

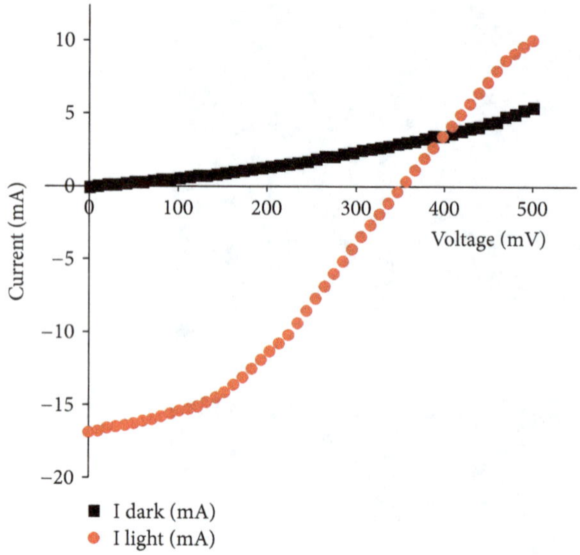

Figure 8: Typical *I-V* curve for the prepared TiO_2/NiO heterojunction thin-film solar cell under illumination and in the dark.

Table 2: Solar cell parameters of the fabricated solar cell.

Sample	J_{sc} (mA)	V_{oc} (mV)	J_{max} (mV)	V_{max} (mV)	FF (%)	η (%)
TiO_2/NiO	16.8	350.0	13.10	173.57	39	2.30

4.2. Short-Circuit Current Density (J_{sc}). J_{sc} is the current that flows through the junction under illumination at zero applied voltage, that is, $J_{sc} = J(V = 0)$. In the ideal case, it equals the photogenerated current density (J_L) and is proportional to the incident number of photons or alternatively the intensity of illumination.

J_{sc} is represented as the intersection of the *J-V* curve under illumination at the current axis. For an ideal solar cell ($R_S = 0$ and $R_{SH} = \infty$), the short-circuit current is given by

$$J_{sc} = J_o\left[\exp\left(\frac{q(0)}{AkT}\right) - 1\right] - J_L, \quad V = 0. \tag{3}$$

4.3. Fill Factor (FF). The fill factor is defined as the inverse of the ratio of the ideal power to the maximum power in operating conditions. It can be defined also as the area of the maximum power rectangle to the product of the short-circuit current and the open-circuit voltage. This is shown as

$$\mathrm{FF} = \frac{V_{max}J_{max}}{V_{oc}J_{sc}}. \tag{4}$$

4.4. Efficiency (η). The most important parameter of a solar cell in terms of its ultimate function is the photovoltaic conversion efficiency. This is defined as the ratio of the output power (electricity) to the input power (light) and can be calculated as

$$\eta = \frac{P_{max}}{P_{in}} = \frac{\mathrm{FF}\,(V_{oc}J_{sc})}{P_{in}}. \tag{5}$$

5. Conclusion

In this study, TiO_2 and NiO thin films were used to fabricate ITO/TiO_2/NiO heterojunction solar cells. It shows that NiO can be used in thin-film solar cells. The conversion efficiency, open-circuit voltage, short-circuit current, and fill factor were 2.30%, 350 mV, 16.8 mA, and 0.39 under 100 mW/cm^2 illumination, respectively. This is an improvement on existing values. This will open up frontiers in affordable and sustainable solar cell fabrication in developing and low-income countries.

Nomenclature

NiO: Nickel oxide
Ni: Nickel
Ti: Titanium
O: Oxygen
TiO_2: Titanium oxide
ZnO: Zinc oxide
CuO: Copper oxide
In_2O_3: Indium oxide
ITO: Indium tin oxide
Au: Gold
SPT: Spray pyrolysis technique
FEGSEM: Field emission gun scanning electron microscope
EDX: Energy dispersive X-ray spectrometer
XRD: X-ray diffractometer
J-V: Current density-voltage
I-V: Current-voltage
V_{oc}: Open-circuit voltage
J_{sc}: Short-circuit current density
FF: Fill factor
P_{in}: Power in
P_{max}: Maximum power.

Conflicts of Interest

The authors declare that there are no conflicts of interest regarding the publication of this paper.

Acknowledgments

The financial assistance of the National Research Foundation and The World Academy of Science (NRF/TWAS) of South Africa under Grant number 105492 towards this research is acknowledged. The first author also wishes to thank EMDI Akure, Adeoye Eyitayo, and Remy Bucher of iThemba Lab for their assistance in the experiment analysis.

References

[1] O. Ellabban, H. Abu-Rub, and F. Blaabjerg, "Renewable energy resources: current status, future prospects and their enabling technology," *Renewable and Sustainable Energy Reviews*, vol. 39, pp. 748–764, 2014.

[2] A. K. Hussein, "Applications of nanotechnology in renewable energies—a comprehensive overview and understanding," *Renewable and Sustainable Energy Reviews*, vol. 42, pp. 460–476, 2015.

[3] N. S. Lewis, "Research opportunities to advance solar energy utilization," *Science*, vol. 351, no. 6271, article aad1920, 2016.

[4] G.-N. Lee, M. P. Machaiah, W.-H. Park, and J. Kim, "Enhanced optical and electrical properties of ITO/Ag/AZO transparent conductors for photoelectric applications," *International Journal of Photoenergy*, vol. 2017, Article ID 8315802, 9 pages, 2017.

[5] S.-H. Hong, J.-H. Yun, H.-H. Park, and J. Kim, "Nanodome-patterned transparent conductor for highly responsive photoelectric device," *Applied Physics Letters*, vol. 103, no. 15, article 153504, 2013.

[6] M. A. Green, "Third generation photovoltaics: ultra-high conversion efficiency at low cost," *Progress in Photovoltaics: Research and Applications*, vol. 9, no. 2, pp. 123–135, 2001.

[7] S. Joshi, M. Mudigere, L. Krishnamurthy, and G. Shekar, "Growth and morphological studies of NiO/CuO/ZnO based nanostructured thin films for photovoltaic applications," *Chemical Papers*, vol. 68, no. 11, pp. 1584–1592, 2014.

[8] A. Czapla, E. Kusior, and M. Bucko, "Optical properties of non-stoichiometric tin oxide films obtained by reactive sputtering," *Thin Solid Films*, vol. 182, no. 1-2, pp. 15–22, 1989.

[9] A. Rahal, A. Benhaoua, M. Jlassi, and B. Benhaoua, "Structural, optical and electrical properties studies of ultrasonically deposited tin oxide (SnO_2) thin films with different substrate temperatures," *Superlattices and Microstructures*, vol. 86, pp. 403–411, 2015.

[10] W. Zhang, S. Ding, Z. Yang et al., "Growth of novel nanostructured copper oxide (CuO) films on copper foil," *Journal of Crystal Growth*, vol. 291, no. 2, pp. 479–484, 2006.

[11] M. Sharon and B. Prasad, "Preparation and characterization of iron oxide thin film electrodes," *Solar Energy Materials*, vol. 8, no. 4, pp. 457–469, 1983.

[12] S. K. Shaikh, S. I. Inamdar, V. V. Ganbavle, and K. Y. Rajpure, "Chemical bath deposited ZnO thin film based UV photoconductive detector," *Journal of Alloys and Compounds*, vol. 664, pp. 242–249, 2016.

[13] M. Jlassi, I. Sta, M. Hajji, and H. Ezzaouia, "Optical and electrical properties of nickel oxide thin films synthesized by sol–gel spin coating," *Materials Science in Semiconductor Processing*, vol. 21, pp. 7–13, 2014.

[14] R. Drevet, C. Legros, D. Bérardan et al., "Metal organic precursor effect on the properties of SnO_2 thin films deposited by MOCVD technique for electrochemical applications," *Surface and Coatings Technology*, vol. 271, pp. 234–241, 2015.

[15] B. Siwach, S. Sharma, and D. Mohan, "Structural, optical and morphological properties of ZnO/MWCNTs nanocomposite photoanodes for dye sensitized solar cells (DSSCs) application," *Journal of Integrated Science and Technology*, vol. 5, no. 1, pp. 1–4, 2017.

[16] O. Langmar, C. R. Ganivet, G. de la Torre, T. Torres, R. D. Costa, and D. M. Guldi, "Optimizing CuO p-type dye-sensitized solar cells by using a comprehensive electrochemical impedance spectroscopic study," *Nanoscale*, vol. 8, no. 41, pp. 17963–17975, 2016.

[17] Z. Yu, I. R. Perera, T. Daeneke et al., "Indium tin oxide as a semiconductor material in efficient p-type dye-sensitized solar cells," *NPG Asia Materials*, vol. 8, no. 9, article e305, 2016.

[18] M. Rokhmat, E. Wibowo, and M. Abdullah, "Performance improvement of TiO_2/CuO solar cell by growing copper particle using fix current electroplating method," *Procedia Engineering*, vol. 170, pp. 72–77, 2017.

[19] Z. Yuan, H. Tai, X. Bao, C. Liu, Z. Ye, and Y. Jiang, "Enhanced humidity-sensing properties of novel graphene oxide/zinc oxide nanoparticles layered thin film QCM sensor," *Materials Letters*, vol. 174, pp. 28–31, 2016.

[20] A. Kathalingam, D. Vikraman, H.-S. Kim, and H. J. Park, "Facile fabrication of n-ZnO nanorods/p-Cu_2O heterojunction and its photodiode property," *Optical Materials*, vol. 66, pp. 122–130, 2017.

[21] T. Shinagawa, M. Chigane, J. Tani, and M. Izaki, "Effect of oxide intermediate layers on pyramidally textured Cu2O/ZnO solar cells prepared by electrodeposition," *ECS Meeting Abstracts*, vol. 2, no. 21, p. 1636, 2016.

[22] A. Samad, M. Furukawa, H. Katsumata, T. Suzuki, and S. Kaneco, "Photocatalytic oxidation and simultaneous removal of arsenite with CuO/ZnO photocatalyst," *Journal of Photochemistry and Photobiology A: Chemistry*, vol. 325, pp. 97–103, 2016.

[23] G. Boschloo and A. Hagfeldt, "Spectroelectrochemistry of nanostructured NiO," *The Journal of Physical Chemistry B*, vol. 105, no. 15, pp. 3039–3044, 2001.

[24] S. Kerli and Ü. Alver, "Preparation and characterisation of ZnO/NiO nanocomposite particles for solar cell applications," *Journal of Nanotechnology*, vol. 2016, Article ID 4028062, 5 pages, 2016.

[25] S. A. Rawool, M. R. Pai, A. M. Banerjee et al., "*pn* Heterojunctions in NiO:TiO_2 composites with type-II band alignment assisting sunlight driven photocatalytic H_2 generation," *Applied Catalysis B: Environmental*, vol. 221, pp. 443–458, 2018.

[26] A. Chrissanthopoulos, S. Baskoutas, N. Bouropoulos, V. Dracopoulos, P. Poulopoulos, and S. N. Yannopoulos, "Synthesis and characterization of ZnO/NiO p–n heterojunctions: ZnO nanorods grown on NiO thin film by thermal evaporation," *Photonics and Nanostructures - Fundamentals and Applications*, vol. 9, no. 2, pp. 132–139, 2011.

[27] M. Boroujerdnia and A. Obeydavi, "Synthesis and characterization of NiO/$MgAl_2O_4$ nanocrystals with high surface area by modified sol-gel method," *Microporous and Mesoporous Materials*, vol. 228, pp. 289–296, 2016.

[28] M. El-Nahass, M. Emam-Ismail, and M. El-Hagary, "Structural, optical and dispersion energy parameters of nickel oxide nanocrystalline thin films prepared by electron beam deposition technique," *Journal of Alloys and Compounds*, vol. 646, pp. 937–945, 2015.

[29] H. Wang, Y. Wang, and X. Wang, "Pulsed laser deposition of the porous nickel oxide thin film at room temperature for high-rate pseudocapacitive energy storage," *Electrochemistry Communications*, vol. 18, pp. 92–95, 2012.

[30] C. E. Morosanu and G. Siddall, *Thin Films by Chemical Vapour Deposition*, Elsevier, Amsterdam, 2016.

[31] I. Sta, M. Jlassi, M. Hajji, and H. Ezzaouia, "Structural, optical and electrical properties of undoped and Li-doped NiO thin films prepared by sol–gel spin coating method," *Thin Solid Films*, vol. 555, pp. 131–137, 2014.

[32] F. Garcés, N. Budini, J. Schmidt, and R. Arce, "Highly doped ZnO films deposited by spray-pyrolysis. Design parameters for optoelectronic applications," *Thin Solid Films*, vol. 605, pp. 149–156, 2016.

[33] M. G. Faraj, "Effect of aqueous solution molarity on the structural and electrical properties of spray pyrolysed lead sulfide (PbS) thin films," *International Letters of Chemistry, Physics and Astronomy*, vol. 57, pp. 122–125, 2015.

[34] A. Enigochitra, P. Perumal, C. Sanjeeviraja, D. Deivamani, and M. Boomashri, "Influence of substrate temperature on structural and optical properties of ZnO thin films prepared by cost-effective chemical spray pyrolysis technique," *Superlattices and Microstructures*, vol. 90, pp. 313–320, 2016.

[35] A. E. Adeoye, E. Ajenifuja, B. A. Taleatu, and A. Fasasi, "Rutherford backscattering spectrometry analysis and structural properties $Zn_xPb_{1-x}S$ of thin films deposited by chemical spray pyrolysis," *Journal of Materials*, vol. 2015, Article ID 215210, 8 pages, 2015.

[36] D. Chen, X. Zou, H. Yang et al., "Effect of annealing process on $CH_3NH_3PbI_3$-$_XCl_X$ film morphology of planar heterojunction perovskite solar cells with optimal compact TiO_2 layer," *International Journal of Photoenergy*, vol. 2017, Article ID 7190801, 9 pages, 2017.

[37] O. K. Ukoba, A. C. Eloka-Eboka, and F. L. Inambao, "Review of nanostructured NiO thin film deposition using the spray pyrolysis technique," *Renewable and Sustainable Energy Reviews*, vol. 82, pp. 2900–2915, 2018.

[38] E. Adeoye Abiodun and A. Salau, "Effect of annealing on the structural and photovoltaic properties of cadmium sulphide: copper sulphide (Cds:Cuxs) heterojunction," *International Journal of Scientific and Research Publications*, vol. 5, no. 8, pp. 1–5, 2015.

[39] S. Sriram and A. Thayumanavan, "Structural, optical and electrical properties of NiO thin films prepared by low cost spray pyrolysis technique," *International Journal of Material Science Engineering*, vol. 1, pp. 118–121, 2013.

[40] M. Gabal, "Non-isothermal decomposition of NiC_2O_4–FeC_2O_4 mixture aiming at the production of $NiFe_2O_4$," *Journal of Physics and Chemistry of Solids*, vol. 64, no. 8, pp. 1375–1385, 2003.

[41] N. A. Bakr, S. A. Salman, and A. M. Shano, "Effect of co doping on structural and optical properties of NiO thin films prepared by chemical spray pyrolysis method," *International Letters of Chemistry, Physics and Astronomy*, vol. 41, pp. 15–30, 2015.

[42] V. Georgieva and A. Tanusevski, "Characterization of p-Cu_2O/n-ZnO heterojunction solar cells," *AIP Conference Proceedings*, vol. 1203, pp. 1074–1078, 2010.

[43] S. Noda, H. Shima, and H. Akinaga, "Cu_2O/ZnO heterojunction solar cells fabricated by magnetron-sputter deposition method films using sintered ceramics targets," *Journal of Physics: Conference Series*, vol. 433, pp. 1–10, 2013.

4

Silicon Powder-Based Wafers for Low-Cost Photovoltaics: Laser Treatments and Nanowire Etching

G. Jia,[1] J. Plentz,[1] A. Gawlik,[1] A. S. Azar,[2] G. Stokkan,[3] M. Syvertsen,[3] P. A. Carvalho,[2] J. Dellith,[1] A. Dellith,[1] G. Andrä,[1] and A. Ulyashin[2]

[1]*Leibniz Institute of Photonic Technology (IPHT), Albert-Einstein-Str. 9, 07745 Jena, Germany*
[2]*SINTEF, Oslo, Norway*
[3]*SINTEF, Trondheim, Norway*

Correspondence should be addressed to G. Jia; guobin.jia@leibniz-ipht.de

Academic Editor: Giulia Grancini

In this study, laser-treated polycrystalline Si (pc-Si) wafers, fabricated by wire sawing of hot-pressed ingots sintered from Si powder, have been investigated. As-cut wafers and those with high-quality thin Si layers deposited on top of them by e-beam have been subjected to laser irradiation to clarify typical trends of structural modifications caused by laser treatments. Moreover, possibility to use laser-treated Si powder-based substrates for fabrication of advanced Si structures has been analysed. It is established that (i) Si powder-based wafers with thicknesses ~180 μm can be fully (from the front to back side) or partly (subsurface region) remelted by a diode laser and grain sizes in laser-treated regions can be increased; (ii) a high-quality top layer can be fabricated by crystallization of an additional a-Si layer deposited by e-beam evaporation on top of the pc-Si; and (iii) silicon nanowires can be formed by metal-assisted wet chemical etching (MAWCE) of polished Si powder-based wafers and as-cut wafers irradiated with medium laser power, while a surface texturing on the as-cut pc-Si wafers occur, and no nanowires can form in the region subject to a liquid phase crystallization (LPC) caused by high-power laser treatments.

1. Introduction

Crystalline silicon (c-Si) wafer-based solar cells contribute to over 90% of commercial photovoltaic (PV) devices at present [1] and will be dominant in the foreseeable future due to the mature technology. Most of the silicon-based solar cells are made from single crystalline or multicrystalline wafers produced by Cz-growth and block casting from costly purified solar grade silicon. The ingots must be subsequently cut into wafers by wire sawing, and the damaged surface layer has to be removed by polishing.

As a result, around 50% of purified solar grade silicon will be lost due to the wafer sawing and subsequent processing [2], in the form of Si kerf (powders) in the sawing waste or in the slurry during the polishing of the wafers, that corresponds to approximately 200,000 tons out of the 400,000 tons of solar grade silicon consumed for solar cell production in 2016 [1, 3].

Therefore, several dicing-free approaches to overcome problems related to Si kerf losses have been proposed [4]. Among them, fabrication of silicon wafers by sintering of silicon powders using hot pressing [5] followed by zone melting recrystallization [6, 7], which is similar to earlier proposed direct growth of silicon sheets (SSP) from compacted silicon powders with a halogen heat source [8], can be considered as one of the promising approaches. However, the disadvantage of the above-mentioned Si powder-based concept is that sintered wafers are quite thick (~300-500 μm), and as a result, consumption of Si is still quite high and comparable with that of conventional Si wafers fabricated by wire sawing of Si ingots. An alternative approach, which is based on ingots fabricated from sintering of Si powders, followed by slicing of wafers using a wire sawing process has been proposed recently [9] and provides a possibility to process Si powder-based wafers with thicknesses below 200 μm. However, production of Si kerf, similar to that of a conventional Si wafer-based approach, occurs in this case as well. Taking into account that purification of Si kerf allows to use purified material for a conventional block casting process [10, 11], it can be expected that Si powders processed from the purified Si kerf by milling can be used for the cost-effective fabrication of sintered Si powder-based wafers. Such approach can be considered as a cost-effective, circular economical solution for an advanced low-cost Si-based PV.

Moreover, another source of silicon feedstock, which can be used to produce high-quality Si powders by milling, comes from microelectronics, which discharge several thousand tons per annum (tpa) of silicon IC waste [12]. In addition, considering that already installed solar modules will reach their end-of-life stage in approximately 30 years after the installation, it is expected that silicon waste from the recycled solar modules is a continuously growing source of solar grade Si feedstock and will alone provide 88,000 and 207,000 tpa by 2040 and 2050, respectively [2]. Important to note that Si extracted from the end-of-life modules or broken solar cells has a doping level adjusted for the processing of good-quality solar cells. Although some efforts to extract such Si feedstock are still required, this process is feasible and can be realised in a cost-effective way [13].

It is therefore of huge economic and environmental advantages, if the above-mentioned Si feedstock from different sources of Si-based wastes can be used for the needs of an advanced Si-based PV after appropriate purification of such feedstock and its conversion into Si powders.

It can be concluded that Si powder-based concepts targeting on utilisation of solar grade Si powders, fabricated from different sources of Si feedstock, have strong background and reasons to be developed further, taking into account the above-mentioned arguments.

Si powder-based wafers can be used for the needs of an advanced Si-based PV by three main routes: (i) to produce Si powder-sintered wafer-based solar cells if the quality is sufficient; (ii) to serve as supporting substrates for solar cells either bonded on top of them [14, 15] or fabricated directly on such polycrystalline Si (pc-Si) wafers; and (iii) to fabricate Si-based structures in the frame of an advanced concept which allow to use low-grade (or defective) silicon wafers to process radial junction (core-shell) nanowire solar cells, in which the generated minority carriers have to be drifted to a very short path (tens of nm in the range of the radii of the nanowires) to be collected by the emitter. Moreover, the core is mostly depleted so that the defects are empty (no carriers are trapped at the defects), so that almost no Shockley-Read-Hall recombination takes place in the core even if it is very defective. As a result, very low-grade silicon materials can be used to produce solar cells [16–18].

It should be noted that apart from the compacting of Si powders, the common issue in the above-mentioned Si powder-based approaches (SSP and hot pressing) is a need for the recrystallization of such substrates. In the case of Si powder-based wafers processed by wire sawing of sintered ingots, such recrystallization is required as well, since the size of Si grains is defined by the size of initial Si powders (1-100 μm), and therefore, grains should be enlarged to get better quality of the final polycrystalline (pc) material.

In this study, following the above-mentioned three routes, we have demonstrated some progress towards fabrication of relevant advanced Si powder-based structures using laser treatments to modify their properties.

In particular, pc-Si wafers, fabricated by wire sawing of the hot-pressed ingots from Si powders, as well as Si powder-based structures processed by e-beam deposition of high-quality Si layers on top of such wafers, were subjected to laser irradiation to clarify typical trends of structural modifications of such structures. Moreover, possibility to use laser-treated Si powder-based substrates for fabrication of advanced Si structures has been analysed.

All the pc-Si powder-based wafers were produced from solar grade p-type silicon powders with a doping level of ~10^{16} cm^{-3} in this study, which have been prepared by ball milling of wafers and sieving techniques.

2. Experimental Results and Discussion

2.1. Laser Recrystallization to Improve the Material Quality. The wire-sawn pc-Si wafer produced from silicon powders was fixed between two glass substrates by Ag paste, so that the to-be-treated area is free-standing bridging between the glass substrates as shown in Figure 1(a), and this setup is advantageous to minimize heat transfer onto the glass substrates. In this case, the glass keeps solid even if the temperature at the laser-treated area is above the melting point of silicon (1414°C). The thickness of the wafer is $170 \pm 10\,\mu$m. In this work, we use a diode laser working at a wavelength of 808 nm, which is commonly used to crystallize Si thin films on glass [19]. It is possible to melt the whole piece of the pc-Si wafer fully from the front to the back side, and a substrate temperature of 600°C was used to reduce the thermal stress during the solidification. Figure 1(a) demonstrates pc-Si wafers treated with a laser power density of 15.0 (upper case) and 13.5 kW cm^{-2} W (lower case) at a scan speed of 10 mm s^{-1}. A complete melt of the whole pc-wafer can be achieved at 15.0 kW cm^{-2}, and at some positions, the liquid Si even flows down on the sample stage as indicated by the arrow of the upper wafer piece. By lowering the laser power density to 13.5 kW cm^{-2}, a melt of the pc-Si wafer occurs while no downflow of the liquid Si takes place. The surface of pc-Si wafer pieces exhibits mirror-like properties at the irradiated regions as shown in Figure 1(a) while the nonirradiated regions appear greyish. A smoothening of the surface can also be seen in a tilted (70°) SEM image in Figure 1(b), which was taken at a cross section of a sample irradiated at a laser power density of 13.5 kW cm^{-2}. However, some microdefects such as grown-up protrusions and microvoids on the surface and in the volume have been detected by SEM measurements in Figures 1(c) and 1(d). Although some microvoids are found in the volume, they are not the same as observed on the surface. The recrystallization shows clear characteristic of liquid phase crystallization (LPC) [20] at the power density level of 13.5-15.0 kW cm^{-2}, which have been monitored by the optical reflection using a control laser measured at the scan front, where a highly reflecting surface is the characteristic for the materials above the melting temperature [20].

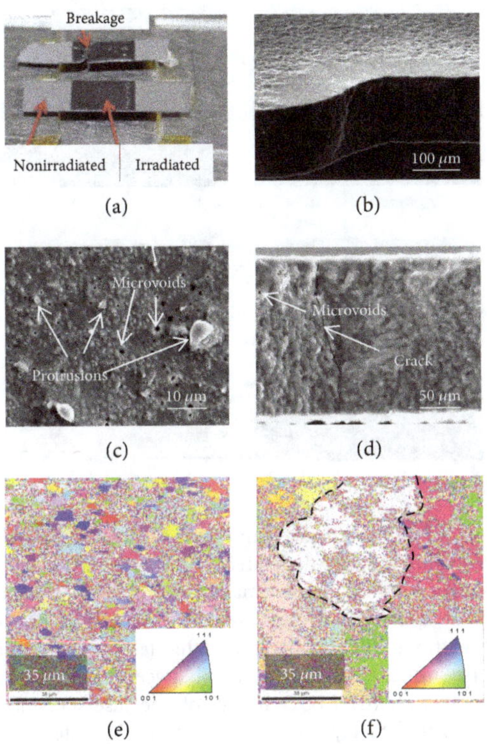

FIGURE 1: (a) Photograph of the samples irradiated by the laser at 15.0 (upper case) and 13.5 kW cm^{-2} (lower case) at a scan speed of 10 mm s^{-1}; (b) SEM image (70° tiled) at the cross section with a rough surface of the nonirradiated region and the smooth laser-recrystallized region; (c) surface morphology of the laser-recrystallized region showing examples of protrusions and microvoids on the surface; (d) a crack induced by the laser recrystallization process and microvoid in the volume; (e) top-view EBSD image of the as-cut pc-Si sample; (f) top-view EBSD image of the laser-recrystallized (at a power density of 13.5 kW cm^{-2}) pc-Si sample.

Crack formation is an issue for the LPC process due to high stress originating from the high temperature gradient during the solidification [21]; it starts within the top layer and spreads to the volume of the pc-wafer as indicated by the arrow in Figure 1(d). Such crack formation sometimes leads to breakage of the wafers; therefore, a substrate temperature of 600°C was used to reduce the thermal stress.

A significant difference between the nonirradiated and laser-recrystallized (with a laser power density of 13.5 kW cm^{-2} at a scan speed of 10 mm s^{-1}) regions can be clearly detected by the top-view electron backscatter diffraction (EBSD) measurements. In particular, the nonirradiated pc-Si wafer shows few grains with a size of about 10 μm embedded in a large amount of small grains as shown in Figure 1(e), whereas the recrystallized region shows substantial growth of large grains up to 50 μm (Figure 1(f)). For the visualization, in Figure 1(f), an enlarged grain is circled by the dashed line (the dashed line is just a guide and not an accurate grain boundary). It can be seen that enlarged grains are in contact with each other and densely packed. The image noise within single enlarged grains may probably originate from the microdefects (protrusions and microvoids) on the surface.

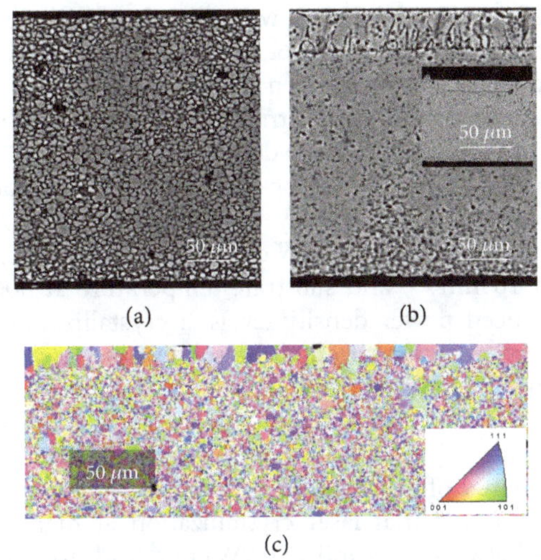

FIGURE 2: (a) SEM cross section image of the as-cut pc-Si sample subject to a polishing; (b) SEM cross section image of the laser-crystallized pc-Si sample (at a power density of 13.5 kW cm^{-2}) and the inset showing another region; (c) cross-sectional EBSD mapping of the laser-recrystallized pc-wafer (at a power density of 13.5 kW cm^{-2}).

A cross-sectional morphological study of the as-cut pc-Si sample as well as of another sample subjected to a laser scan at 13.5 kW cm^{-2} has been performed. Both samples were embedded in an epoxy resin and subsequently polished and ion-milled (in order to prevent charging of the epoxy) prior to the EBSD measurements. The cross-sectional SEM image of the as-cut pc-Si sample in Figure 2(a) shows similar crystal size distribution as indicated by the top-view EBSD image in Figure 1(e). However, the cross section of the laser-recrystallized pc-Si sample (in Figure 1(f)) clearly shows a top layer with larger grains with a thickness of 25-30 μm, which extends up to 60 μm into the bulk in some positions (see the inset in Figure 2(b)). EBSD measurement has been performed at the cross section of the laser-recrystallized sample as shown in Figure 2(b), and large grains with sizes in the range of 20-30 μm can be clearly seen in the subsurface area of the laser-treated side (Figure 2(c)). This result is in a good agreement with the results of the cross-sectional SEM (Figure 2(b)) and the top-view EBSD (Figure 1(f)) measurements.

It is demonstrated that the wire-sawed pc-Si wafers produced from silicon powder can be recrystallized by a high-power diode laser scan, the grain size in the top layer can be significantly enlarged through the LPC mechanism, and the conductivity of the laser-treated pc-Si-based structures can be improved, which can be attributed to the growth of a larger grain as well as to an improved crystallinity.

2.2. Laser Crystallization of an Additional a-Si Layer Deposited on Top of the pc-Si Wafer. The second route, mentioned above, supposes to use low-cost pc-Si wafers as supporting substrates as well as a conductive back contact. In the frame of such approach, a high-quality Si layer (solar cell grade) has to be fabricated on top of the supporting pc-Si wafers. In this study, a high purity 10 μm a-Si layer was deposited on top of the pc-Si wafers by the e-beam evaporation. An *n*-type doping of ~6e16 cm^{-3} of the a-Si layer was achieved by heating an effusion cell with a phosphorus source during the deposition, as determined by our previous systematic investigation at the same deposition condition.

In order to prevent the pc-Si wafers from breakage, the crystallization of the a-Si layer was performed at reduced power of 11.0 up to 12.0 kW cm^{-2} while keeping the scan speed of 10 mm s^{-1} and substrate temperature at 600°C. At these reduced power density levels, a crystallization of the top a-Si layer occurs through the solid phase crystallization (SPC) instead of the LPC mechanism, as confirmed by the optical reflection measurement through a control laser at the scan front [20], because the SPC does not show substantial change of the surface reflection during the crystallization. It is noted that laser crystallization at higher power densities between 13.5 and 15.0 kW cm^{-2} leads to immediate crack formation and breakage of the wafers, and this is much frequently happening than those laser recrystallization processes performed on the as-cut pc-Si wafers. This is probably because of the stress that was already induced by the additional a-Si layer.

The crystallized top layer got slightly smoothened, though, not like the one subject to laser recrystallization of the whole wafer. Figure 3(a) shows a SEM image (sample is 70° tilted) at the cross section. The corresponding higher magnification SEM image in Figure 3(b) shows the morphology of the surface after the laser-induced SPC process. No substantial protrusions have been detected on top of the crystallized a-Si layer. Figure 3(c) shows the boundary (step) between two regions—the region with deposited 10 μm a-Si (left side) and that without a-Si (right side) after laser treatment.

The EBSD image in Figure 3(d) shows that most of the grains have size around 10 μm, in the range of the a-Si layer thickness, which is typical for the SPC process.

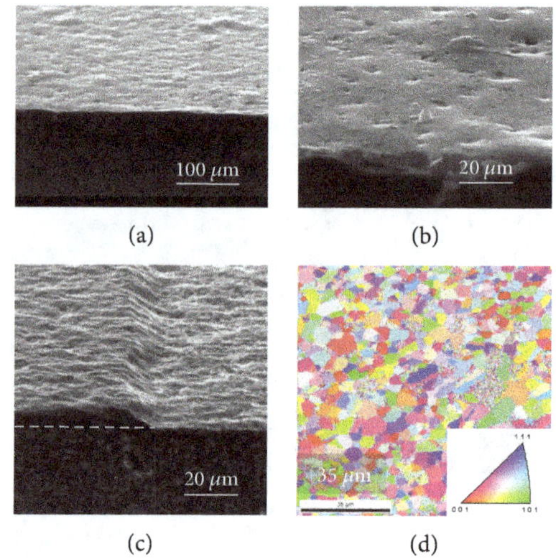

Figure 3: (a) SEM image (sample is 70° tilted) of the laser-crystallized thin e-beam-deposited Si layer at the cross section; (b) the corresponding higher magnification SEM image; (c) the boundary at the region deposited with 10 μm a-Si (left side) and that without a-Si (right side) after laser crystallization, the dashed line indicated the boundary between the additional layer and the pc-wafer; (d) EBSD image of the laser-crystallized thin e-beam-deposited Si layer on top of the pc-Si wafer.

2.3. Nanowire Etching for Core-Shell Solar Cells. Nanowire core-shell solar cells, the so-called radial junction, have the advantages that allow to produce high-efficiency solar cells using low-grade silicon material [16, 22]. This is due to the fact that light absorption [23] will be significantly enhanced by light trapping of the densely packed nanowire array [24]. Additionally, the carrier recombination in the bulk will be greatly reduced because of the short collection path of the generated carriers [16, 22, 25], which is only in the range of the radii of the nanowires. Moreover, the core is mostly depleted so that the defects are empty (no carriers are trapped in the defects); therefore, almost no Shockley-Read-Hall recombination takes place in the core even if it is very defective. As a result, very low-grade silicon materials can be used to produce solar cells.

The core-shell nanowire solar cell concept fits very well with the low-cost pc-Si wafers produced from silicon powder, so the nanowire etching on these wafers is a prerequisite for this advanced concept.

In this study, one-step and two-step metal-assisted wet chemical etching (MAWCE) processes have been performed on the pc-Si wafers. All the wafers have been cleaned by acetone and subsequent isopropanol to remove surface contamination prior to the etching. All the etching has been done at room temperature. The cleaned wafers were etched in a $AgNO_3$ (0.02 M) : HF (5 M) solution (one-step etching) for the etching of nanowires corresponding to an etching rate of about 30 nm min^{-1}. The two-step etching was done by immersion of the wafers in the $AgNO_3$ (0.02 M) : HF (5 M) solution for 30 s to deposit Ag nanoparticles on the surface as the first step. Then the second step was performed in a HF (5 M) : H_2O_2 (30%) solution [26, 27] with a volume ratio of 20 : 1. Here, the etching rate is much faster at about 300 nm min^{-1} and one order of magnitude higher than the one-step etching. In both cases, the Ag nanoparticles were etched away afterwards by concentrated HNO_3 (65%) for 10 min followed by deionized water rinsing.

The etching of nanowires on the as-cut pc-wafers using the above-described two methods has failed independently of the etching time. The etching produces rather a surface texturing as shown in the SEM image in Figure 4(a) and the corresponding higher magnification SEM image in Figure 4(b). Some nanofeatures can be seen between the textures as shown in the insets of Figure 4(b) (regions marked with red rectangles).

On polished pc-Si wafers, a very homogeneous nanowire array can be etched. The vertically densely packed nanowire array can be obtained despite the different crystal orientation. This result is comparable with the etching on high-quality pc-Si wafers and thin films. Figure 4(c) and the corresponding higher magnification SEM images show an example of nanowires etched on the polished pc-Si wafers by the two-step process. The duration of the second step was 5 min, and the length of the nanowires is about 1.5 μm.

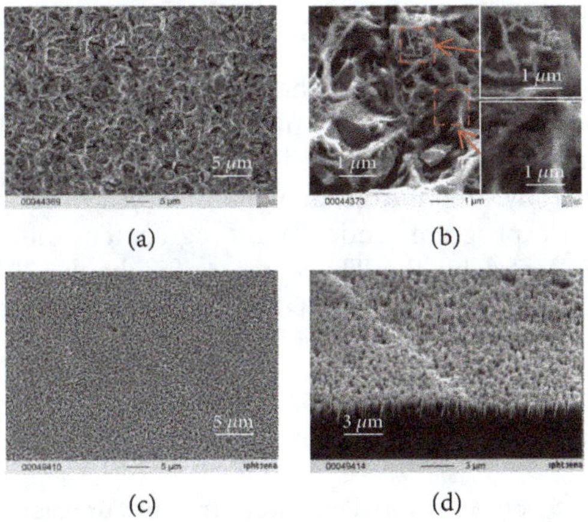

FIGURE 4: (a) Surface texturing of the as-cut pc-Si wafers subject to a MAWCE; (b) higher magnification SEM image of the surface texture showing nanofeatures (examples in the insets are shown in the regions marked with red rectangles); (c) densely packed nanowires formed after the two-step MAWCE on polished wafers; (d) vertically orientated silicon nanowires etched by the two-step MAWCE process.

Both nanowire etching processes have been applied on the pc-Si wafers after the laser recrystallization process. Here, it shows different behaviors in regions treated with different laser powers. pc-Si-based structures treated with different laser powers at 12.0 and 13.5 kW cm^{-2} have been tested for the nanowire fabrication process. Moreover, a reference pc-Si sample is placed on the heater along with the samples to be laser treated (the pc-Si samples were heated up to 600°C in about 2 hours, but no laser treatment has been performed on the reference). The two-step MAWCE was performed on the pc-Si samples, in which the duration of the first etching keeps unchanged at 30 s, and the second etching took 10 min.

It has been found that nanowires could not be etched as before for the reference sample without laser treatment. Instead, a surface texturing occurs as shown in the SEM image in Figure 5(a), (A) and in the corresponding higher magnification SEM image in Figure 5(a), (B). However, the textures are quite different with those generated on the as-cut pc-Si wafers shown in Figures 4(a) and 4(b). No nanofeatures and only grains with sharp edges can be seen in this case. This change in the structure compared to the as-cut pc-Si wafer may closely relate to the thermal process during the laser crystallization, indicating that some solid-state transformation has happened at the process temperature even without laser scanning.

It was established that Si nanowires can be etched on the region treated with medium laser power of 12.0 kW cm^{-2}, and this region appears greyish similar to the as-cut wafer, though, with a laser track, which underwent a SPC crystallization. The SEM image in Figure 5(b), (A) and the corresponding higher magnification SEM image in Figure 5(b), (B) show deep and dense nanowires prepared in this region by a two-step MAWCE, where the first step is kept 30 s and the second step for 10 min.

On the region treated with a higher laser power of 13.5 kW cm^{-2}, the surface is mirror smooth after the LPC process initiated by the laser scan, and no nanowire can be etched by both methods. In this case, the surface looks visually untouched after the etching. The typical surface structure of the laser-recrystallized regions after the two-step etching (duration of second step is 5 min) is shown in Figure 5(c), (A) and the corresponding higher magnification SEM images in Figure 5(c), (B). It can be clearly seen that some small grains are embedded in large grains, and the "grain boundaries" were etched away completely, leaving behind hollow "boundaries" between neighboring grains. This behavior is quite different compared to previous results obtained for a multicrystalline Si (mc-Si) thin film deposited on glass, EFG-grown thin mc-Si wafers and low-cost mc-Si wafers [28], where the grain boundaries remain after the nanowire etching. It can be assumed that silicon oxide may exist on the surface of Si powders which have been used for the hot pressing. Moreover, it is supposed that an oxide layer is sandwiched between the neighboring grains after the hot pressing sintering as well as after the laser recrystallization processes. In this case, such oxide layer can be etched away by the HF during the nanowire etching, which results in the formation of a hollow "grain boundary" structure. To verify the above-discussed assumptions, more detailed analysis of interfaces of Si powder-based structures, as well as an optimization of the hot pressing process, is required. The observed small grains embedded in the larger

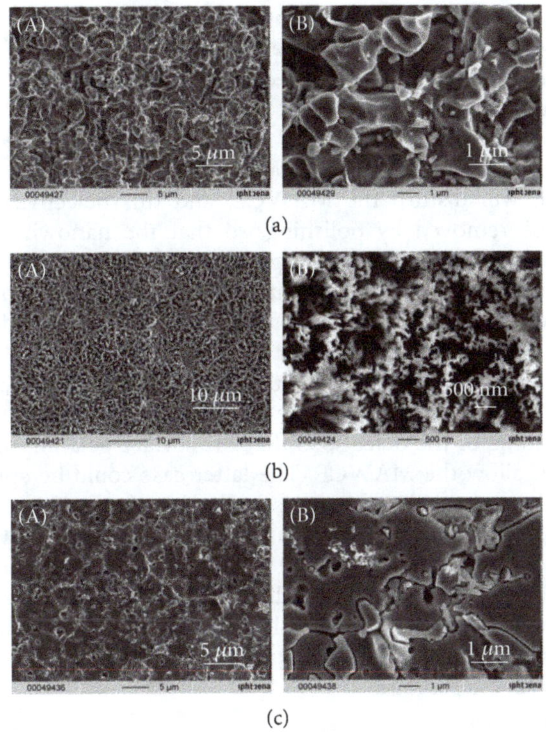

FIGURE 5: (a), (A) SEM image showing the surface textures of the nonirradiated pc-Si sample (only heat up to 600°C for about 2 h without laser treatment) subject to a MAWCE process and (a), (B) the corresponding higher resolution SEM image; (b), (A) nanowires formed by MAWCE in the region treated with medium laser power density of 12.0 kW cm^{-2} and (b), (B) the corresponding higher resolution SEM image; (c), (A) no nanowires being detected in the region treated with a laser power density of 13.5 kW cm^{-2} and (c), (B) the corresponding higher resolution SEM image.

grains could be a result of the oxide shell, which does not allow for Si powders to be merged.

It is still unclear if what causes the difference between the as-cut and the polished pc-wafers for the results of nanowire etching, supposed that there is a defective surface layer on the pc-Si wafers after the wire sawing, which will disturb the MAWCE, and this defective layer will be removed by polishing, so that the nanowires can be etched on the polished pc-Si wafer. This explanation is consistent with the results observed on the samples treated with laser, where a defective surface layer is induced after a LPC process at laser power at 13.5 kW cm^{-2} due to the rapid quenching at the surface region, while the SPC process treated at laser power at 12.0 kW cm^{-2} will heal some of the defects, so that the surface gets improved, which allow the MAWCE. The latter case could be applied for the annealed reference sample, in which some solid phase transformation happens at the surface region, which will heal the defects at the surface region to a certain extent but not enough to allow triggering the MAWCE. However, this explanation should be tested with further experiments including structure analysis by transmission electron microscope.

3. Conclusion

The following trends of the structural modifications of pc-Si-based structures caused by laser treatments have been established:

(i) Wire-sawed pc-Si wafers produced from silicon powder can be recrystallized by a high-power diode laser scan, and the grain size in the top layer can be enlarged through the LPC mechanism

(ii) The a-Si layer deposited by the e-beam evaporation on top of the pc-Si wafers can be crystallized by the diode laser scan. However, a breakage-free crystallization can only be achieved by a SPC-based process

(iii) No nanowires can be etched on the as-cut pc-Si wafers. Instead, a surface texturing of such pc-Si occurs

(iv) Nanowires can be etched on polished pc-wafers as well as in the regions treated with the medium level laser power

(v) Crack formation of the pc-Si wafers occurs during the laser scan and has to be investigated in order to obtain large-area crack-free wafers, and the laser parameters such as power, scan speed, focus, and the substrate temperatures require careful optimizations

Conflicts of Interest

The authors declare that there are no conflicts of interest regarding the publication of this paper.

Acknowledgments

The authors acknowledge the experimental assistance by Dr. Thomas Schmidt, Dr. Ingmar Höger at IPHT, and Mr. Wilhelm Dall at SINTEF. This work was financially supported by the European Commission within the 7th framework program under grant agreement no. 246331 (NanoPV) and by the Thüringer Aufbaubank (2015 FGR 0078) and the European Social Fund (ESF) within the project "Bi-PV".

References

[1] "Photovoltaics report-Fraunhofer-Institut Für Solare Energiesysteme ISE," December 2017, https://www.ise.fraunhofer.de/content/dam/ise/.../Photovoltaics-Report.pdf.

[2] Y. K. Yi, H. S. Kim, T. Tran, S. K. Hong, and M. J. Kim, "Recovering valuable metals from recycled photovoltaic modules," *Journal of the Air & Waste Management Association*, vol. 64, no. 7, pp. 797–807, 2014.

[3] P. Sonntag, N. Preissler, M. Bokalič et al., "Silicon solar cells on glass with power conversion efficiency above 13% at thickness below 15 micrometer," *Scientific Reports*, vol. 7, no. 1, article 873, 2017.

[4] M. J. McCann, K. R. Catchpole, K. J. Weber, and A. W. Blakers, "A review of thin-film crystalline silicon for solar cell applications. Part 1: native substrates," *Solar Energy Materials and Solar Cells*, vol. 68, no. 2, pp. 135–171, 2001.

[5] S. Duboisa, E. Bere, P. Grosbras, and A. Straboni, "Correlation between compacity distributions in compressed Si powders and in Si sintered pellets," *The European Physical Journal B - Condensed Matter and Complex Systems*, vol. 26, no. 1, pp. 35–39, 2002.

[6] P. Bellanger, A. Sow, M. Grau et al., "New method of fabricating silicon wafer for the photovoltaic application based on sintering and recrystallization steps," *Journal of Crystal Growth*, vol. 359, pp. 92–98, 2012.

[7] A. Sow, P. Bellanger, F. Dupont, J. Serra, N. Alonso-Vante, and A. Straboni, "Characterization of recrystallized sintered silicon substrates for photovoltaic's solar cells," *Energy Procedia*, vol. 27, pp. 13–20, 2012.

[8] A. Eyer, N. Schillinger, I. Reis, and A. Rauber, "Silicon sheets for solar cells grown from silicon powder by the SSP technique," *Journal of Crystal Growth*, vol. 104, no. 1, pp. 119–125, 1990.

[9] R. Buchwald, S. Würzner, H. J. Möller et al., "Microstructural characterization of Si wafers processed by multi-wire sawing of hot pressed silicon powder based ingots," *Physica Status Solidi A: Applications and Materials Science*, vol. 211, p. 25, 2015.

[10] T. Y. Wang, Y. C. Lin, C. Y. Tai, R. Sivakumar, D. K. Rai, and C. W. Lan, "A novel approach for recycling of kerf loss silicon from cutting slurry waste for solar cell applications," *Journal of Crystal Growth*, vol. 310, no. 15, pp. 3403–3406, 2008.

[11] M. Syvertsen, T. Halvorsen, K. Mork, A. Nordmark, T. Kaden, and A. Ulyashin, "Remelting and purification of Si-kerf for PV-wafers," in *33rd European Photovoltaic Solar Energy Conference and Exhibition*, pp. 1882–1885, Amsterdam, the Netherlands, 2017.

[12] Z. Liang, H. Shen, H. Xu, Q. Li, and Z. Li, "Study on the recycle of solar grade silicon from waste IC wafers," in *Proceedings of*

ISES World Congress 2007 (Vol. I – Vol. V), D. Y. Goswami and Y. Zhao, Eds., pp. 1189–1193, Springer, Berlin, Heidelberg, 2007.

[13] M. Syvertsen, B. Ryningen, M. Di Sabatino et al., "Recycling of broken Si based structures and solar cells," in *32nd European Photovoltaic Solar Energy Conference and Exhibition*, pp. 1768–1774, Munich, Germany, June 2016.

[14] H. Sivaramakrishnan Radhakrishnan, K. Van Nieuwenhuysen, J. Govaerts et al., "Solar cells and mini-modules based on 40 Mμm-thick epitaxial Si foils: towards conductive bonding onto low-cost Si powder sintered conductive carriers from the recycling of silicon waste," in *32nd European Photovoltaic Solar Energy Conference and Exhibition*, pp. 1268–1271, Munich, Germany, 2016.

[15] P. Bellanger, A. Slaoui, S. Roques et al., "Silicon foil solar cells on low cost supports," *Journal of Renewable and Sustainable Energy*, vol. 10, no. 2, article 023502, 2018.

[16] B. M. Kayes, H. A. Atwater, and N. S. Lewis, "Comparison of the device physics principles of planar and radial p-n junction nanorod solar cells," *Journal of Applied Physics*, vol. 97, no. 11, article 114302, 2005.

[17] B. Tian, X. Zheng, T. J. Kempa et al., "Coaxial silicon nanowires as solar cells and nanoelectronic power sources," *Nature*, vol. 449, no. 7164, pp. 885–889, 2007.

[18] G. Jia, M. Steglich, I. Sill, and F. Falk, "Core–shell heterojunction solar cells on silicon nanowire arrays," *Solar Energy Materials and Solar Cells*, vol. 96, pp. 226–230, 2012.

[19] J. Plentz, T. Schmidt, A. Gawlik et al., "Applicability of an economic diode laser emitting at 980 nm for preparation of polycrystalline silicon thin film solar cells on glass," *Physica Status Solidi A*, vol. 214, no. 10, article 1600882, 2017.

[20] T. Schmidt, A. Gawlik, H. Schneidewind, A. Ihring, G. Andrä, and F. Falk, "Experimental setup for investigating silicon solid phase crystallization at high temperatures," *Optics Express*, vol. 21, no. 14, pp. 16296–16304, 2013.

[21] T. Pliewischkies, T. Schmidt, I. Höger et al., "Thermal stresses and cracking behavior during laser crystallization of silicon on glass for thin film solar cells," *Physica status solidi (a)*, vol. 212, no. 2, pp. 317–322, 2013.

[22] G. Jia, G. Andrä, A. Gawlik et al., "Nanotechnology enhanced solar cells prepared on laser-crystallized polycrystalline thin films (<10 μm)," *Solar Energy Materials and Solar Cells*, vol. 126, pp. 62–67, 2014.

[23] G. Jia, A. Gawlik, J. Bergmann et al., "Silicon nanowire solar cells with radial p-n heterojunction on crystalline silicon thin films: light trapping properties," *IEEE Journal of Photovoltaics*, vol. 4, no. 1, pp. 28–32, 2014.

[24] M. Vetter, G. Jia, A. Sanei, A. Gawlik, J. Plentz, and G. Andrä, "Evaluation of light trapping structures for liquid-phase crystallized silicon on glass (LPCSG)," *Physica Status Solidi A*, vol. 214, no. 8, article 1600859, 2017.

[25] G. Jia, J. Plentz, I. Höger, J. Dellith, A. Dellith, and F. Falk, "Core–shell diodes for particle detectors," *Journal of Physics D: Applied Physics*, vol. 49, no. 6, article 065106, 2016.

[26] X. Li and P. W. Bohn, "Metal-assisted chemical etching in HF/H2O2 produces porous silicon," *Applied Physics Letters*, vol. 77, no. 16, pp. 2572–2574, 2000.

[27] G. Jia, J. Westphalen, J. Drexler et al., "Ordered silicon nanowire arrays prepared by an improved nanospheres self-assembly in combination with Ag-assisted wet chemical etching," *Photonics and Nanostructures - Fundamentals and Applications*, vol. 19, pp. 64–70, 2016.

[28] G. Jia, I. Höger, A. Gawlik et al., "Wet chemically prepared silicon nanowire arrays on low-cost substrates for photovoltaic applications," *Physica status solidi (a)*, vol. 210, no. 4, pp. 728–731, 2013.

5

MPPT for Photovoltaic System Using Nonlinear Controller

Ramsha Iftikhar,[1] **Iftikhar Ahmad**,[1] **Muhammad Arsalan,**[1] **Neelma Naz,**[1] **Naghmash Ali,**[2] **and Hammad Armghan**[2]

[1]*SEECS, National University of Science and Technology, Islamabad, Pakistan*
[2]*School of Electrical Engineering, The University of Faisalabad, Faisalabad, Pakistan*

Correspondence should be addressed to Iftikhar Ahmad; iftikhar.rana@seecs.edu.pk

Academic Editor: Nimrod Vazquez

Photovoltaic (PV) system generates energy that varies with the variation in environmental conditions such as temperature and solar radiation. To cope up with the ever increasing demand of energy, the PV system must operate at maximum power point (MPP), which changes with load as well as weather conditions. This paper proposes a nonlinear backstepping controller to harvest maximum power from a PV array using DC-DC buck converter. A regression plane is formulated after collecting the data of the PV array from its characteristic curves to provide the reference voltage to track MPP. Asymptotic stability of the system is proved using Lyapunov stability criteria. The simulation results validate the rapid tracking and efficient performance of the controller. For further validation of the results, it also provides a comparison of the proposed controller with conventional perturb and observe (P&O) and fuzzy logic-based controller (FLBC) under abrupt changes in environmental conditions.

1. Introduction

Human dependence on fossil fuels, for the generation of energy, has created numerous environmental catastrophes across the planet. Increased carbon emission, global warming, and ozone depletion are the direct consequences of this ill use of fossil fuels. This dire environmental situation is demanding us to utilize renewable energy resources to restore the damage done by fuel consumption. Renewable energy sources are not only ecofriendly but are also conveniently available to everyone and everywhere. The most eminent among these renewable sources for energy generation is solar energy [1]. Energy expenditure on earth is almost ten thousand times lesser than the energy bestowed upon us by the sun. Therefore, there is a dire need to develop instruments to utilize this unrestricted energy source. Solar cell is one such promising device that converts solar energy into electrical energy that can be used directly in a number of ways. Although solar or PV cells are quite promising, yet they are unable to convert all the solar energy into electricity. The percentage of the solar energy shining on a PV device that is converted into usable electricity is termed as conversion efficiency [2]. Hence, different techniques have been devised to extract maximum power from PV cells, so that they can operate at their maximum operating efficiency [3].

The power characteristics of photovoltaic cells are nonlinear, that vary with the variation in the environmental conditions [4]. Variation in temperature and irradiance, for instance, changes the voltage produced, as well as, the generated current by the PV module [5]. As a result, the generated power also varies. Consequently, the operating point of PV array for maximum power generation changes. This operating point is called maximum power point, and the voltage at which PV module can produce maximum power is called maximum power voltage (or peak power voltage). As this point varies by varying environmental conditions, it makes the maximum power extraction a complex task. The power characteristic curve of a PV module at different irradiance levels is shown in Figure 1. Since the temperature and irradiance changes all the time, so a procedure is required to track this maximum power point.

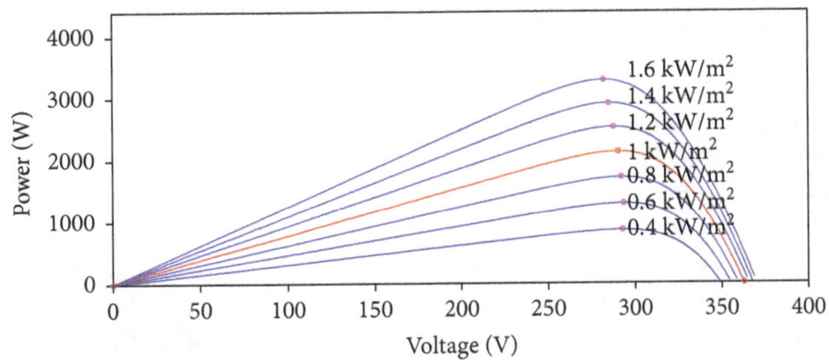

FIGURE 1: Power characteristics of a PV module.

Numerous methods have been proposed in the literature to accomplish the goal of maximum power point tracking (MPPT). They can be categorized into three families of techniques, each having distinct approach to reach MPP. They are

(i) conventional algorithms,

(ii) bioinspired algorithms,

(iii) artificial intelligence- (AI-) based algorithms.

Conventional algorithms mainly constitute a number of variants of two basic techniques, namely, perturb and observe (P&O) and incremental conductance [6]. In P&O-based algorithms, the output voltage of PV module is *perturbed* and output power is *observed.* If $\Delta Power > 0$, then the voltage will be further perturbed in the same direction, that is, voltage will be increased if it was previously increased and vice versa. But if it is less than 0, then the voltage will be perturbed in the opposite direction. These perturbations are introduced periodically and the whole process keeps on repeating itself to eventually reach the maximum power point [7]. As the perturbations are periodic in nature, they result in oscillations of the operating point about the MPP. The downside of this algorithm is its slow convergence to the MPP, resulting in degraded efficiency, especially in conditions where environmental variables are varying rapidly. Perturbations in the output are also an eventual outcome of this algorithm.

Incremental conductance (IC) is more expeditious, as well as efficient in comparison with P&O [8]. This algorithm works on the principle that the $\Delta I_{PV}/\Delta V_{PV}$ is equal to $-I_{PV}/V_{PV}$ at MPP. So, if the PV module is being operated at the left of MPP in the power VS voltage curve, then $\Delta I_{PV}/\Delta V_{PV} > -I_{PV}/V_{PV}$. But, if it is operating on the right of MPP, then $\Delta I_{PV}/\Delta V_{PV} < -I_{PV}/V_{PV}$ [9]. The algorithm is capable of tracking the MPP, even when the environmental conditions are varying swiftly. Once the system reaches MPP, it will eventually stop the iterations and will result in much better efficiency in comparison with that of P&O. The cost of better performance is increased complexity and the execution of larger number of instructions to accurately perform the necessary calculations [10].

Both P&O and IC are also categorized as hill-climbing methods, since their principle of operation is based on the assumption of existence of global maxima. In the event of partial shading, that is, when a number of PV modules are connected with each other and some of them are under shadow while rest are under sunlight, then the whole system will experience multiple local maxima. The conventional algorithms are viable to converge at these local maxima, since they are unable to distinguish between a global and a local peak [11]. Similarly, both the methods continuously oscillate about the MPP, thus introducing oscillations in the system and power loss.

Bioinspired algorithms are much efficient when compared to the conventional ones. They are capable enough to quickly converge to a global maxima and hence can save power loss even in a partially shaded environment. These algorithms generate a population of individuals and each individual represents a distinguished solution. Depending on the type of algorithm, they interact with each other to converge at the maxima. Since the population is initialized randomly, the chances of reaching a global maxima becomes very high. Genetic algorithm (GA) is one such algorithm that solved the obstacle of partial shading [12]. Hardware implementation of GA-based MPPT using FLBC verified its effectiveness under partial shading [13]. Particle swarm optimization (PSO) is another bioinspired algorithm, which is employed successfully in [14]. The particles or solutions swarm independently and evaluate their respective positions using a cost function to estimate their closeness to MPP. The particles eventually converge on a solution that will provide MPP.

Despite their usefulness in varying environmental conditions, these techniques are inefficient because of their slow response. These algorithms continuously evaluate and compare the outcomes of a large number of possible solutions, which also introduce oscillations in the output of the PV array. To reduce these oscillations, an improved variant of PSO was suggested in [15], which increased the efficiency of the system. Similarly, PSO was combined with P&O in [16] to achieve results that are better in comparison with either of the two parent techniques. Ant colony optimization (ACO) is another population-based algorithm which was integrated with P&O in [17] to reduce oscillations. Although there are several advantages of bioinspired algorithms, but their difficult encoding schemes, too many parameter assignments, slow convergence under rapidly varying conditions, and difficult theoretical analysis inhibit their practical usage.

There are two manifestations of AI-based MPPT algorithms, which are different in nature from one another. *Fuzzy logic-based controllers (FLBCs)* or algorithms incorporate the human knowledge and information of a particular system in determining a fuzzy rule base to control it. It does not require any mathematical model of the system but it maps the inputs to the output using fuzzy *If Else* rules. Hence, its performance completely depends on the designer's information about the behavior of the system and its working in varying environment. Due to this property of fuzzy systems, these controllers are relatively simple to design and are robust in performance, since they are also nonlinear in nature [18]. Their only disadvantage is computational complexity especially during implementation. The second type of AI algorithm is *artificial neural network- (ANN-)* based MPPT. It is computationally less costly and improves its performance with time on its own. It does require a training data set in the beginning to train the input output relation, but once deployed, they become robust in operation in response to rapid variation in input parameters [19]. A variant of both FLBC- and ANN-based controller is developed in [20], which outperforms its predecessors in performance, robustness, and efficiency. The resultant artificial neuro fuzzy interference system (ANFIS) shows less overshoot, less settling time, and few oscillations about the MPP.

In this paper, data points were collected using characteristic curves of a PV module. These points map a particular irradiance and temperature to the *peak power voltage*. Linear regression is then executed over these data points to generate a *regression plane*, which provides the reference peak power voltage under varying temperature and irradiance levels. The generation of reference is the first step in achieving MPPT. To extract actual power, we require a DC-DC converter to operate in succession with the PV array. Sometimes, the operating voltage for loads is different than the output voltage of a PV module. For instance, the nominal voltage of a battery is usually much lower than the panel's output voltage. In this scenario, it is obvious to use some kind of interface between the input power and the output load [21]. DC-DC buck converter is used in the proposed study to interface loads that require low input voltage [22]. Being the simplest among all the converters, it has the advantage of lowest part count [23]. For the same output power, the size of inductor is much smaller than that of a boost converter, which makes buck more efficient [24]. Buck converter can be operated at full range of duty cycle, that is, [0.1], because it is inherently stable [25]. Converters are usually modelled with the assumption that they depict linear behavior, which is wrong. Abrupt changes in duty cycle introduces abrupt transients in the output that depicts the nonlinear behavior of converters. Hence, it is unwise to use a linear controller for a tracking problem with the converters [5].

The paper is organized in the following manner. The model of buck converter is established in Section 2. Section 3 describes the generation of regression plane and the reference voltage to extract maximum power from the PV array. A nonlinear backstepping controller is designed in Section 4, and the analysis of global asymptotic stability using Lyapunov stability criteria is given in the same section. Results obtained after simulation are revealed in Section 5. This section also includes results obtained after

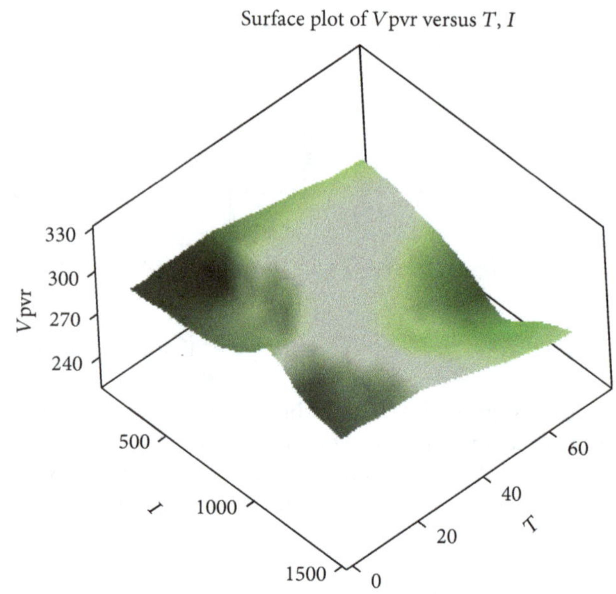

FIGURE 2: Regression plane.

comparing the proposed controller with the conventional P&O and FLBC. The advantages and disadvantages of the abovementioned techniques are presented in Section 6. Finally the conclusion is presented in Section 7.

2. Reference Voltage Generation by Regression Plane

PV characteristic curves are generated by varying temperature from 5°C to 75°C at constant irradiance level of $1000\,W/m^2$. Similarly, more data points were obtained by varying irradiance levels from $200\,W/m^2$–$1400\,W/m^2$ at constant temperature of 25°C. The data set obtained by these characteristic curves is used for generation of regression plane that provides us the required peak power voltage (v_{PVR}). The generated regression plane is shown in Figure 2 and is given by the following equation:

$$v_{PVR} = 322 - 1.31 * T - 0.00037 * I, \quad (1)$$

where T is temperature and I is irradiance.

3. Modeling of Buck Converter

Buck is a switched mode DC-DC converter, whose output voltage has lesser magnitude than the input voltage. It is also termed as a step-down converter. Its circuit diagram is shown in Figure 3. It is assumed to be operated in continuous conduction mode (CCM) throughout this paper. It has two modes of operation. In *model*, *Switch S* is on and *Diode D* is off. By Kirchoff's current and voltage law, we can write

$$\begin{aligned} i_{C1} &= i_{PV} - i_L, \\ v_L &= v_{C1} - v_{C2}, \\ i_{C2} &= i_L - \frac{v_{C2}}{R}. \end{aligned} \quad (2)$$

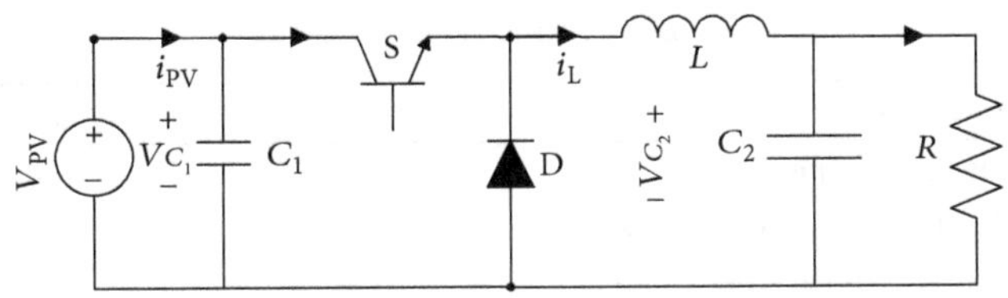

FIGURE 3: Buck converter.

In mode 2, *Switch S* is off and *Diode D* is on. Using Kirchoff's current and voltage laws, we get

$$\begin{aligned} i_{C1} &= i_{PV}, \\ v_L &= -v_{C2}, \\ i_{C2} &= i_L - \frac{v_{C2}}{R}. \end{aligned} \tag{3}$$

By utilizing inductor's volt second balance and capacitor's charge balance, we can write:

$$\begin{aligned} \frac{d_{vC1}}{dt} &= \frac{i_{PV}}{C_1} - \frac{i_L}{C_1}u, \\ \frac{di_L}{dt} &= \frac{v_{C1}}{L}u - \frac{v_{C2}}{L}, \\ \frac{dv_{C2}}{dt} &= \frac{i_L}{C_2} - \frac{v_{C2}}{RC_2}. \end{aligned} \tag{4}$$

After averaging the model for one switching period and assuming x_1, x_2, x_3, and μ to be the average value of v_{C1}, i_L, v_{C2}, and u, respectively, we can write them as

$$\begin{aligned} x_1 &= \langle v_{C1} \rangle, \\ x_2 &= \langle i_L \rangle, \\ x_3 &= \langle v_{C2} \rangle, \\ \mu &= \langle u \rangle. \end{aligned} \tag{5}$$

Evaluating the time derivative of (5) using (4), we get

$$\begin{aligned} \dot{x}_1 &= \frac{i_{PV}}{C_1} - \frac{x_2}{C_1}\mu, \\ \dot{x}_2 &= \frac{x_1}{L}\mu - \frac{x_3}{L}, \\ \dot{x}_3 &= \frac{x_2}{C_2} - \frac{x_3}{RC_2}. \end{aligned} \tag{6}$$

This averaged state space model is then used to track the reference peak power voltage.

4. Backstepping Control

In order to effectively track the reference generated by the regression plane, a nonlinear controller based on backstepping approach is designed. The controller provides the input μ that will determine the duty ratio to be supplied to the switch in buck converter. The reference v_{PVR} generated in Section 2 is termed here as x_{1ref} to avoid any confusion while deriving the controller.

Assuming e_1 to be the error between actual and required PV array output voltage

$$e_1 = x_1 - x_{1ref}. \tag{7}$$

The goal is to converge the error signal e_1 to zero. Derivative of (7) with respect to time gives

$$\dot{e}_1 = \dot{x}_1 - \dot{x}_{1ref}. \tag{8}$$

Inserting (6) in (8) gives

$$\dot{e}_1 = \frac{i_{PV}}{C_1} - \frac{x_2}{C_1}\mu - \dot{x}_{1ref}. \tag{9}$$

Let V_1 be a positive definite Lyapnuov candidate function for checking the convergence of e_1 to 0.

$$V_1 = \frac{1}{2}{e_1}^2. \tag{10}$$

To ensure asymptotic stability, derivative of the Lyapnuov function must be negative definite. Taking time derivative of (10), we have

$$\dot{V}_1 = e_1\dot{e}_1. \tag{11}$$

Using (9), we get

$$\dot{V}_1 = e_1\left(\frac{i_{PV}}{C_1} - \frac{x_2}{C_1}\mu - \dot{x}_{1ref}\right). \tag{12}$$

For $\dot{V}_1$ to be negative definite, let

$$\frac{i_{PV}}{C_1} - \frac{x_2}{C_1}\mu - \dot{x}_{1ref} = -K_1e_1, \tag{13}$$

so that $\dot{V}_1$ becomes

$$\dot{V}_1 = -K_1{e_1}^2. \tag{14}$$

Rewriting (13) as

$$x_2 = \frac{C_1}{\mu}\left(K_1e_1 + \frac{i_{PV}}{C_1} - \dot{x}_{1ref}\right). \tag{15}$$

Let (15) be the reference current for inductor, given by

$$\beta = \frac{C_1}{\mu}\left(K_1e_1 + \frac{i_{PV}}{C_1} - \dot{x}_{1ref}\right). \tag{16}$$

Let us define the error e_2 to track x_2 to β

$$e_2 = x_2 - \beta. \tag{17}$$

Rewriting (17) as

$$x_2 = e_2 + \beta. \tag{18}$$

Putting (18) in (9) gives

$$\dot{e}_1 = \frac{i_{PV}}{C_1} - \frac{e_2 + \beta}{C_1}\mu - \dot{x}_{1ref}. \tag{19}$$

Putting β from (16) in (19). After simplification, we get

$$\dot{e}_1 = -K_1 e_1 - \frac{e_2}{C_1}\mu. \quad (20)$$

Hence, (11) becomes

$$\dot{V}_1 = e_1\dot{e}_1 = e_1\left(-K_1 e_1 - \frac{e_2}{C_1}\mu\right), \quad (21)$$

$$\dot{V}_1 = -K_1 {e_1}^2 - \frac{e_1 e_2}{C_1}\mu. \quad (22)$$

Here, the first term in (22) is negative definite, but we are not sure about the second term. By taking the derivative of (16) and (17) and simplifying the expressions

$$\dot{e}_2 = \dot{x}_2 - \dot{\beta}, \quad (23)$$

and

$$\dot{\beta} = \frac{C_1}{\mu}\left(K_1\dot{e}_1 + \frac{\dot{i}_{PV}}{C_1} - \ddot{x}_{1ref}\right) - \frac{\dot{\mu}}{\mu^2}C_1\left(K_1 e_1 + \frac{i_{PV}}{C_1} - \dot{x}_1 ref\right). \quad (24)$$

Simplifying using (16) and (20)

$$\dot{\beta} = \frac{C_1}{\mu}\left(K_1\left(-K_1 e_1 - \frac{e_2}{C_1}\mu\right) + \frac{i_{PV}}{C_1} - \ddot{x}_{1ref}\right) - \frac{\dot{\mu}}{\mu}\beta. \quad (25)$$

Inserting (25) in (23), $\dot{e}_2$ becomes

$$\dot{e} = \dot{x}_2 - \frac{C_1}{\mu}\left(-K_1^2 e_1 - \frac{K_1 e_2}{C_1}\mu\right) - \frac{C_1}{\mu}\left(\frac{\dot{i}_{PV}}{C_1} - \ddot{x}_{1ref}\right) + \frac{\dot{\mu}}{u}\beta. \quad (26)$$

Now, to guarantee convergence of both e_1 and e_2 to zero, a composite Lyapunov function V_c is defined as follows:

$$V_C = V_1 + \frac{1}{2}e_2^2. \quad (27)$$

If the time derivative of V_C is negative definite, then according to Lyapunov stability criteria, both the errors e_1 and e_2 will converge to 0. In other words, it will ensure that x_1 will converge to x_{1ref}, so that our system can reach to MPP. Taking the time derivative of (27), we get

$$\dot{V}_C = \dot{V}_1 + e_2\dot{e}_2 = -K_1{e_1}^2 - \frac{e_1 e_2}{C_1}\mu + e_2\dot{e}_2, \quad (28)$$

or

$$\dot{V}_C = -K_1 e_1^2 + e_2\left(\dot{e}_2 - \frac{e_1}{C_1}\mu\right). \quad (29)$$

For $\dot{V}_C$ to be negative definite, take

$$\dot{e}_2 - \frac{e_1}{C_1}\mu = -K_2 e_2, \quad (30)$$

where K_2 is a positive constant. So that the $\dot{V}_C$ becomes

$$\dot{V}_C = -K_1 e_1^2 - K_2 e_2^2. \quad (31)$$

Using (6), (26), and (30), we get

$$\begin{aligned} -K_2 e_2 = {} & \frac{x_1}{L}\mu - \frac{x_3}{L} + \frac{K_1^2 C_1 e_1}{\mu} + K_1 e_2 \\ & - \frac{\dot{i}_{PV}}{\mu} + \frac{C_1\ddot{x}_{1ref}}{\mu} + \frac{\dot{\mu}}{\mu}\beta - \frac{e_1}{C_1}\mu. \end{aligned} \quad (32)$$

Solving (32) for $\dot{\mu}$

$$\begin{aligned} \dot{\mu} = {} & \frac{\mu}{\beta}\left(-K_2 e_2 - \frac{x_1}{L}\mu + \frac{x_3}{L} - \frac{K_1^2 C_1 e_1}{\mu}\right) \\ & \cdot \frac{\mu}{\beta}\left(-K_1 e_2 + \frac{\dot{i}_{PV}}{\mu} - \frac{C_1\ddot{x}_{1ref}}{\mu} + \frac{e_1}{C_1}\mu\right), \end{aligned} \quad (33)$$

where $0 < \mu < 1$ and $\beta \neq 0$. Using μ obtained by integrating (33), $\dot{V}_C$ becomes negative definite, proving the asymptotic stability of the system, which is evident from (31) as $\dot{V}_C \leq 0$. Moreover, the convergence of e_1 to 0 or PV array input voltage to v_{PVR} is also ensured.

Table 1: Parameters of PV array.

Parameter	Value
PV module per string	10
Parallel connected strings	1
Number of cells per module	72
Open circuit voltage	363
Short circuit current	7.84
Voltage at MPP	290
Current at MPP	7.35
Maximum power per module	213.15

Table 2: Parameters of controller and converter.

Parameter	Value
K_1	8
K_2	26,000
Input capacitor, C_1	39 uF
Inductor, L	7 mH
Output capacitor, C_2	39 uF
Load resistor, R	10 ohms
Switching frequency, f_s	100 KHz

5. Simulation and Results

The parameters of PV array that are used in this work are mentioned in Table 1. Similarly, the parameters of controller and converter are mentioned in Table 2. Simulations of the proposed controller are performed in MATLAB/SIMULINK to verify its performance. The section is divided into four

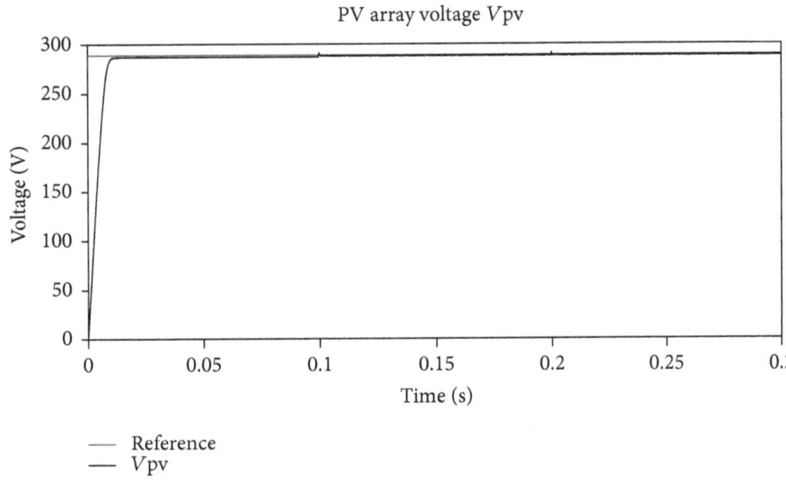

FIGURE 4: Tracking of PV module voltage.

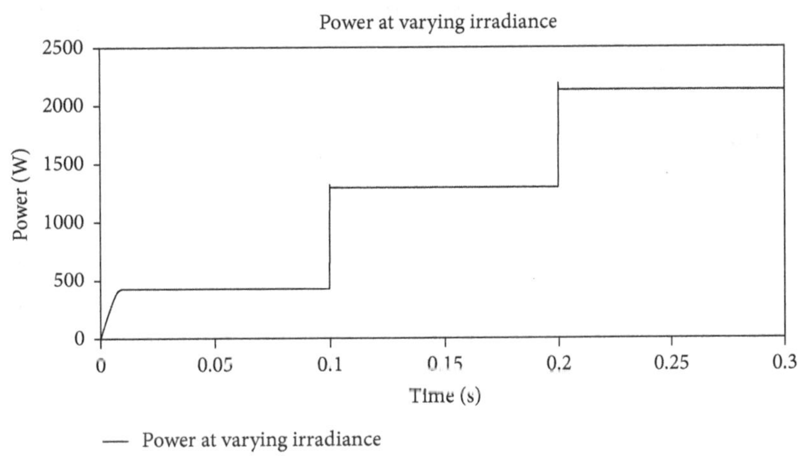

FIGURE 5: Generated power under varying irradiance.

subsections. The first two subsections critically analyzes the performance of the proposed controller under sudden changes in temperature and irradiance. Whereas, the latter two compare the proposed controller with P&O and FLBC-based MPPT algorithms.

5.1. Test under Varying Irradiance. To test the proposed controller in harsh environment, the initial irradiance is kept at 200 W/m^2, which is abruptly changed to 600 W/m^2 after 0.1 s. Similarly, after 0.2 s, it is changed to 1000 W/m^2. The whole experiment is performed while keeping the temperature of PV module equal to 25°C. The regression plane successfully generates the tracking peak power voltage which is successfully tracked by the controller, as shown in Figure 4. Similarly, Figure 5 depicts the change in generated power by the system as a result of abrupt variation in irradiance. Again, the PV module reaches at maximum power within 0.002 seconds with almost negligible ripple.

5.2. Test under Varying Temperature. In this case, the initial temperature of the PV cell is first maintained at 25°C, which is then increased to 40°C after an interval of 0.1 s. Similarly, after 0.2 s, the temperature is sharply increased to 55°C. Throughout this experiment, the irradiance is kept 1000 W/m^2, so that the system's performance can be verified only under varying temperature condition. The proposed controller yet again successfully tracks the reference voltage, as shown in Figure 6. Similarly, the controller is robust enough to maximize the power by reaching MPP in less than 0.001 seconds. The generated power under varying temperature is shown in Figure 7.

5.3. Comparison with P&O. Conventional P&O and the proposed controller are first compared under varying irradiance while keeping the temperature constant and then under varying temperature while keeping the irradiance constant. The conditions of both the tests are kept same as before in the previous respective experiments. The proposed controller clearly outperforms the P&O algorithm. Here, in Figure 8, the generated power under varying irradiance is shown. The proposed controller is not only robust, but the ripples are also negligible. The efficiency of the system is greatly enhanced when the proposed controller is used. The power generated under varying temperature conditions, shown in Figure 9, also verifies the abovementioned results. The P&O

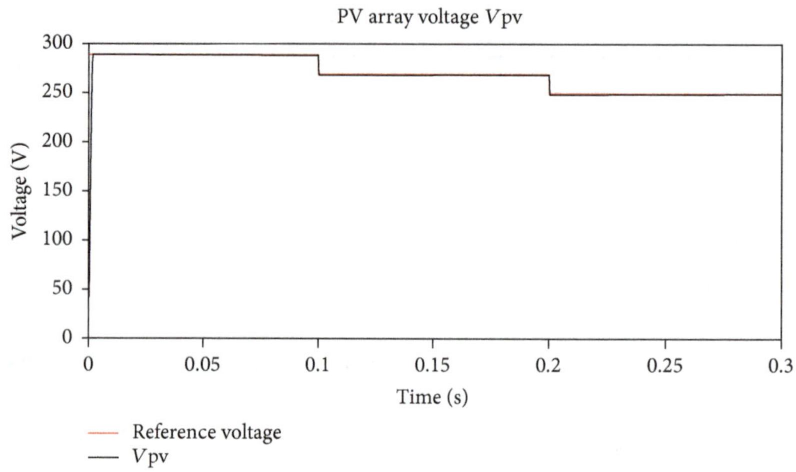

Figure 6: Tracking of PV module voltage.

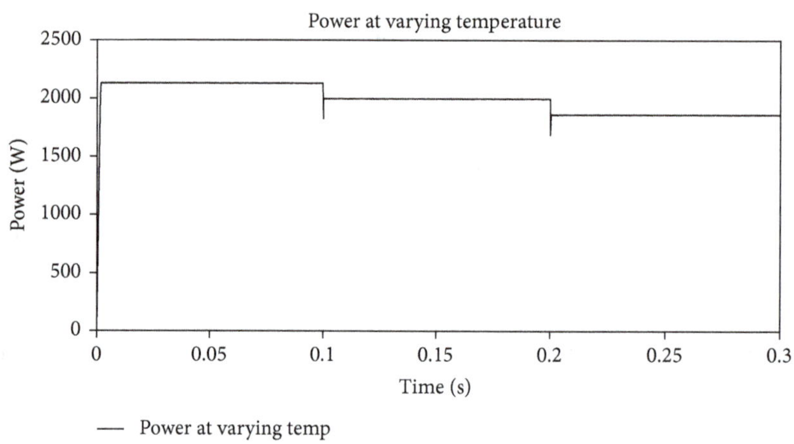

Figure 7: Generated power under varying temperature.

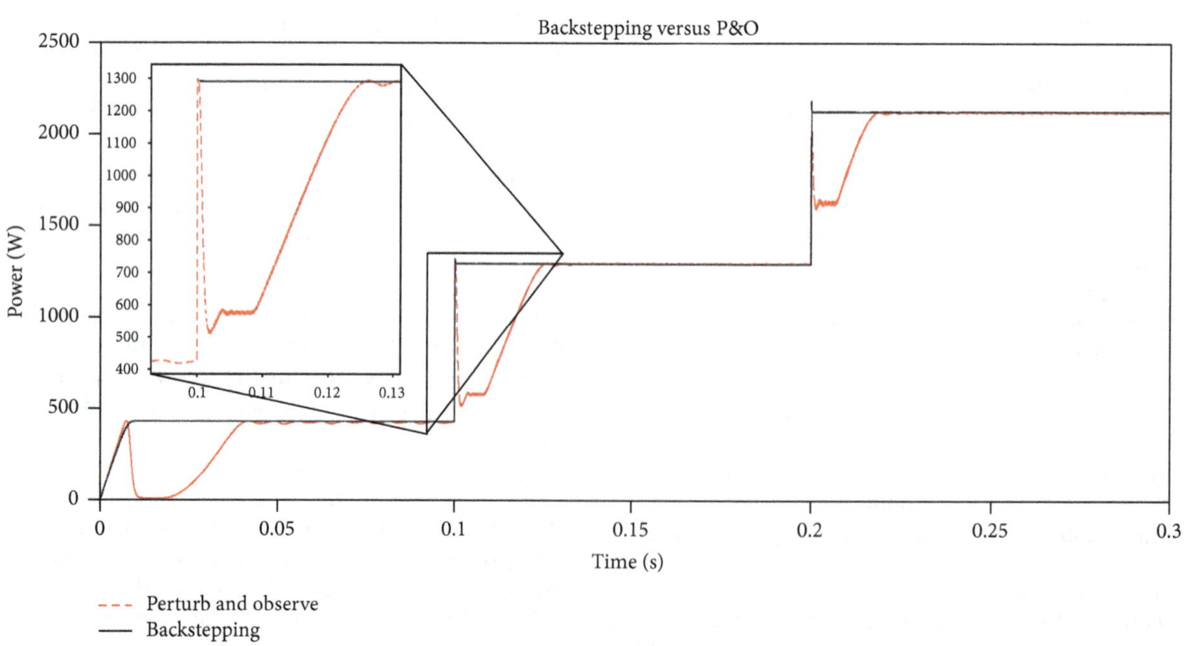

Figure 8: Power comparison under varying irradiance.

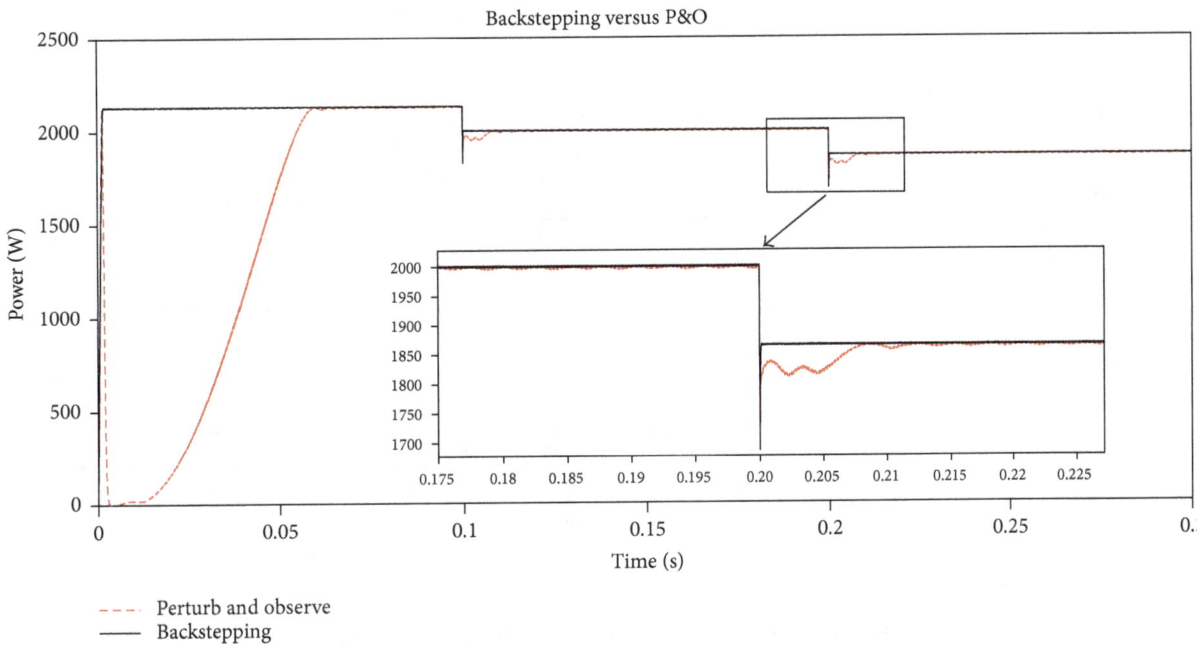

FIGURE 9: Power comparison under varying temperature.

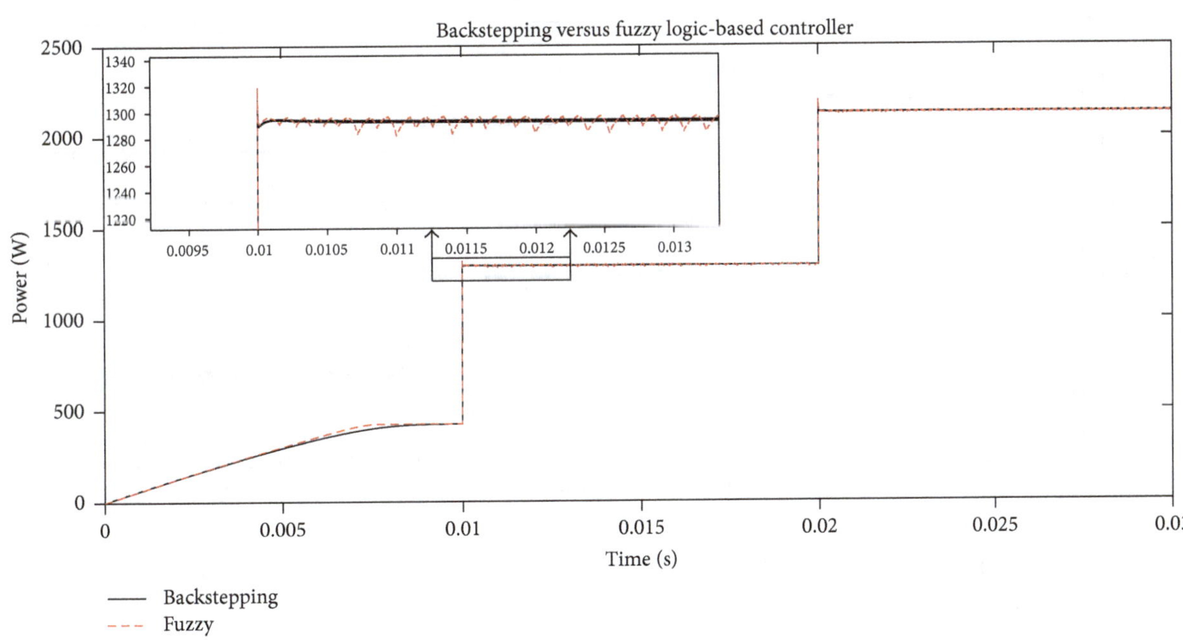

FIGURE 10: Power comparison under varying irradiance.

algorithm also takes considerably much more time to reach the MPP when initially the conditions were kept constant.

5.4. Comparison with FLBC. Once again, the same two tests were performed to study the comparison between backstepping and fuzzy logic-based controller, except for one change. Since both the controllers showed very rapid response to variations, the time between the successive variations was reduced 10 times. So the changes in temperature and irradiance are introduced after every 0.01 s. The comparison of generated power under varying irradiance is shown in Figure 10, and the comparison under temperature variation is shown in Figure 11. The results that are obtained by using backstepping controller are free of ripples and overshoot, but with FLBS, both of them are easily visible. To compare the two techniques further, a comparison between the voltage tracking of the two controllers under varying temperature is shown in Figure 12. Although both the controllers successfully track the reference, but still the FLBC displays large ripples in the voltage waveform along with an overshoot. If

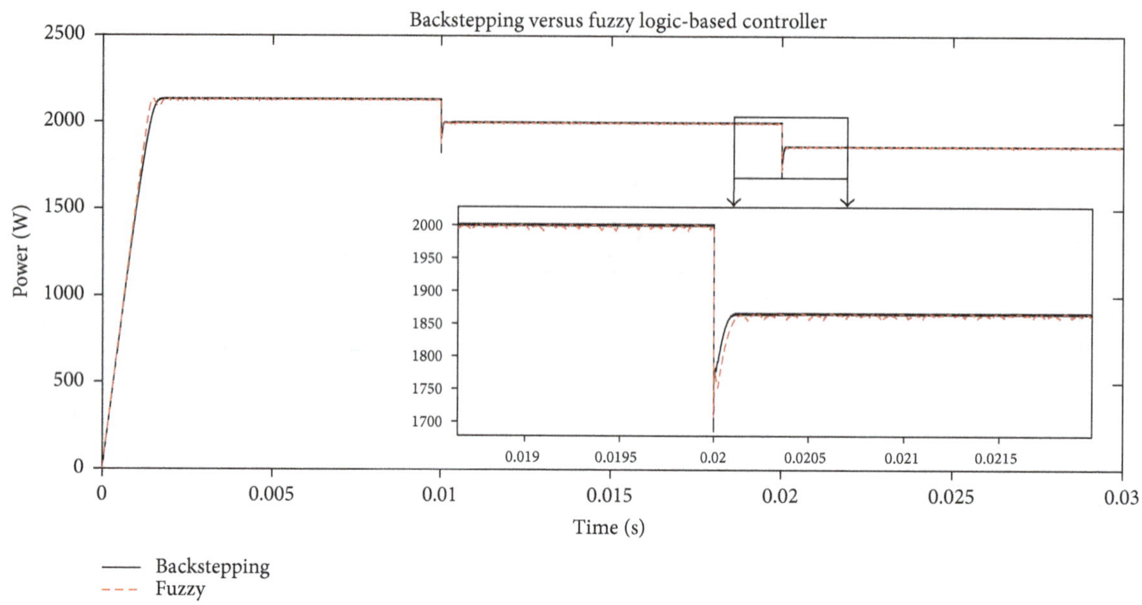

FIGURE 11: Power comparison under varying temperature.

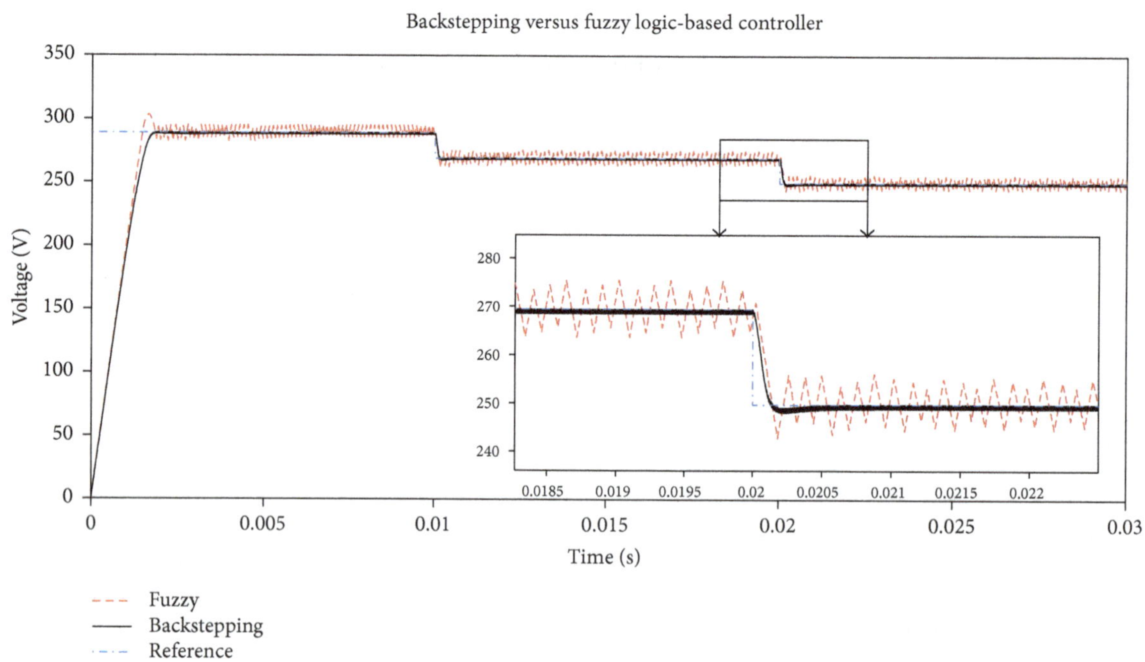

FIGURE 12: v_{PVR} tracking, backstepping versus FLBC.

we take into account the computational complexity and the results unveiled in this document, we can easily state that the proposed backstepping controller successfully outperforms the FLBC-based MPPT.

6. Comparison between Analyzed Techniques

All the techniques analyzed in the previous section vary the output voltage of PV array by varying the duty cycle of the converter. Hence, the PV array output voltage waveforms obtained using all the techniques are compared and the results are presented in Table 3. The results are compared on the basis of rise time (RT), settling time (ST) (2% and 5% criteria), steady-state error (SSE), and overshoot and ripples in the output voltage of PV array, measured in voltage from peak to peak. Both P&O and FLBC show large oscillations about the reference, so their output voltage never reached within 2% of the steady-state value. Their value is shown as not applicable (NA) in Table 3.

Both the backstepping-based control and FLBC with regression plane require three sensors in total; one voltage sensor, one temperature sensor, and one sensor to measure

Table 3: Comparison.

Method	RT (ms)	ST 5% criteria (ms)	ST 2% criteria (ms)	SSE (%)	Overshoot (V)	Ripples (V)
Backstepping	1.4	1.6	1.81	0.16	0.6	0.9
P&O	1.3	59	NA	0.3	70	8.9
FLBC	1.3	1.9	NA	0.29	12.8	9

irradiance. However, P&O algorithm requires a voltage and a current sensor to measure PV array output voltage and current for its operation. The regression plane, used in the proposed technique and FLBC, requires regular maintenance to accurately generate the reference. FLBC is also computationally complex and can cause unwanted delays in MPPT, which will result in wastage of useful energy. However, P&O is the simplest of all. But when we analyze the data presented in Table 3, the superiority of the proposed technique becomes evident. Robustness of controller along with negligible steady-state error validates its exceptional performance. Similarly, least overshoot and ripples have been recorded for backstepping-based approach. Consequently, electrical components with small sizes, such as inductor, capacitor, switches, and diodes, can be selected, when used with the proposed controller, which will increase the efficiency of the overall system.

7. Conclusion

In this paper, nonlinear backstepping controller is proposed to be used for MPPT using buck converter. To extract maximum power, the duty cycle of buck converter is controlled to track the reference generated by the regression plane using the proposed controller. The performance of the proposed controller outclassed the conventional P&O and FLBC and it also proves the global asymptotic stability using Lyapunov stability criteria, whereas the previous two techniques are unable to do so. Regression plane does require some maintenance, because in real world, the PV arrays are subjected to wear and tear. The work can be further extended by successfully implementing the proposed converter in experimental setup. Similarly, robustness of bioinspired algorithms can be improved to generate the reference voltage swiftly and it should replace the regression plane.

Conflicts of Interest

The authors declare that they have no conflicts of interest.

References

[1] G. K. Singh, "Solar power generation by PV (photovoltaic) technology: a review," *Energy*, vol. 53, pp. 1–13, 2013.

[2] F. Dincer, "The analysis on photovoltaic electricity generation status, potential and policies of the leading countries in solar energy," *Renewable and Sustainable Energy Reviews*, vol. 15, no. 1, pp. 713–720, 2011.

[3] M. Metry, M. B. Shadmand, R. S. Balog, and H. Abu-Rub, "MPPT of photovoltaic systems using sensorless current-based model predictive control," *IEEE Transactions on Industry Applications*, vol. 53, no. 2, pp. 1157–1167, 2017.

[4] Z. Wang, Y. Li, K. Wang, and Z. Huang, "Environment-adjusted operational performance evaluation of solar photovoltaic power plants: a three stage efficiency analysis," *Renewable and Sustainable Energy Reviews*, vol. 76, pp. 1153–1162, 2017.

[5] A. D. Martin, J. M. Cano, J. F. A. Silva, and J. R. Vazquez, "Backstepping control of smart grid-connected distributed photovoltaic power supplies for telecom equipment," *IEEE Transactions on Energy Conversion*, vol. 30, no. 4, pp. 1496–1504, 2015.

[6] D. Sera, L. Mathe, T. Kerekes, S. V. Spataru, and R. Teodorescu, "On the perturb-and-observe and incremental conductance MPPT methods for PV systems," *IEEE Journal of Photovoltaics*, vol. 3, no. 3, pp. 1070–1078, 2013.

[7] M. A. Elgendy, B. Zahawi, and D. J. Atkinson, "Evaluation of perturb and observe MPPT algorithm implementation techniques," in *6th IET International Conference on Power Electronics, Machines and Drives (PEMD 2012)*, pp. 1–6, Bristol UK, 2012.

[8] M. A. Elgendy, B. Zahawi, and D. J. Atkinson, "Assessment of the incremental conductance maximum power point tracking algorithm," *IEEE Transactions on Sustainable Energy*, vol. 4, no. 1, pp. 108–117, 2013.

[9] M. W. Rahman, C. Bathina, V. Karthikeyan, and R. Prasanth, "Comparative analysis of developed incremental conductance (IC) and perturb & observe (P&O) MPPT algorithm for photovoltaic applications," in *2016 10th International Conference on Intelligent Systems and Control (ISCO)*, pp. 1–6, Coimbatore, India, 2016.

[10] G. J. Kish, J. J. Lee, and P. W. Lehn, "Modelling and control of photovoltaic panels utilising the incremental conductance method for maximum power point tracking," *IET Renewable Power Generation*, vol. 6, no. 4, p. 259, 2012.

[11] D. F. Teshome, C. H. Lee, Y. W. Lin, and K. L. Lian, "A modified firefly algorithm for photovoltaic maximum power point tracking control under partial shading," *IEEE Journal of Emerging and Selected Topics in Power Electronics*, vol. 5, no. 2, pp. 661–671, 2017.

[12] M. B. Smida and A. Sakly, "Genetic based algorithm for maximum power point tracking (MPPT) for grid connected PV systems operating under partial shaded conditions," in *2015 7th International Conference on Modelling, Identification and Control (ICMIC)*, pp. 1–6, Sousse, Tunisia, 2015.

[13] A. A. S. Mohamed, A. Berzoy, and O. A. Mohammed, "Design and hardware implementation of FL-MPPT control of PV systems based on GA and small-signal analysis," *IEEE Transactions on Sustainable Energy*, vol. 8, no. 1, pp. 279–290, 2017.

[14] F. M. de Oliveira, F. R. Durand, V. D. Bacon, S. A. O. da Silva, L. P. Sampaio, and L. B. G. Campanhol, "Grid-tied photovoltaic system based on PSO MPPT technique with active power line conditioning," *IET Power Electronics*, vol. 9, no. 6, pp. 1180–1191, 2016.

[15] K. Ishaque, Z. Salam, M. Amjad, and S. Mekhilef, "An improved particle swarm optimization (PSO)–based MPPT

for PV with reduced steady-state oscillation," *IEEE Transactions on Power Electronics*, vol. 27, no. 8, pp. 3627–3638, 2012.

[16] K. L. Lian, J. H. Jhang, and I. S. Tian, "A maximum power point tracking method based on perturb-and-observe combined with particle swarm optimization," *IEEE Journal of Photovoltaics*, vol. 4, no. 2, pp. 626–633, 2014.

[17] K. Sundareswaran, V. Vigneshkumar, P. Sankar, S. P. Simon, P. Srinivasa Rao Nayak, and S. Palani, "Development of an improved P&O algorithm assisted through a colony of foraging ants for MPPT in PV system," *IEEE Transactions on Industrial Informatics*, vol. 12, no. 1, pp. 187–200, 2016.

[18] A. El Khateb, N. A. Rahim, J. Selvaraj, and M. N. Uddin, "Fuzzy-logic-controller-based SEPIC converter for maximum power point tracking," *IEEE Transactions on Industry Applications*, vol. 50, no. 4, pp. 2349–2358, 2014.

[19] L. M. Elobaid, A. K. Abdelsalam, and E. E. Zakzouk, "Artificial neural network-based photovoltaic maximum power point tracking techniques: a survey," *IET Renewable Power Generation*, vol. 9, no. 8, pp. 1043–1063, 2015.

[20] R. K. Kharb, S. L. Shimi, S. Chatterji, and M. F. Ansari, "Modeling of solar PV module and maximum power point tracking using ANFIS," *Renewable and Sustainable Energy Reviews*, vol. 33, pp. 602–612, 2014.

[21] R. Abid, F. Masmoudi, B. S. Fatma, and D. Nabil, "Modeling and simulation of conventional DC-DC converters deticated to photovoltaic applications," in *2016 7th International Renewable Energy Congress (IREC)*, pp. 1–6, Hammamet, Tunisia, July 2016.

[22] L. An and D. D. C. Lu, "Design of a single-switch DC/DC converter for a PV-battery-powered pump system with PFM+PWM control," *IEEE Transactions on Industrial Electronics*, vol. 62, no. 2, pp. 910–921, 2015.

[23] J. Chauhan, P. Chauhan, T. Maniar, and A. Joshi, "Comparison of MPPT algorithms for DC-DC converters based photovoltaic systems," in *2013 International Conference on Energy Efficient Technologies for Sustainability*, pp. 476–481, Nagercoil, India, July 2013.

[24] L. Corradini, D. Maksimovic, P. Mattavelli, and R. Zane, *Continuous-Time Averaged Modeling of DC–DC Converters*, John Wiley & Sons, Inc, Hoboken, NJ, USA, 2015.

[25] T. Laagoubi, M. Bouzi, and M. Benchagra, "Analysis and comparison of MPPT nonlinear controllers for PV system," in *2015 3rd International Renewable and Sustainable Energy Conference (IRSEC)*, pp. 1–5, Marrakech, Morocco, 2015.

6

Investigation on Potential-Induced Degradation in a 50MWp Crystalline Silicon Photovoltaic Power Plant

Jingsheng Huang,[1,2] Hongtao Li,[2] Yaojie Sun,[1] He Wang,[3] and Hong Yang[3]

[1]*School of Information Science and Technology, Fudan University, Shanghai 200433, China*
[2]*State Key Laboratory of Operation and Control of Renewable Energy & Storage Systems, China Electric Power Research Institute, Nanjing 210003, China*
[3]*MOE Key Laboratory for Nonequilibrium Synthesis and Modulation of Condensed Matter, School of Science, Xi'an Jiaotong University, Xi'an 710049, China*

Correspondence should be addressed to Yaojie Sun; yjsun@fudan.edu.cn and Hong Yang; hongy126@126.com

Academic Editor: Chun-Sheng Jiang

The regular performance deterioration of P-type crystalline silicon solar modules and module strings caused by potential-induced degradation in a photovoltaic power plant was found in the field. The PID-affected solar modules dismounted from the photovoltaic power plant were further investigated systematically in the laboratory. For the first time, we found that the neutral point of voltage in a module string moved forward to the positive pole for a PID-affected module string as time goes on. Even if low positive voltage is applied to a PID-prone module, it could cause PID. The thermographic and electroluminescence (EL) images of a PID-affected module string also exhibit a regular degradation pattern. This is in good agreement with the measured power loss of the dismounted solar modules under standard test conditions. The results obtained in this paper show that the maximum power degradation rate of solar modules was as high as 53.26% after only one year of operation because of PID in the field. Due to the vast amount of solar modules and incomplete recovery, this is a terrible catastrophe for the owner of a power plant and module producer.

1. Introduction

The crystalline silicon solar module has been considered to possess high reliability over 25 years with about a 0.8% power degradation rate per year [1, 2]. However, in recent years, a catastrophic power degradation caused by potential induced degradation (PID) has attracted considerable attention from power plant owners, module producers, and researchers. Some investigators thought that several module components, such as the front cover substrate, encapsulate, silicon nitride (SiNx:H) film, rain, and soiling are related with PID, and they suggested preventing PID on the level of solar cells and module manufacturing [3–6]. But other authors think that the metal ions from the soda lime front cover glass are a key reason for PID. These sodium decorated stacking faults are identified as the primary cause of potential-induced degradation [7–10].

Up to now, there always exist some debates about the mechanism of PID, and there are also no effective measures to prevent PID. Yang et al. even reported the performance deterioration of substrings for PID-affected modules [11]. In this paper, the regular performance deterioration of P-type crystalline silicon solar modules and module strings caused by potential-induced degradation in a photovoltaic power plant was found in the field. The PID-affected solar modules dismounted from the photovoltaic power plant were further investigated systematically in the laboratory. A detailed analysis on the PID that occurred in the 50 MWp crystalline silicon solar power plant is presented in this contribution. We find that the neutral point of voltage in a module string moved forward to the positive pole for a PID-affected module string as time goes on for the first time. Even if low positive voltage is applied to the PID-prone module, it could cause PID. The regular degradation patterns of

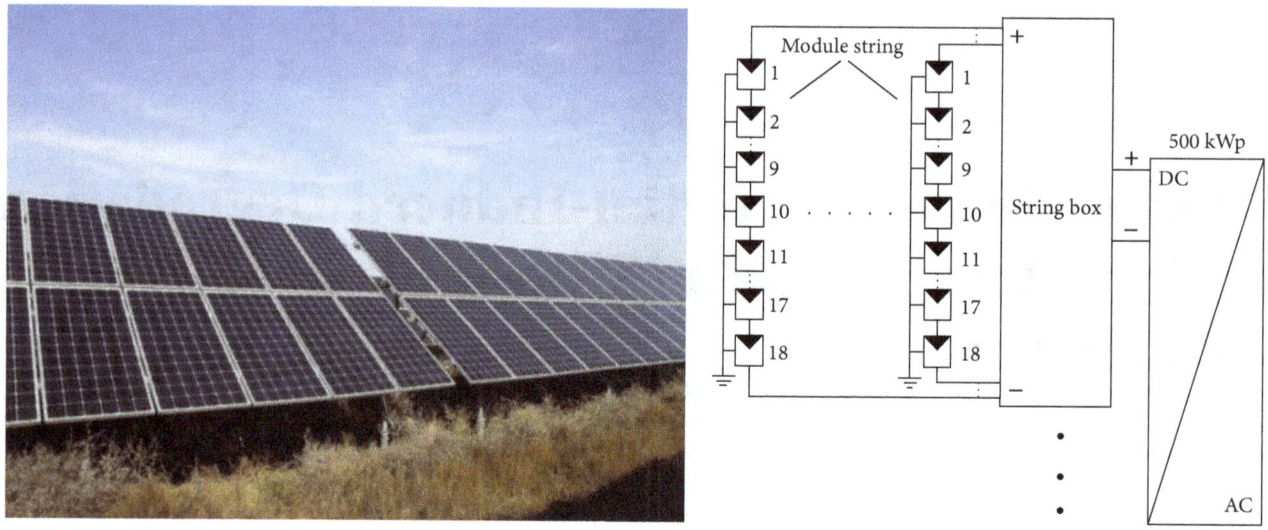

(a) (b)

FIGURE 1: The scene of the 50 MWp grid-connected photovoltaic power plant (a) and the circuit diagram of the subarray in the power plant (b).

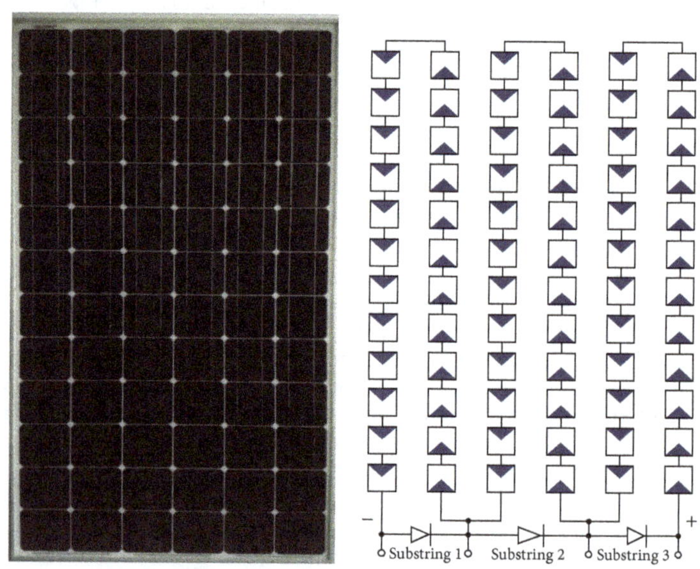

(a) (b)

FIGURE 2: A photograph of the PID-affected solar module (a) and the circuit diagram of the solar module with 72 cells (b).

the thermographic and electroluminescence (EL) images for the PID-affected module string were revealed. The electrical performance of PID-affected modules also shows that the power degradation rate of these solar modules is related with the position of module installation in a module string. The new findings obtained in this paper can help identify and uncover the mystery of PID.

2. Experiments and Methods

In this work, the PID-affected solar modules come from the 50 MWp crystalline silicon solar power plant located in China. Figure 1(a) shows the scene of the 50 MWp grid-connected power plant, and Figure 1(b) shows the circuit diagram of the subarray in the power plant. Figure 2(a) shows a photograph of the PID-affected solar module, and Figure 2(b) shows the circuit diagram of the solar module with 72 solar cells. In the photovoltaic power plant, the eighteen solar modules are interconnected serially into a module string which is grounded by the aluminum frame of every module. Every solar module consists of 72 pieces of P-type monocrystalline silicon solar cells connected in a series, respectively. The cells feature an unpassivated full-area screen-printed aluminium back surface field with dimensions of $125 \times 125\ mm^2$ and 70 nm SiNx:H antireflective coating [12].

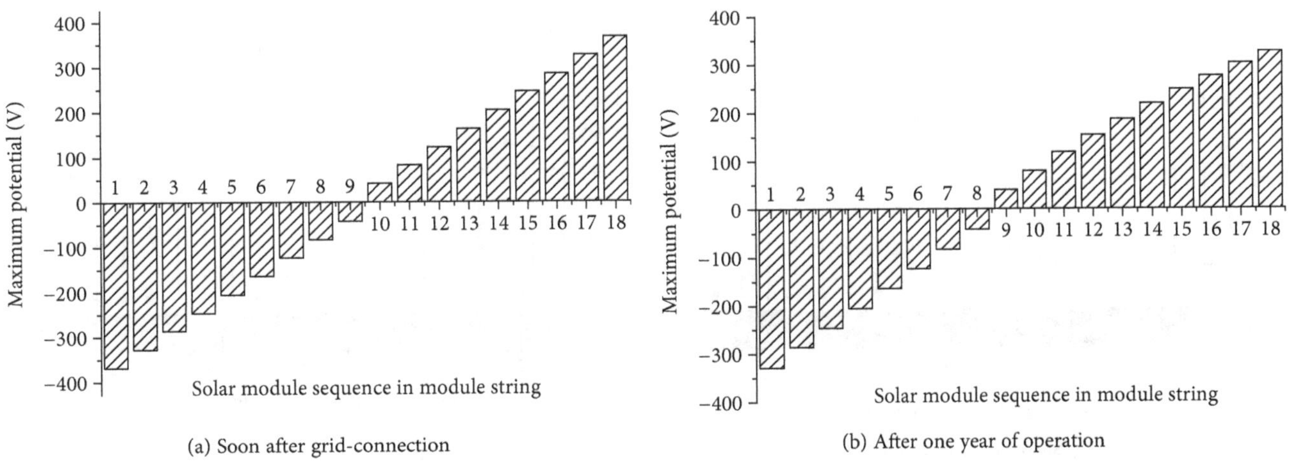

FIGURE 3: The measured potential difference of 18 modules from the same module string between the ground and the pole of module under an open circuit.

In order to get stable results, we chose an ~25°C ambient temperature, more than 800 W/m^2 solar irradiation, and ~60% relative humidity as test conditions in the field. Thermographic images of the solar modules were obtained by a T420 infrared camera with a resolution of 320×240 pixels and a spectral range of 7.5–13 μm (FLIR Systems Inc., T420). The sunlight exposure dose was measured by a YFJ-10 solar irradiation meter. The operation temperature of the solar module was measured by a thermocouple. The solar modules, which are from the PID-affected module string, were dismounted and marked in sequence. The performance of the solar modules was measured by a solar simulator (PASAN Sunsim 3C) under standard test conditions (STC: 1 kW/m^2 irradiance, 25°C module temperature and AM1.5 global spectrum). The solar simulator was calibrated by the standard module. The electroluminescence (EL) images of the solar modules were measured by an EL measurement system in the laboratory.

3. Results and Discussion

3.1. The Potential and the Neutral Point Shift of Module String. Figure 3 shows the maximum potential differences of the 18 modules from the same module string between the ground and cells of the solar module, which were measured in the field under an open circuit. The potential difference between the ground and cells depends on the number of solar modules, solar irradiation, and module temperature. Figure 3(a) shows the measured results of the maximum potential differences of the 18 modules from the same module string soon after the grid for the power plant was finished. Figure 3(b) shows the measured results of the maximum potential differences of the 18 modules from the same module string after one year of operation. From Figure 3, we find that the potential difference between the ground and cells is negative from module 1 to module 9, and that the potential difference between the ground and cells is positive from module 10 to module 18 soon after the power plant was connected with the grid. After one year of operation, the neutral point of voltage in the module string moved forward to the positive pole. The potential difference between the ground and cells in module 9 evolved from the negative to positive. From module 9 to module 18, the closer the module gets to the negative pole of system, the more serious the degradation of open circuit voltage for module is [11]. Because of a transformerless inverter topology, the neutral point of the voltage in the module string moved forward to the positive pole when the aluminium frame of the solar modules is grounded. When the direction of the electric field of module 9 reverses after one year of operation, even if low positive voltage is applied to the PID-prone module, it could cause PID.

Comparing Figure 3(a) with Figure 3(b), the open circuit voltage of the module string becomes smaller for the same module string under nearly the same test conditions mentioned above. This is caused by PID of the module string. The negative potential versus the ground leads to an occurrence of PID.

The potential difference between the ground and cells creates an electric field on the surface of solar cells. Figure 4(a) shows the electric fields of 18 modules from the same module string soon after grid connection, and Figure 4(b) shows the electric fields of 18 modules after one year of operation.

From Figure 4, we find that only the electric field produced by the positive potential difference between the ground and cells could drive the metal ions, such as sodium ions driven into the PN junction which has existing stacking faults. This leads to damage of the PN junction and reduction of the open circuit voltage. This conclusion agrees with [13].

Figure 5 shows the direction of the electric field between the ground and solar cells of modules from the same module string. Soon after the plant was grid-connected, the direction of the electric field is directed from the solar cells to the Al frame for module 1 to module 9, as shown in Figure 5(a), while the direction of the electric field is reversed for module 10 to module 18, as shown in Figure 5(b). After one year of operation, the direction of the electric field for module 1 to module 8 is directed from the solar cells to the Al frame, while the direction of the electric field for module 9 to module 18 is reversed because of PID. This shows that the direction of the electric field of module 9 reverses after one year of

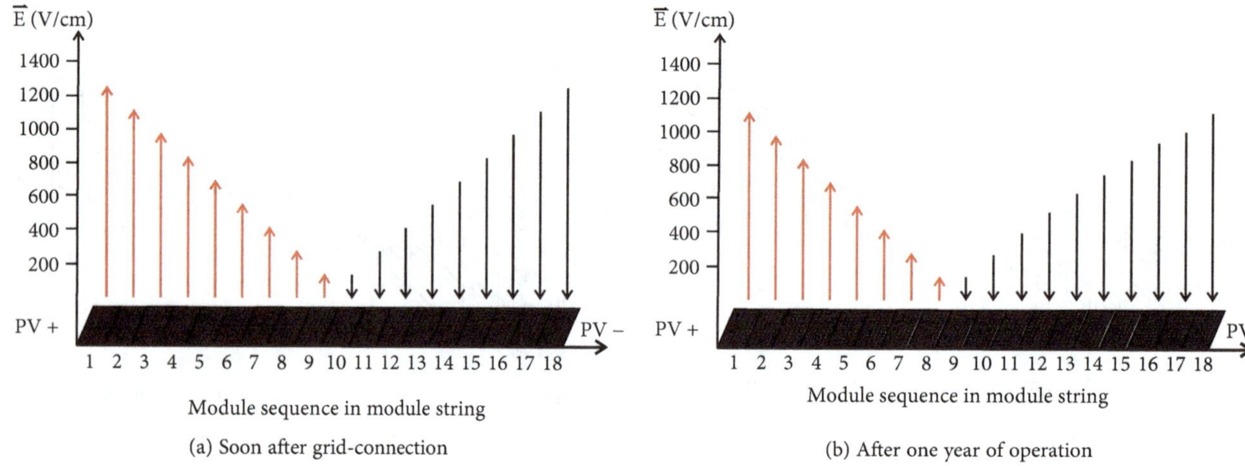

FIGURE 4: The electric fields of 18 modules from the same module string.

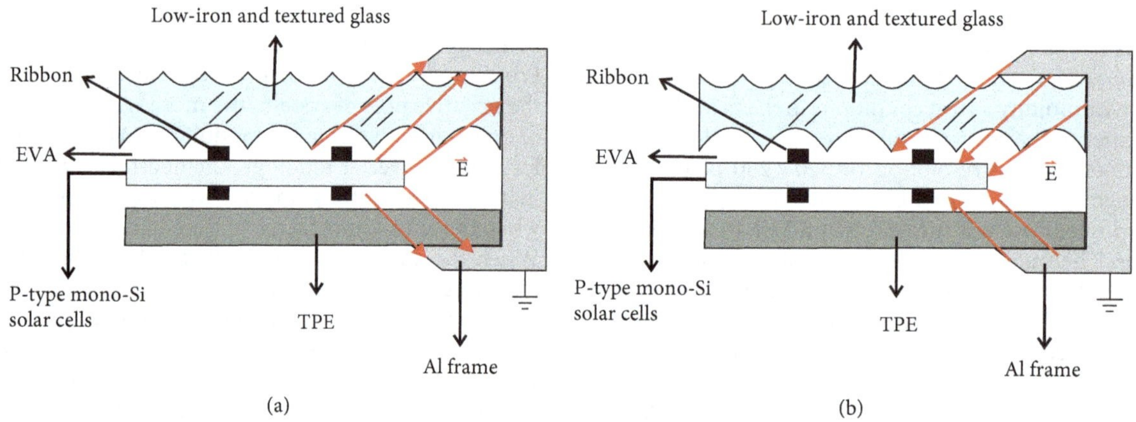

FIGURE 5: The direction of the electric field between the Al frame and solar cells of the module. (a) The direction of the electric field is directed from the solar cells to the Al frame. (b) The direction of the electric field is directed from the Al frame to the solar cells.

operation. This phenomenon means that the neutral point of voltage in a module string moved forward to the positive pole for a PID-affected module string as time goes on. If there does not exist an electric field produced by positive voltage applied to the cells, PID does not happen.

Based on the above analysis, we find that the positive potential difference between the ground and solar cells is a necessary condition for PID. To eliminate the electric field produced by the positive potential difference between the ground and cells is an effective measure to prevent PID.

3.2. Thermographic Images of Solar Modules from the Same PID-Affected Module String. The thermographic images of the solar modules from the same PID-affected module string are displayed in Figure 6. The photo was taken in the field under an ambient temperature of ~20°C, global irradiation of ~800 W/m^2, and wind speed of ~1 m/s. During image acquisition, the module strings were operated at maximum power point. The thermographic images of 18 solar modules had the highest temperature spot, where the junction box was built on the back. This is because the junction box affects the heat dissipation of the solar cells. From Figure 6, the thermographic images from module 1 to module 9 exhibit a

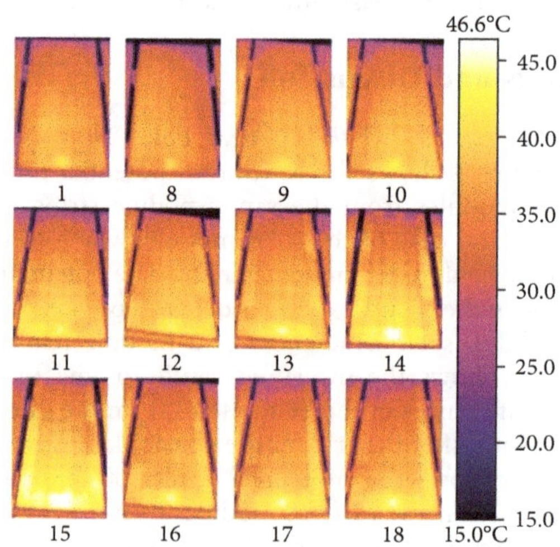

FIGURE 6: The thermographic images of solar modules from the same PID-affected module string.

homogeneous temperature distribution. No solar cells have an elevated temperature from module 1 to module 9. It is seemingly suggested that the modules from 1 to 9 are PID-

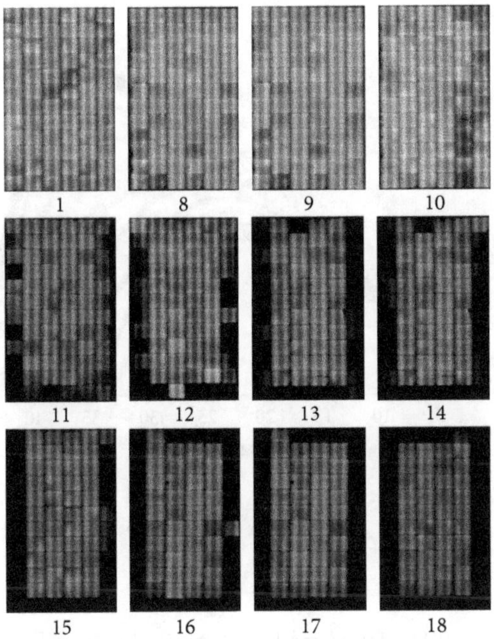

FIGURE 7: The electroluminescence images of solar modules dismounted from the same PID-affected module string.

free. This agrees with [14, 15]. Instead, there are some solar cells with an elevated temperature from module 10 to module 18. The closer a module is to the negative pole, the greater is the number of cells with an elevated temperature. The thermographic images of solar modules from the same PID-affected module string in the field show a regular degradation pattern. This is different with the results obtained in the lab oratory [16, 17].

An interesting finding is that the temperature of the solar cells adjacent to the Al frame is higher than the middle cells in a PID-affected solar module. This is because the electric field adjacent to the Al frame is more intensive than the middle area; that is, the solar cells adjacent to the Al frame suffer more severe PID. The shunt resistance of PID-affected cells decreases dramatically; consequently, the dark current of the cells increases remarkably under illumination. The performance of the solar cells adjacent to the Al frame is inferior to the middle cells for a PID-affected solar module. So, a smaller portion of the sunlight is converted to electrical energy. Thus, PID-affected cells exhibit a higher temperature than sound ones.

3.3. Electroluminescence Images of Solar Modules Dismounted from the Same PID-Affected Module String. Figure 7 shows the electroluminescence images of solar modules dismounted from the same PID-affected module string in the power plant. The photos were taken indoors. From Figure 7, we find that the cells from modules 1 to 9 have the same EL intensity. From module 10 to module 18, some solar cells adjacent to the Al frame appear dark in the EL images, while the middle cells appear bright in a PID-affected solar module. The closer a module is to the negative pole, the greater is the number of dark cells in a PID-affected module. The darker the appearance of the solar cell is, the more serious is the power degradation caused by PID. The EL images of the solar modules dismounted from the same PID-affected module string yield a regular degradation pattern. Comparing Figure 6 with Figure 7, there is a good agreement between the electroluminescence images and the thermographic images, which exhibit a similar degradation pattern.

3.4. The Power Degradation of Solar Modules Dismounted from the Same PID-Affected Module String in the Field. Figure 8 shows the I-V and P-V curves of solar modules dismounted from the same PID-affected module string under standard test conditions. Table 1 shows the electrical parameters of module 8 before shipments and after one year of operation. The electrical parameters of module 9 before shipments and after one year of operation are presented in Table 2. From here, it can be observed that the power degradation rate is nearly the same from module 1 to module 8. Module 9 has a 2.5% power drop after one year of operation, and it is higher than that of module 8. This shows that the low positive potential difference between the ground and cells also leads to PID for a PID-prone solar module. This shows that the solar module slightly affected by PID could not be identified by the thermographic and electroluminescence images because of limitations in accuracy. The peak powers and open circuit voltages (Voc) of modules from 9 to 18 decrease remarkably in sequence. However, the short circuit currents of the PID-affected modules from the same module string have a slight change; this is consistent with some findings in the literature [18]. The closer a module is to the negative pole, the more serious is the power degradation of the PID-affected solar module.

Table 3 shows the electrical parameters for module 18 before shipments and after one year of operation. From Table 3, we found that the Voc of module 18 decreases from 45.01 V to 31.86 V, and the peak power of module 18 decreases from 196.95 W to 92.06 W, which was up to 53.26% power degradation after one year of operation. This shows that the potential-induced degradation leads to a catastrophic impact on the performance and energy yield of photovoltaic power plants.

4. Conclusions

This paper presents an investigation of PID-affected crystalline silicon solar modules from the field. For the first time, we find that low positive voltage could cause PID for PID-prone solar modules. For a PID-affected module string connected with the grid, the neutral point of voltage in a module string moved forward to the positive pole as time goes on. In a PID-affected photovoltaic power plant, the thermographic and electroluminescence images of the solar modules in a module string exhibit a regular degradation pattern simultaneously. This study also indicates that the power degradation rate of solar modules dismounted from the field is related with the position of the module installation in a module string. These new findings in this contribution can help identify and uncover the mystery of PID. The microanalysis of PID-affected cells dismounted from the field must be further

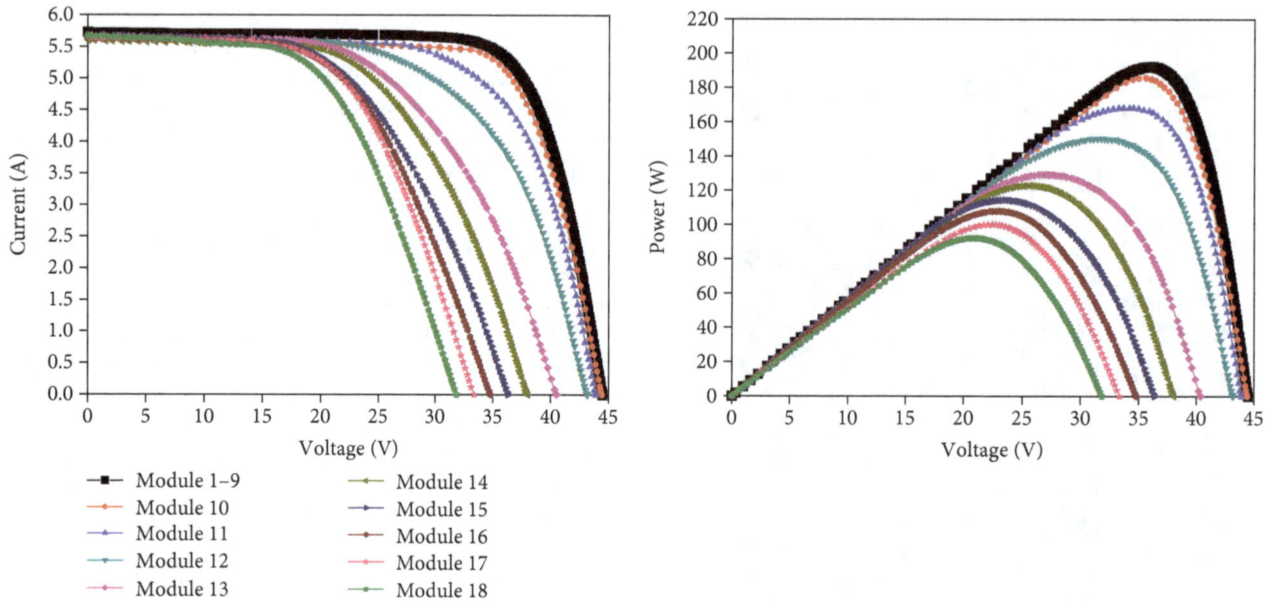

FIGURE 8: The I-V and P-V curves of solar modules disassembled from the same PID-affected module string in the power plant under standard test conditions.

TABLE 1: The electrical parameters for module 8 before shipments and after one year of operation.

Module 8	Voc (V)	Isc (A)	Vm (V)	Im (A)	Rsh (Ω)	P_{max} (W)
Before shipments	44.96	5.73	36.58	5.38	398.23	196.81
After one year of operation	44.92	5.62	36.48	5.22	385.45	193.28
Degradation rate (%)	0.09	1.92	0.27	2.97	3.21	1.79

Test conditions: 1 kW/m^2 irradiance, 25°C module temperature and AM1.5 global spectrum.

TABLE 2: The electrical parameters of module 9 before shipments and after one year of operation.

Module 9	Voc (V)	Isc (A)	Vm (V)	Im (A)	Rsh (Ω)	P_{max} (W)
Before shipments	45.02	5.78	36.52	5.40	389.27	197.19
After one year of operation	44.87	5.65	36.32	5.29	360.54	192.27
Degradation rate (%)	0.33	2.25	0.55	2.04	7.38	2.5

Test conditions: 1 kW/m^2 irradiance, 25°C module temperature and AM1.5 global spectrum.

TABLE 3: The electrical parameters for module 18 before shipments and after one year of operation.

Module 18	Voc (V)	Isc (A)	Vm (V)	Im (A)	Rsh (Ω)	P_{max} (W)
Before shipments	45.01	5.75	36.54	5.39	409.13	196.95
After one year of operation	31.86	5.60	19.30	4.77	182.36	92.06
Degradation rate (%)	29.22	2.61	47.18	11.50	55.43	53.26

Test conditions: 1 kW/m^2 irradiance, 25°C module temperature and AM1.5 global spectrum.

developed. An extension of this study is currently underway at our laboratory.

Conflicts of Interest

The authors declare that they have no conflicts of interest.

Acknowledgments

The authors would like to thank the support of the National Key Technology R&D Program (Grant no. 2015BAA09B01). This study was also supported by the Natural Science Foundation of China (Grant nos. 61376067 and 61274050) and the 2016 Grid State Science and Technology Program (NY71-16-027).

References

[1] D. C. Jordan, S. R. Kurtz, K. Vansant, and J. Newmiller, "Compendium of photovoltaic degradation rates," *Progress in Photovoltaics: Research and Applications*, vol. 24, no. 7, pp. 978–989, 2016.

[2] A. Ndiaye, A. Charki, A. Kobi, C. M. F. Kébé, P. A. Ndiaye, and V. Sambou, "Degradations of silicon photovoltaic modules: a literature review," *Solar Energy*, vol. 96, pp. 140–151, 2013.

[3] K. Mishina, A. Ogishi, K. Ueno et al., "Plasma-enhanced chemical-vapor deposition of silicon nitride film for high resistance to potential-induced degradation," *Japanese Journal of Applied Physics*, vol. 54, no. 8S1, article 08KD12, 2015.

[4] S. Jonai, K. Hara, Y. Tsutsui, H. Nakahama, and A. Masuda, "Relationship between cross-linking conditions of ethylene vinyl acetate and potential induced degradation for crystalline silicon photovoltaic modules," *Japanese Journal of Applied Physics*, vol. 54, no. 8S1, article 08KG01, 2015.

[5] S. Pingel, O. Frank, M. Winkler et al., "Potential induced degradation of solar cells and panels," in *2010 35th IEEE Photovoltaic Specialists Conference*, pp. 2817–2822, Honolulu, HI, USA, June 2010.

[6] M. Koehl and S. Hoffmann, "Impact of rain and soiling on potential induced degradation," *Progress in Photovoltaics: Research and Applications*, vol. 24, no. 10, pp. 1304–1309, 2016.

[7] B. Ziebarth, M. Mrovec, C. Elsasser, and P. Gumbsch, "Potential-induced degradation in solar cells: electronic structure and diffusion mechanism of sodium in stacking faults of silicon," *Journal of Applied Physics*, vol. 116, no. 9, article 093510, 2014.

[8] V. Naumann, D. Lausch, A. Hähnel et al., "Explanation of potential-induced degradation of the shunting type by Na decoration of stacking faults in Si solar cells," *Solar Energy Materials and Solar Cells*, vol. 120, pp. 383–389, 2014.

[9] S. P. Harvey, J. Moseley, A. Norman et al., "Investigating PID shunting in polycrystalline silicon modules via multiscale, multitechnique characterization," *Progress in Photovoltaics: Research and Applications*, vol. 26, no. 6, pp. 377–384, 2018.

[10] H. Wang, P. Zhao, H. Yang, J. Chang, D. Song, and S. Sang, "Performance variation of dark current density-voltage characteristics for PID-affected monocrystalline silicon solar modules from the field," *Microelectronics Reliability*, vol. 81, pp. 320–327, 2018.

[11] H. Yang, F. Wang, H. Wang, J. Chang, D. Song, and C. Su, "Performance deterioration of P-type single crystalline silicon solar modules affected by potential induced degradation in photovoltaic power plant," *Microelectronics Reliability*, vol. 72, pp. 18–23, 2017.

[12] H. Yang, H. Wang, D. Cao, D. Sun, and X. Ju, "Analysis of power loss for crystalline silicon solar module during the course of encapsulation," *International Journal of Photoenergy*, vol. 2015, Article ID 251615, 5 pages, 2015.

[13] J. Oh, S. Bowden, and G. S. Tamizhmani, "Potential-induced degradation (PID): incomplete recovery of shunt resistance and quantum efficiency losses," *IEEE Journal of Photovoltaics*, vol. 5, no. 6, pp. 1540–1548, 2015.

[14] C. Buerhop-Lutz, R. Weibmann, H. Scheuerpflug, R. Auer, and C. J. Brabec, "Quality control of PV-modules in the field using a remote-controlled drone with an infrared camera," in *Proceedings of the 27th European Photovoltaic Solar Energy Conference and Exhibition (EU PVSEC 2012) 2012*, pp. 3370–3373, Frankfurt, Germany, 2012.

[15] T. Kaden, K. Lammers, and H. J. Moller, "Power loss prognosis from thermographic images of PID affected silicon solar modules," *Solar Energy Materials and Solar Cells*, vol. 142, pp. 24–28, 2015.

[16] M. B. Koentopp, M. Krober, and C. Taubitz, "Toward a PID test standard: understanding and modeling of laboratory tests and field progression," *IEEE Journal of Photovoltaics*, vol. 6, no. 1, pp. 252–257, 2016.

[17] P. Hacke, S. Spataru, K. Terwilliger et al., "Accelerated testing and modeling of potential-induced degradation as a function of temperature and relative humidity," *IEEE Journal of Photovoltaics*, vol. 5, no. 6, pp. 1549–1553, 2015.

[18] F. Martínez-Moreno, G. Figueiredo, and E. Lorenzo, "In-the-field PID related experiences," *Solar Energy Materials and Solar Cells*, vol. 174, pp. 485–493, 2018.

7

CFD Analysis and Electrical Efficiency Improvement of a Hybrid PV/T Panel Cooled by Forced Air Circulation

Gökhan Ömeroğlu

Mechanical Engineering Department, Engineering Faculty, Atatürk University, 25300 Erzurum, Turkey

Correspondence should be addressed to Gökhan Ömeroğlu; gomeroglu@atauni.edu.tr

Academic Editor: Philippe Poggi

The thermal and electrical efficiency of a custom-designed PV/T panel cooled by forced air circulation was investigated by experimental and computational fluid dynamics (CFD) analysis. Experiments were carried out with four different array configurations, under constant irradiation of $1100\,W/m^2$ and 3 different air velocities (3.3 m/s, 3.9 m/s, and 4.5 m/s). The heat transfer surface area and forced air circulation are known to positively affect the total heat transfer, and therefore, it is foreseeable that an increased number of fins and higher air velocities will help maintain the electrical efficiency of the panel at higher levels. The main objective of this study is to determine the critical threshold for the abovementioned parameters as well as to show how important parameters, such as fin arrangement and consequent turbulent air flows, are for satisfying the heat removal needs. Highest efficiency was achieved with a 108 pc type 1 arrangement at 12.02% as expected. Nevertheless, while the 108 pc type 2 arrangement could maintain the electrical efficiency at 11.81%, a close level of 11.55% could be obtained with a 54 pc type 2 arrangement. Experimental results are compared with ANSYS Fluent program, and the effect of the number and arrangement of the fins on the efficiency of the panel has been shown.

1. Introduction

A photovoltaic/thermal (PV/T) collector is considered to be a solar collector combined with a photovoltaic module and produces electricity and heat at the same time. Different kinds of solar energy systems are used in various engineering fields [1]. There are mainly two types of PV/T collectors depending on the medium used to collect the thermal energy, air-based and water-based. PV/T air collectors are significantly advantageous than are PV/T water collectors, which require a variety of thermal collection materials depending on PV modules. One of the most important components of a solar energy system is the collector. The various applications of the solar energy collectors have been well recorded [2]. Solar collectors can be used in various areas especially for drying, floor heating, and solar desalination and so on Conventional solar air collectors have inherent disadvantages in lower thermal efficiency. Ekechukwu and Norton [3] conducted experiments to demonstrate a typical 40% to 50% efficiency of the conventional solar air collector. For a PV/T air collector with a ventilated system, the heat losses from PV to air flow are generally not very good, so the losses in the environment are large and the average thermal efficiency is around 10% to 20% [4]. Wijeysundera et al. [5] also investigated the thermal performance of single-glazed and double-glazed multipass solar air heaters. Choudhury et al. [6] found that the single-plated double-pass air heater was the most cost effective compared to the other design. Joshi et al. [7] performed a new optimum efficiency of a hybrid PV/T system. To investigate the performance in New Delhi meteorological conditions, they investigated two types of PV modules (glass-to-plastic and glass-to-glass) and the results show that the overall performance of the hybrid thermal collector with a glass-to-glass PV module is better than with a glass-to-plastic module.

In addition, Choudhury et al. [8] studied theoretical models for single- and double-ended three-pass solar air heaters. Ho et al. [9] studied the improvement of device performance of external recirculating multipass solar air heaters under counter flow conditions. Fudholi et al. [10] stated that they had achieved efficiencies as high as 70% of multipass solar collectors with an increased heat transfer surface.

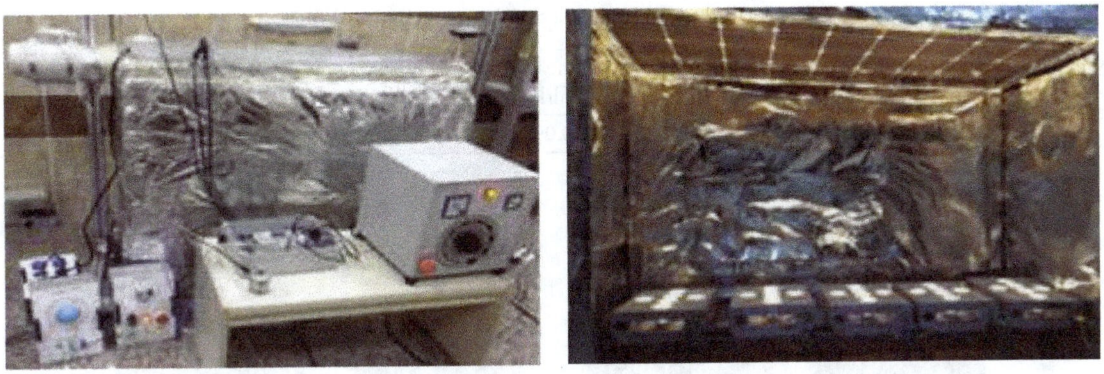

(a) (b)

FIGURE 1: (a) PV/T hybrid test setup. (b) Artificial solar radiation and the position of the photovoltaic panel.

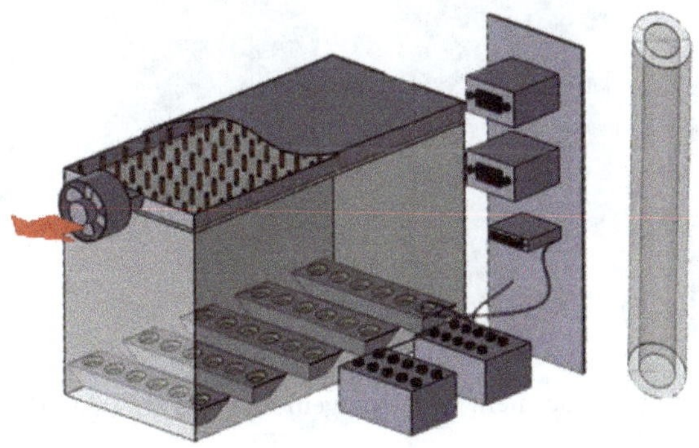

(a) (b)

FIGURE 2: (a) Schematic view of the experimental setup. (b) Fin geometry.

The different pathways leading to significant improvements in collector efficiency are, for example, corrugated surfaces [11–13], matrix-type absorbers [14], composite honeycomb aggregators [15], box-type absorbers [16], porous media [17], and finned absorbers [18–20].

This study shows that the heat transfer coefficient increases with increasing Reynolds number. The thermal efficiency of the PV module increases with increasing flow rate. In this study, investigating the effect of active cooling with different configurations, such as the fin number, arrangement, and air velocity, on the electrical efficiency of the PV module is aimed. Moreover, a CFD model has been used to validate the results of the experiments. The main objective was to investigate the effects of fin arrangement and subsequent formation of turbulent air rather than the heat transfer surface area (the number of fins in this case) and air velocity on the total heat drawn from the PV panel. As the highest insolation on the earth's surface is 1100 W/m^2, this figure was chosen for the experiments.

The fins used in this study are cheap and easy to manufacture. Besides, on the contrary to other heat transfer elements used in other studies, it is the first time that they are being used in PV/T systems. Another reason to choose them is that they are easy to install and offer a quite low friction loss during the air circulation and hence does not require a high fan power.

2. Material and Methodology

A test setup was designed to investigate the thermal and electrical performance of the PV/T air system. This system was studied in the Renewable Energy Resources Laboratory in Ataturk University. A general view of the experimental setup and a schematic diagram is shown in Figures 1 and 2(a), respectively. A fan with 10 W of power has been employed to feed more air through the control volume in order to create forced convection, and ambient air, used as the working fluid, was sucked through the air duct of the test setup, as shown in Figure 2. The fan is attached to the outlet of the control volume rather than the inlet in order to prevent the fan from heating up the feed air entering the control volume over time.

This experimental study was designed to investigate how temperature affects the efficiency and power output of a PV panel during operation and by using various fin arrays. Polycrystalline solar modules were used in the experiment to generate electricity. The electricity generated by the solar modules is stored in batteries. And the fan was powered by the PV per se, and no external power supply is employed for feeding air. In order to increase the active cooling effect, copper fins, the specifications of which are given in Table 1 and illustrated in Figure 2(b), were fitted in different arrangements as shown in Figure 3. The copper fins were glued with thermal paste onto the back of the PV module, and the air

Table 1: Physical properties of copper cylindrical fins.

Specific weight (kg/dm^3)	Thermal conductivity (W/m·K)	Melting point (°C)	Specific heat capacity (kcal/kg·K)	Resistivity ($\mu\Omega$)	Weight (kg/m)	Outer/inner diameter (mm)	Length (mm)
8.9	336	1065–1085	0.092	1.71	0.70	12/10	50

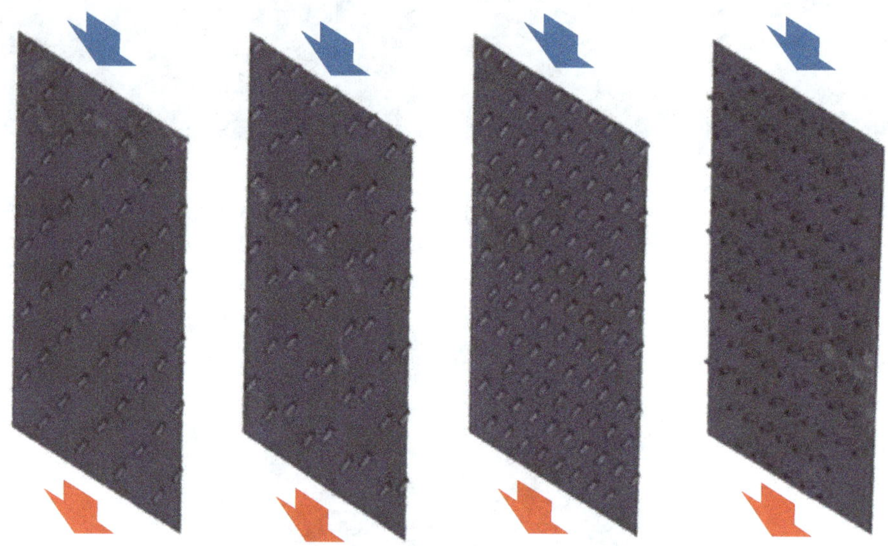

Figure 3: Placement and arrangement of fin configurations.

duct providing air passage was connected to the back of the PV module. The inlet/outlet manifold of the duct uses precision air pressure and velocity gauges to ensure that the air pressure and velocity of the air have been also measured when uniform air flow is being made. Temperature measurements have a crucial role in this experiment and were therefore used to calibrate T-type thermocouples. In this experiment, the PV current, the PV voltage, the temperature of the panels, the air temperatures at the inlet and outlet manifolds, the air velocity, and the solar radiation were measured. Also in this experiment, instead of natural sunlight, artificial solar radiation was used. The panel size is 55 cm wide and 120 cm in length.

The copper fins are attached to the back surface of the PV panel in two different numbers, 54 pcs and 108 pcs, and four different arrangements, 54 pc shifted (type 1), 54 pc staggered (type 2), 108 pc double-shifted (type 1), and 108 pc shifted (type 2). For 54 pc type 1 fins, odd and even rows have 4 and 5 fins, respectively. The vertical spacing between each row is set to be 10.5 cm, with a horizontal spacing of 12.5 cm between fins. In the staggered second arrangement (type 2), each row is staggered one over another with a 5 cm vertical spacing, and spacing between fins is again 12.5 cm. This configuration is a grouped form of type 1, having a vertical spacing of 20 cm between the row groups. As for 108 pc copper fins, in a double-shifted type 1 arrangement, odd and even rows have 8 and 10 fins, respectively. The vertical spacing of each row is 10 cm, and the horizontal spacing between the fins is 5.5 cm. In a shifted type 2 arrangement, odd rows and even rows have 9 and 10 fins, respectively, except for the last two rows missing 3 fins on both sides in order to keep the number of fins the same as in type 1. The vertical spacing between the rows is 5 cm, and the horizontal spacing between fins is 5.5 cm. The amount of radiation was kept constant at $\dot{Q}_{\text{solar}} = 1100\ \text{W/m}^2$ by using artificial solar lamps during the experiments. The intensity of this radiation is the maximum insolation level which can be measured on earth.

As shown in Figure 4, the test setup is mainly composed of two basic sections, namely, the test section and the control section, and 17 parts or auxiliary elements named in Table 2.

The law of conservation of mass (also referred to as the first law), the most basic law of thermodynamics, applies to all thermodynamic calculations. Before performing the thermodynamic analysis of the photovoltaic thermal (PVT) system, which is the subject of this study, it should be noted that the system is a control volume with continuous flow (an open system). A control volume is the general name for systems with a mass flow within their boundaries. Both mass and energy can be transferred through the boundaries of the control volume [21–24].

The mass transfer from any system to the control volume or from the control volume to the system in a Δt time interval is equal to the change of mass in the control volume at that time interval [25, 26].

$$\dot{m}_{\text{inlet}} - \dot{m}_{\text{outlet}} = \Delta_{\text{cv}} \left[\frac{\text{kg}}{\text{s}}\right]. \tag{1}$$

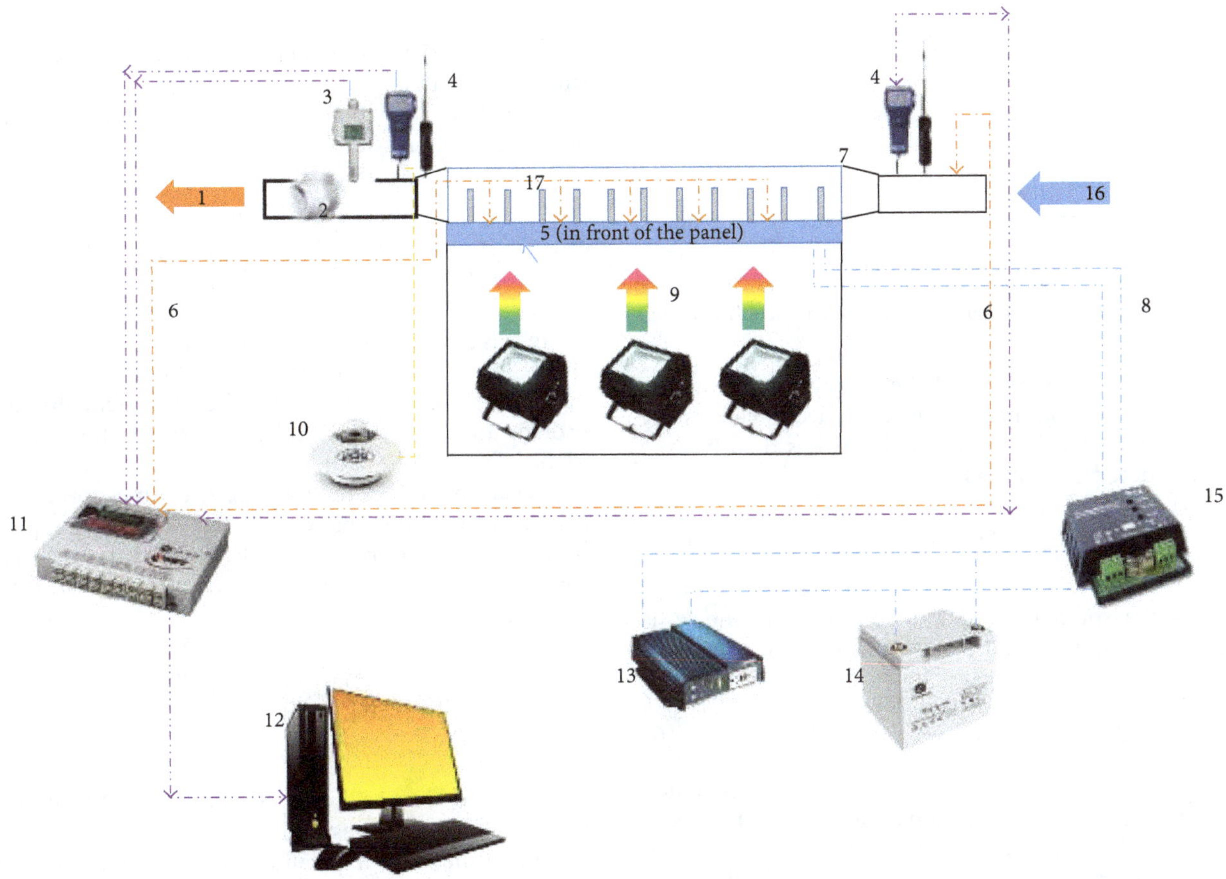

Figure 4: Experimental schematic setup.

Table 2: Parts of the experimental setup.

Test section
(1) Air outlet
(2) Fan
(3) Temperature humidity transmitter
(4) Input velocity anemometer
(5) Photovoltaic panel
(6) Thermocouple
(7) Control volume
(8) Photovoltaic power transmission
(9) Artificial solar radiation
Control section
(10) Pyranometer
(11) Data logger
(12) Computer
(13) DC-AC inverter
(14) Acidic gel battery
(15) Solar charger regulator
(16) Air inlet
(17) Cooling fins

The energy equation for a continuous flow system, as given by Masters [27] and Radziemska [21], is

$$\dot{Q}_U = \dot{m} \cdot c_p (T_{\text{inlet}} - T_{\text{outlet}}) \text{ [kW]}. \tag{2}$$

The total useful solar power from the sun is the sum of the thermal power gain and the electrical power production ($\dot{Q}_e = I \cdot V$) [28–31] and is given as

$$\dot{Q}_{\text{gain}} = \dot{m} \left[c_p (T_{\text{inlet}} - T_{\text{outlet}}) \right] + I \cdot V \text{ [kW]}. \tag{3}$$

The total electrical energy produced is obtained if the energy drawn by the fan ($W_{\text{fan}} = I_{\text{fan}} \cdot V_{\text{fan}}$) is removed, and the first-law efficiency, as given by Zagorska et al. [29] and Zimmermann et al. [31], is

$$\begin{aligned} \eta_I &= \frac{\dot{Q}_u + \dot{Q}_e - W_{\text{fan}}}{\dot{Q}_{\text{solar}}} \\ &= \frac{\dot{Q}_{\text{gain}} - W_{\text{fan}}}{\dot{Q}_{\text{solar}}}. \end{aligned} \tag{4}$$

When there is a flow in the system, regarding the calculations, it is important to determine whether the flow is

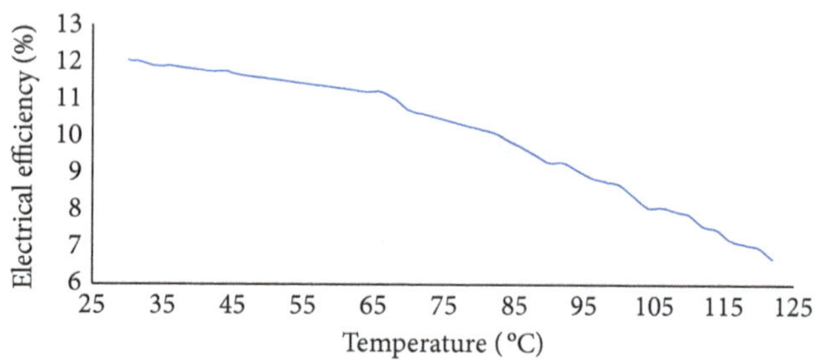

FIGURE 5: Electrical efficiency-temperature of the noncooling panel.

compressible or incompressible. Whether the compressibility effect in a flow is important or not is concluded by calculating the Mach number. If the Mach number is less than 0.3, the flow is considered incompressible.

$$\mathrm{Ma} = \frac{V}{c}, \tag{5}$$

where c is the speed of sound ($c = 340.29$ m/s) and V is the velocity of the fluid, whereby the flow is concluded to be Newtonian.

$$Q = h \cdot A_{\mathrm{pv}} \cdot \left(\Delta T_{\mathrm{air}} - T_{\mathrm{pv}}\right) \,[\mathrm{kW}]. \tag{6}$$

Using (6), with the help of the energy the air has gained within the designed control volume, the total heat transfer coefficient for the system can be calculated. As the total heat transfer coefficient is a function of the total heat drawn from the system, Nusselt number calculation can be made there from

$$\mathrm{Nu} = \frac{h \cdot D_{\mathrm{h}}}{k_{\mathrm{air}}}. \tag{7}$$

Against each Reynolds value for the air acting as the refrigerant in the system, a different Nusselt number will correspond.

The channel friction factor can be computed along the test section by pressure loss, ΔP, as

$$f = \frac{\Delta P}{(L/D_{\mathrm{h}}) \cdot \rho \cdot U_{\mathrm{m}}^2/2}. \tag{8}$$

In this equation, ρ is density of film temperature of the air and U_{m} is the average velocity of the fluid.

In the subject of strategies for heat transfer enhancement, enough attention should be paid towards flow resistance, friction factor value, and performance factor.

Regarding the Nu number and friction factor, the performance factor of the fins is evaluated to have an appropriate comparison of overall thermohydrodynamic performance among test fins. Performance evaluation criteria, η, is written as [32–34]

$$\eta = \left(\frac{\mathrm{Nu}}{\mathrm{Nu}_0}\right)\left(\frac{f_0}{f}\right)^{1/3}, \tag{9}$$

where f_0 and Nu_0 are the friction factor and Nusselt number of the plain channel, respectively.

With the aim of validation, the obtained Nusselt number (Nu) and friction factor (f) of the plain channel have been compared to those correlations proposed in the literature. The empirical correlations for the plain channel are considered to achieve the Nusselt number and friction factor as [35, 36]

$$\begin{aligned} &\mathrm{Nu} = 0.023\mathrm{Re}^{0.8}\,\mathrm{Pr}^{\frac{1}{3}}\text{: Colburn correlation } \left(\text{for } 10^4 \le \mathrm{Re}\right),\\ &\mathrm{Nu} = 0.023\mathrm{Re}^{0.8}\,\mathrm{Pr}^{n}\ (n = 0.3\text{–}0.4)\text{: Dittus-Boelter correlation } \left(\text{for } 10^4 \le \mathrm{Re}\right),\\ &f = 0.316\mathrm{Re}^{-0.25}\text{: Blasius correlation } (\mathrm{Re} \le 20000),\\ &f = (0.790 \ln \mathrm{Re} - 1.64)^{-2}\text{: Petukhov correlation } \left(\text{for } 3000 \le \mathrm{Re} \le 5 \times 10^6\right). \end{aligned} \tag{10}$$

To compare the experimental results, experiments were performed using the identical fins in different numbers and arrangements as shown in Figure 3.

3. Results and Discussion

The surface temperature of a noncooled photovoltaic panel increased to about 120°C over time due to the fact that solar cells convert only a small portion of incident solar radiation into electricity and accumulates the rest as waste heat in the unit unless it is recovered and utilized through another thermal system [37]. The ratio of generated electrical energy to total incoming energy is called the electrical efficiency, whereas the thermal efficiency is the ratio of net utilized thermal energy to total incident solar energy received on the PV panel.

The increase in cell temperature reduces the electrical efficiency of the photovoltaic panel, which is plotted versus the surface temperature in Figure 5.

In order to keep the electrical efficiency of the photovoltaic panel at an appropriate level, the optimum cell temperature should be around 50–60 degrees. In order to achieve the necessary cooling in the cell temperature, the heat transfer surface area was increased using fins, convection heat transfer was increased using forced circulation, and different fin arrangements were tried in order to create turbulence within the air flow.

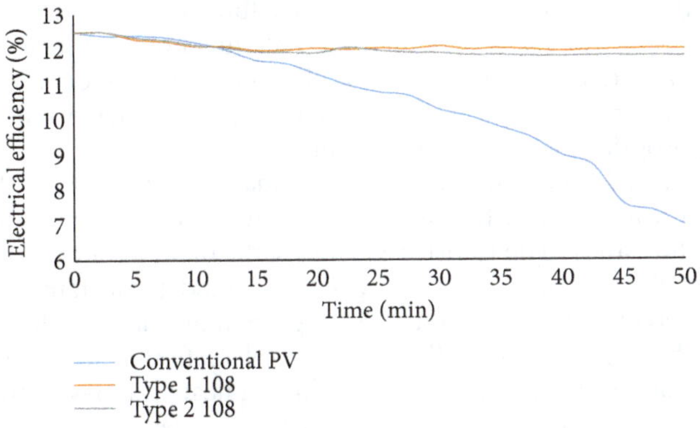

FIGURE 6: Change of electrical efficiency with 108 pc fin arrangements at 4.5 m/s.

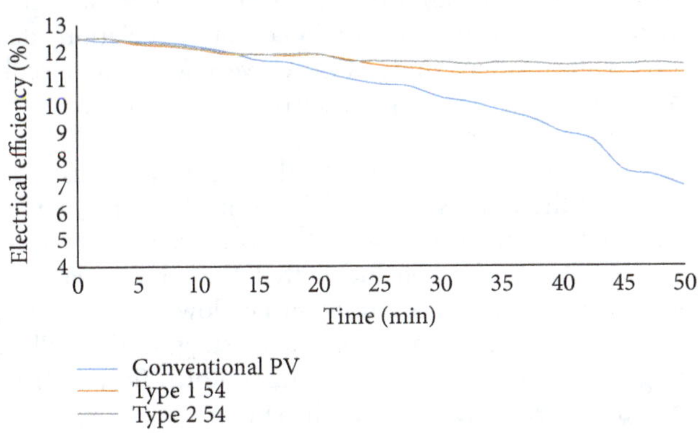

FIGURE 7: Change of electrical efficiency with 54 pc fin arrangements at 4.5 m/s.

Consequently, the total heat drawn from the PV panel has been increased and the cell temperature decreased to near optimum operating temperature; hence, the electrical efficiency could be maintained at higher levels.

The improvement in the electrical efficiency of the photovoltaic panel is shown in Figures 6 and 7 per fin arrangement at 4.5 m/s. In the experiments, it was observed that electrical efficiency did not change during the first 15-minute time period, but electrical efficiency decreased drastically in the next time period. As a result, while the electrical efficiency of the uncooled photovoltaic panel reduced from 12.5% to approximately 7%, thanks to the heat transfer enhancement provided by 54 pc fin arrangements and 108 pc fin arrangements used in active cooling, the reduction in the electrical efficiency has been prevented at a significant extent and could be kept at fairly high levels. Efficiencies for type 1 and type 2 arrangements of 108 pc fins are 12.02% and 11.81%, respectively (Figure 6). As for 54 pc fin type 1 and type 2 arrangements, these figures were 11.2% and 11.55%, respectively (Figure 7).

The rate of change in the thermal efficiency of the PV/T panel is shown in Figures 8 and 9 for different arrangements. The best thermal efficiency rate increased to 74% at type 108 and to 60% at type 54. The improvement of thermal efficiency increased the electrical efficiency as well.

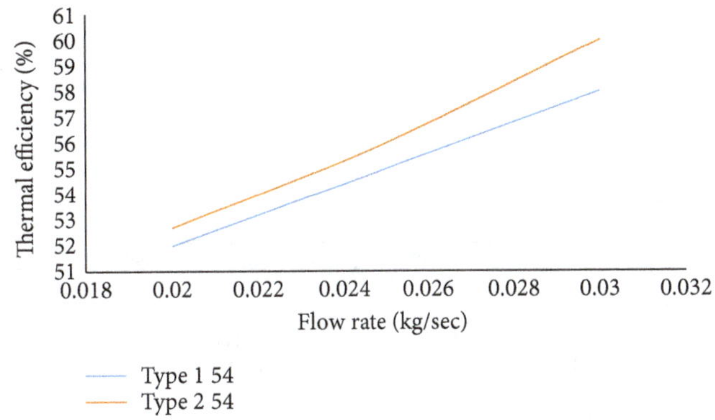

FIGURE 8: Variation of thermal efficiency versus flow rate (kg/s) for 54 pc arrangements.

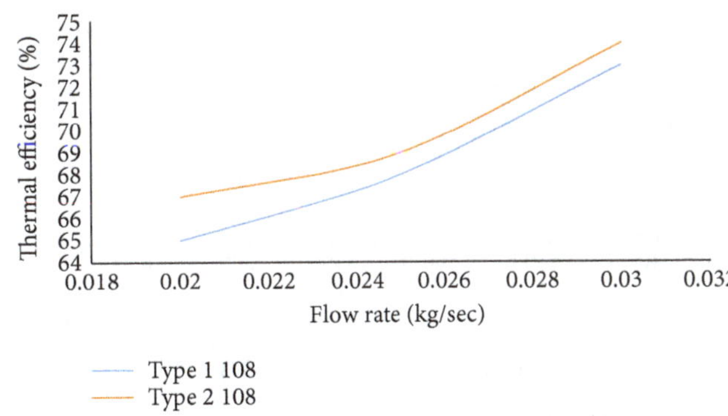

FIGURE 9: Variation of thermal efficiency versus flow rate (kg/s) for 108 pc arrangements.

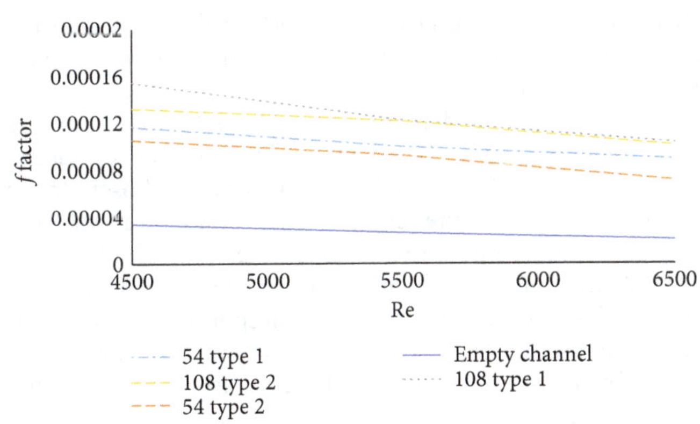

FIGURE 10: Friction factor—Reynolds number.

The heat transfer friction factor and heat transfer coefficient of fins used in the experiments are given in Figures 10 and 11. As seen from Figures 10 and 11, while the rate of heat transfer increased with increasing air velocity, the effectiveness of the friction coefficient decreased.

Figure 12 shows the change in Reynolds number and thermal enhancement factor (η) for all fin arrangements. Nusselt number and friction factor values were compared for the same Reynolds number. It is seen that all thermal enhancement factor values are above 1. With increased Reynolds numbers, there is generally a decrease in the

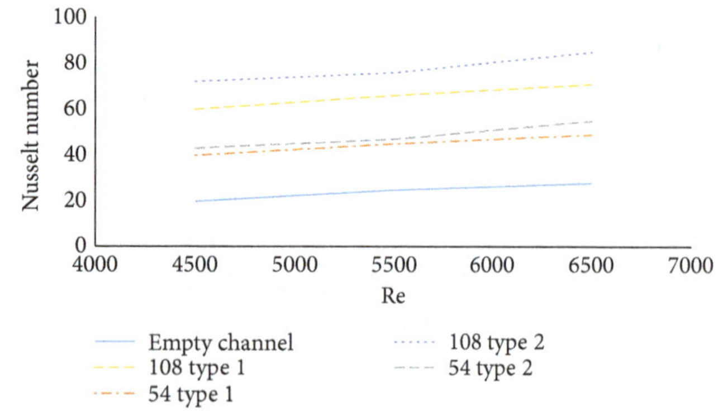

Figure 11: Nusselt representations of different fin configurations in the photovoltaic system.

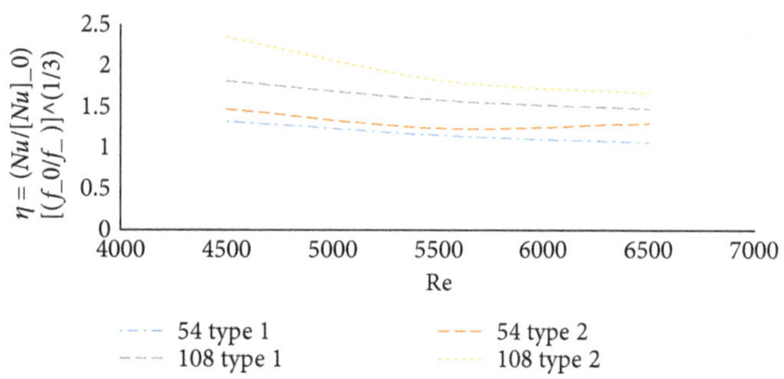

Figure 12: Thermal enhancement factor η for varying Reynolds numbers.

thermal enhancement factor. The best result for the thermal enhancement factor was 108 type 2, while 54 type 1 gave the lowest result.

In order to validate these results obtained from the experiments, a CFD model was prepared. The boundary conditions used for the CFD analysis were chosen to be those in effect in the experimental evaluation, wherein the input air temperature is 23°C, the solar irradiance is 1100 W/m^2, and air velocities are 3.3, 3.9, and 4.5 m/s. In the model preparation, the most important phase prior to the solution is the phase in which the material properties are defined, the geometric model is established, and meshing is performed.

A solution validation study was performed according to the CFD verification and validation standard to evaluate the mesh uncertainty and appropriate mesh parameters for the mesh simulation.

For the model presented in Figure 13, nonstructural four-sided cages were formed which were refined with prismatic near-wall elements (inflated). Stage grating treatments with a factor of 1.3 were applied to the edge separation bar and bar. The ratio between the height of the last six-sided layer and the first prism was the same for all cages. Also, different manifold designs were modeled using the ANSYS Fluent program, the CFD of generic calculated fluids, to provide for the distribution of flows in channels.

The standard k-ε model solution does not require a wall model for the solution of the problem. In the k-ε turbulence models, "enhanced wall function" was used as the wall model. The enhanced wall function is the most advanced mode among the wall models. This model also incorporates the viscous effects in the walls to provide a more appropriate solution to the problem.

In addition, by forming a boundary layer effect on the hulls, it was seen that the flow structure gave better results in this study. The transfer equations, the method of calculating the turbulence viscosity, and the model constants are different for each model. The generation of turbulence, turbulence due to the lift force, the calculation of the effects of compressibility, and the modeling of heat and mass transfer are essentially common within each model.

In Figures 14–16 are images obtained from the simulations showing velocity contours for three different air velocities per configuration, along with those for the empty control volume. It can easily be seen from these figures that, with the contribution of forced air circulation, dead volume zones, where there is no or very low flow circulation occurs, have been significantly removed, especially in the air outlet region.

As it is seen in Figure 16, the air flow distribution throughout the cross section of the control volume is more intense and temperature distribution is more homogeneous. Moreover, it may be concluded that the air circulation is more effective due to the turbulent air flows in the vicinity of fins. The simulation results also suggest that all fin arrangements, in terms of providing uniform flow distribution with no recirculation, are efficient.

No doubt, the designed system will help draw more heat from the PV module and reduce the temperature during operation, which in turn will help maintain the electrical efficiency at higher levels. But the point is whether another satisfactory configuration with less fins and lower air velocity is possible with just changing the fin arrangement or not. Temperature contours suggest that it is.

Figures 17–19 show the temperature distribution for three different air velocities in all configuration types. As seen in Figures 17–19, the blown air could draw almost no heat from the panel surface at any air velocities when the control volume has no fins installed within. Again, for this configuration, the intensity of dead volume where the air gets stuck and heated up is significant. For 54-piece and 108-piece type 1 arrangements, the temperature distribution is far better, and the removal of dead volume is clearly seen in Figures 17–19.

The best temperature distribution, hence the most heat drawing as well as minimum dead volume zones, is achieved with a 108-piece type 2 arrangement at 4.5 m/s air velocity, which is quite foreseeable, but the most important finding here is that the resulting 54-piece fins in type 1 arrangement offers is very close to those of 108-piece type 1 and type 2. This shows that the arrangement is as much important a parameter as the heat transfer surface area for the total heat transfer rate. Taking into account the cost of fins and the friction loss, and the pressure drop, hence the increased fan power, the 54-piece type 1 arrangement can be preferred rather than any arrangement of 108-piece fins.

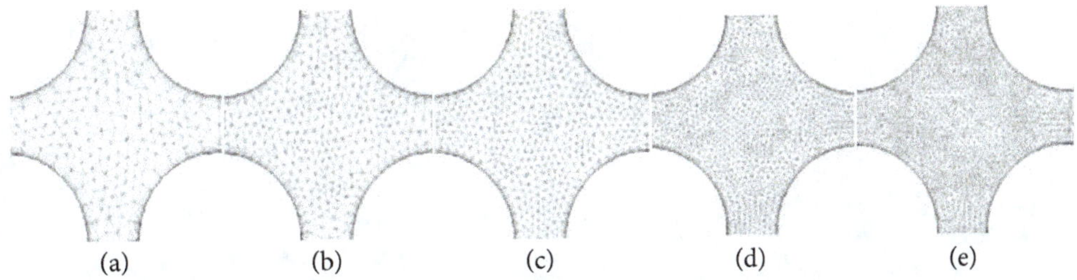

FIGURE 13: Mesh section in the areas where there are no fins.

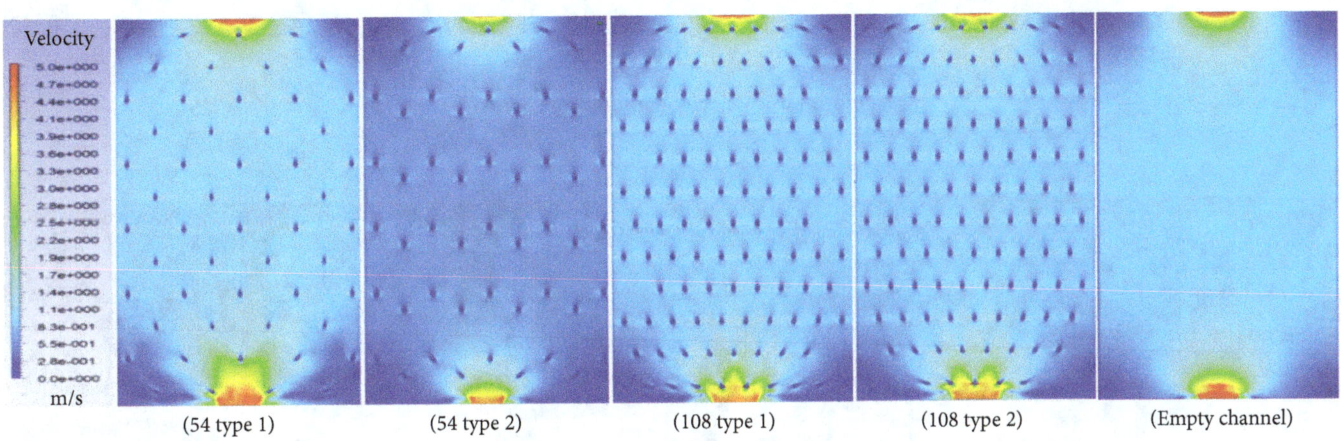

FIGURE 14: Velocity contours at 4.5 m/s.

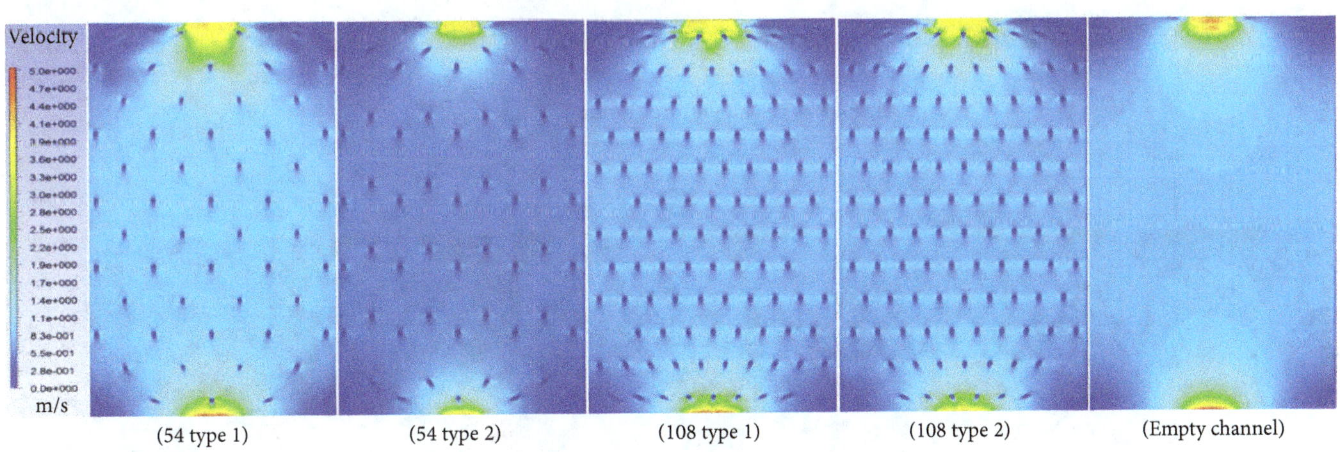

FIGURE 15: Velocity contours at 3.9 m/s.

4. Conclusion

In this work, both electrical and thermal energy is produced by the hybrid PV/T system. The electrical efficiencies of the PV/T panels are severely reduced as the surface temperatures increase. Cooling the surface of the panel is an effective way to avoid this reduction in efficiency. If no cooling is used, the operating temperature of the PV module reaches a value as high as 120°C and the electrical efficiency drops about to 7% from 12.5%. As the efficiency of the PV panel without active cooling decreases in relation to increasing surface temperature when the panel is not cooled, the forced convection heat transfer zone is formed in the control volume in the back of the PV panel. As a result, a cooling of up to 60–65 degrees at the surface temperature of 120 degrees has been achieved, so the decrease in the electrical current is prevented to a great extent and the efficiency could be maintained at 12.02%.

Consequent turbulent air flows occurring within the control volume, especially in the vicinity of the fins, contributes to heat removal from the panel; therefore, with the 108 pc type 1 arrangement, the highest efficiency was achieved

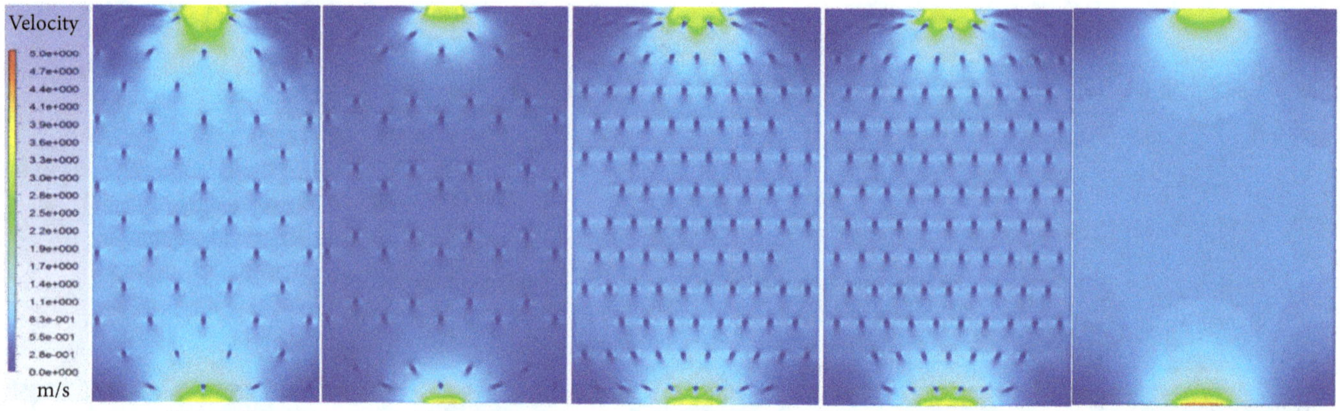

(54 type 1) (54 type 2) (108 type 1) (108 type 2) (Empty channel)

FIGURE 16: Velocity contours at 3.3 m/s.

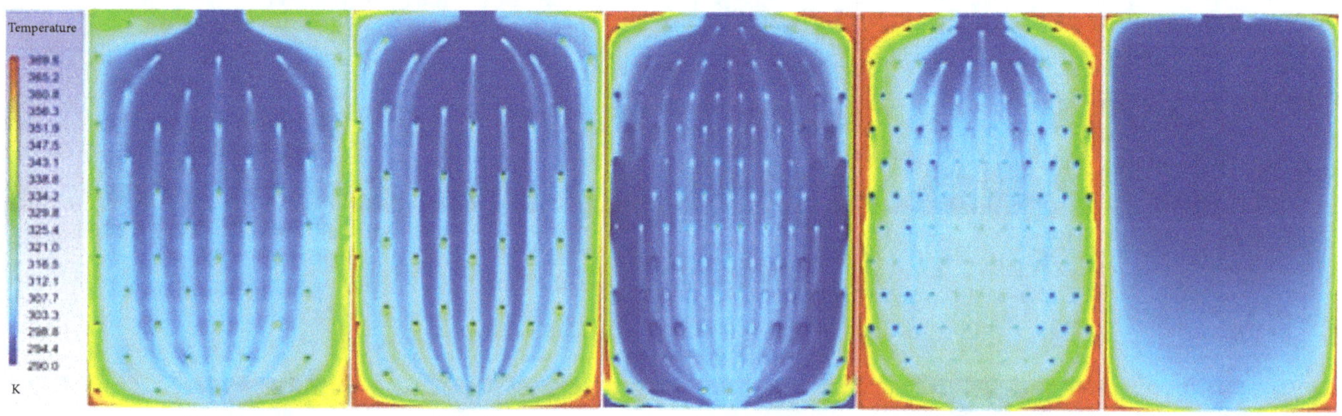

(54 type 1) (54 type 2) (108 type 1) (108 type 2) (Empty channel)

FIGURE 17: Temperature contours at 4.5 m/s.

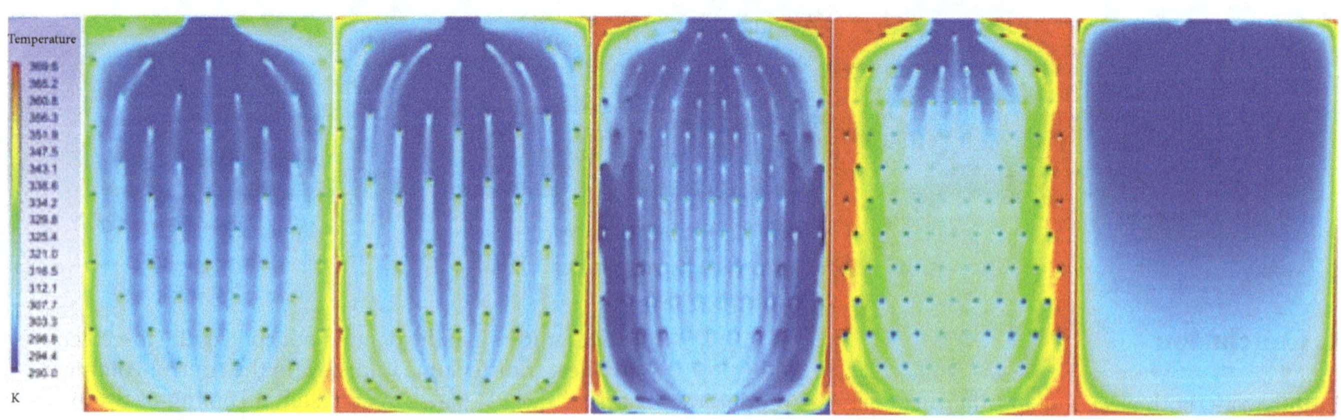

(54 type 1) (54 type 2) (108 type 1) (108 type 2) (Empty channel)

FIGURE 18: Temperature contours at 3.9 m/s.

at 12.02% as expected. Nevertheless, while the 108 pc type 2 arrangement could maintain the electrical efficiency at 11.81%, a close level of 11.55% could be obtained with the 54 pc type 2 arrangement, whereby we can see the impact of arrangement on heat transfer and could conclude that there is a threshold value for the heat transfer surface area somewhere between 50 and 60 pc fins. Similar efficiency gains can be achieved by using fewer fin elements and by

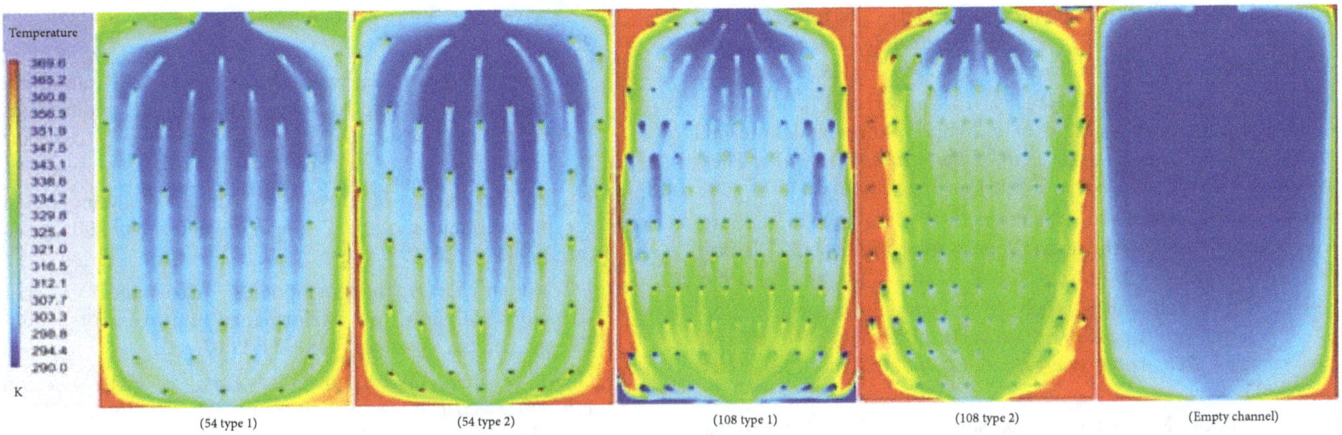

FIGURE 19: Temperature contours at 3.3 m/s.

creating the correct turbulence model compared to the cost of the cooling unit.

The best panel efficiency was with 108-piece type 1 at 4.5 m/s air velocity, but the efficiency achieved with the 54-piece type 2 arrangement was very close with 11.55%. This suggests that, thanks to the turbulence effect created by the fin arrangement, the heat transfer rate with 54 fins has been very close to that of twice as much fins.

In addition to the heat transfer surface area and air velocity, it is seen that the fin arrangement is an important parameter in the heat transfer rate. It can be concluded that with better designed fin arrangements, the required fan speed could be lowered for any number of fins or similar results could be obtained with fewer fins. On the other hand, since the power required by the PV-powered fan will also decrease with fewer fins and at lower air velocities, the total efficiency obtained from the PV system will also increase.

Nomenclature

c: Velocity of sound (m/s)
$\dot{Q}_e$: Electrical power production (W)
$\dot{Q}_u$: Usable thermal power (W)
$\dot{Q}_{solar}$: Solar radiation (W/m^2)
$\Delta\dot{m}_{cv}$: Difference of mass in control volume (kg/s)
Ma: Mach number
$\dot{m}_{inlet}$: Inlet mass of air (kg/s)
$\dot{m}_{outlet}$: Outlet mass of air (kg/s)
I: Direct current (A)
V: Voltage (volts)
η: Performance evaluation criteria
ρ: Density of film temperature of the air
U_m: Average fluid velocity
Re: Reynolds number
Nu: Nusselt number
Pr: Prandtl number
T_{inlet}: Inlet air temperature (°C)
T_{outlet}: Outlet air temperature (°C)
v: Velocity of air (m/s)
w_{fan}: Work of fan (W)
c_p: Specific heat capacity (J/kg·K)
h: Heat transfer coefficient in air duct (W/m^2·K)
A_{pv}: Aperture area of PV module (m^2)
T_{air}: Air temperature (°C)
T_{pv}: Photovoltaic cell temperature (°C)
Nu: Nusselt number
$\dot{Q}_{gain}$: Gained total power (W)
k: Heat transfer coefficient for conduction (W/m·K)
f: Friction factor
Dh: Effective diameter of the channel
ΔP: Pressure loss
L: Channel length.

Conflicts of Interest

The author declares that he has no conflicts of interest.

Acknowledgments

The author would like to thank the laboratory staff of Ataturk University Engineering Faculty Mechanical Engineering for their time and support.

References

[1] A. Fudholi, K. Sopian, M. H. Ruslan, M. A. Alghoul, and M. Y. Sulaiman, "Review of solar dryers for agricultural and marine products," *Renewable and Sustainable Energy Reviews*, vol. 14, no. 1, pp. 1–30, 2010.

[2] M. Grupp, H. Bergler, J. P. Bertrand, B. Kromer, and J. Cieslok, ""Convective" flat plate collectors and their applications," *Solar Energy*, vol. 55, no. 3, pp. 195–207, 1995.

[3] O. Ekechukwu and B. Norton, "Review of solar-energy drying systems III: low temperature air-heating solar collectors for crop drying applications," *Energy Conversion and Management*, vol. 40, no. 6, pp. 657–667, 1999.

[4] H. A. Zondag, W. G. J. Van Helden, M. Bakker et al., *PVT roadmap. A European guide for the development and market introduction of PVT Technology*, Energy research Centre of the Netherlands ECN, Petten (Netherlands), 2006.

[5] N. Wijeysundera, L. L. Ah, and L. E. Tjioe, "Thermal performance study of two-pass solar air heaters," *Solar Energy*, vol. 28, no. 5, pp. 363–370, 1982.

[6] C. Choudhury, P. Chauhan, and H. Garg, "Performance and cost analysis of two-pass solar air heaters," *Heat Recovery Systems and CHP*, vol. 15, no. 8, pp. 755–773, 1995.

[7] A. Joshi, A. Tiwari, G. N. Tiwari, I. Dincer, and B. V. Reddy, "Performance evaluation of a hybrid photovoltaic thermal (PV/T) (glass-to-glass) system," *International Journal of Thermal Sciences*, vol. 48, no. 1, pp. 154–164, 2009.

[8] C. Choudhury, P. M. Chauhan, H. P. Garg, and S. N. Garg, "Cost—benefit ratio of triple pass solar air heaters," *Energy Conversion and Management*, vol. 37, no. 1, pp. 95–116, 1996.

[9] C. Ho, C. Yeh, and S. Hsieh, "Improvement in device performance of multi-pass flat-plate solar air heaters with external recycle," *Renewable Energy*, vol. 30, no. 10, pp. 1601–1621, 2005.

[10] A. Fudholi, M. Ruslan, and K. Sopian, "Review of multi-pass solar air collectors with heat transfer surface enhancement," in *Proceedings of the 4th Scientific Conference PPI-UKM*, pp. 3–11, Supranto, Malaysia, 2009.

[11] A. Fudholi, K. Sopian, M. Y. Othman et al., *Heat transfer correlation for the V-groove solar collector*, Solar Energy Research Institute, University Kebangsaan Malaysia, 43600 Bangi Selangor Malaysia, 2008.

[12] M. Hafiz, M. Y. Othman, M. M. Salleh, and K. Sopian, "Design and indoor testing of the V-groove back pass solar collector," *Renewable Energy*, vol. 16, pp. 2118–2122, 1999.

[13] W. Gao, W. Lin, T. Liu, and C. Xia, "Analytical and experimental studies on the thermal performance of cross-corrugated and flat-plate solar air heaters," *Applied Energy*, vol. 84, no. 4, pp. 425–441, 2007.

[14] A. Kolb, E. Winter, and R. Viskanta, "Experimental studies on a solar air collector with metal matrix absorber," *Solar Energy*, vol. 65, no. 2, pp. 91–98, 1999.

[15] A. Abdullah, H. Abou-Ziyan, and A. Ghoneim, "Thermal performance of flat plate solar collector using various arrangements of compound honeycomb," *Energy Conversion and Management*, vol. 44, no. 19, pp. 3093–3112, 2003.

[16] K. Matrawy, "Theoretical analysis for an air heater with a box-type absorber," *Solar Energy*, vol. 63, no. 3, pp. 191–198, 1998.

[17] K. Sopian, M. A. Alghoul, E. M. Alfegi, M. Y. Sulaiman, and E. A. Musa, "Evaluation of thermal efficiency of double-pass solar collector with porous–nonporous media," *Renewable Energy*, vol. 34, no. 3, pp. 640–645, 2009.

[18] S. Youcef-Ali, "Study and optimization of the thermal performances of the offset rectangular plate fin absorber plates, with various glazing," *Renewable Energy*, vol. 30, no. 2, pp. 271–280, 2005.

[19] M. A. Karim and M. Hawlader, "Performance investigation of flat plate, v-corrugated and finned air collectors," *Energy*, vol. 31, no. 4, pp. 452–470, 2006.

[20] P. Naphon, "On the performance and entropy generation of the double-pass solar air heater with longitudinal fins," *Renewable Energy*, vol. 30, no. 9, pp. 1345–1357, 2005.

[21] E. Radziemska, "The effect of temperature on the power drop in crystalline silicon solar cells," *Renewable Energy*, vol. 28, no. 1, pp. 1–12, 2003.

[22] M. Rahman, M. Hasanuzzaman, and N. Rahim, "Effects of various parameters on PV-module power and efficiency," *Energy Conversion and Management*, vol. 103, pp. 348–358, 2015.

[23] E. Saloux, A. Teyssedou, and M. Sorin, "Analysis of photovoltaic (PV) and photovoltaic/thermal (PV/T) systems using the exergy method," *Energy and Buildings*, vol. 67, pp. 275–285, 2013.

[24] F. Sobhnamayan, F. Sarhaddi, M. A. Alavi, S. Farahat, and J. Yazdanpanahi, "Optimization of a solar photovoltaic thermal (PV/T) water collector based on exergy concept," *Renewable Energy*, vol. 68, pp. 356–365, 2014.

[25] H. Teo, P. Lee, and M. Hawlader, "An active cooling system for photovoltaic modules," *Applied Energy*, vol. 90, no. 1, pp. 309–315, 2012.

[26] J. Tonui and Y. Tripanagnostopoulos, "Air-cooled PV/T solar collectors with low cost performance improvements," *Solar Energy*, vol. 81, no. 4, pp. 498–511, 2007.

[27] G. M. Masters, "Wind power systems," in *Renewable and Efficient Electric Power Systems*, pp. 307–383, John Wiley & Sons, Inc., Hoboken, NJ, USA, 2005.

[28] J. Tonui and Y. Tripanagnostopoulos, "Performance improvement of PV/T solar collectors with natural air flow operation," *Solar Energy*, vol. 82, no. 1, pp. 1–12, 2008.

[29] V. Zagorska, I. Ziemelis, L. Kancevica, and H. Putans, "Experimental investigation of photovoltaic-thermal hybrid solar collector," *Agronomy Research Biosystem Engineering Special*, vol. 1, pp. 227–234, 2012.

[30] W. Harry, *Recent facts about photovoltaics in Germany*, Fraunhofer Institute for Solar Energy Systems (ISE), Germany, 2014.

[31] S. Zimmermann, H. Helmers, M. K. Tiwari et al., "A high-efficiency hybrid high-concentration photovoltaic system," *International Journal of Heat and Mass Transfer*, vol. 89, pp. 514–521, 2015.

[32] P. Promvonge, "Thermal augmentation in circular tube with twisted tape and wire coil turbulators," *Energy Conversion and Management*, vol. 49, no. 11, pp. 2949–2955, 2008.

[33] S. Eiamsa-ard and P. Promvonge, "Experimental investigation of heat transfer and friction characteristics in a circular tube fitted with V-nozzle turbulators," *International Communications in Heat and Mass Transfer*, vol. 33, no. 5, pp. 591–600, 2006.

[34] G. Ömeroğlu, "Experimental and computational fluid dynamics analysis of a photovoltaic/thermal system with active cooling using aluminum fins," *Journal of Photonics for Energy*, vol. 7, no. 4, article 045503, 2017.

[35] K. Yakut and B. Sahin, "The effects of vortex characteristics on performance of coiled wire turbulators used for heat transfer augmentation," *Applied Thermal Engineering*, vol. 24, no. 16, pp. 2427–2438, 2004.

[36] S. Eiamsa-ard and P. Promvonge, "Enhancement of heat transfer in a tube with regularly-spaced helical tape swirl generators," *Solar Energy*, vol. 78, no. 4, pp. 483–494, 2005.

[37] Ö. İlhan Volkan, M. K. Yeşilyurt, E. Ç. Yilmaz, and G. Ömeroğlu, "Photovoltaic thermal (PVT) solar panels," *International Journal of New Technology and Research*, vol. 2, no. 12, pp. 13–16, 2016.

8

Short-Term Photovoltaic Power Generation Combination Forecasting Method Based on Similar Day and Cross Entropy Theory

Qi Wang,[1,2] Shunxiang Ji,[1,2] Minqiang Hu,[1,2] Wei Li,[3] Fusuo Liu,[3] and Ling Zhu[3]

[1]*School of Electrical & Automation Engineering, Nanjing Normal University, Nanjing 210042, China*
[2]*Jiangsu Province Gas-Electricity Integrated Energy Engineering Laboratory, Nanjing 210046, China*
[3]*State Key Laboratory of Smart Grid Protection and Control, Nanjing 211106, China*

Correspondence should be addressed to Qi Wang; wangqi@njnu.edu.cn

Academic Editor: Mohammed Ashraf Gondal

The forecast for photovoltaic (PV) power generation is of great significance for the operation and control of power system. In this paper, a short-term combination forecasting model for PV power based on similar day and cross entropy theory is proposed. The main influencing factors of PV power are analyzed. From the perspective of entropy theory, considering distance entropy and grey relation entropy, a comprehensive index is proposed to select similar days. Then, the least square support vector machine (LSSVM), autoregressive and moving average (ARMA), and back propagation (BP) neural network are used to forecast PV power, respectively. The weights of three single forecasting methods are dynamically set by the cross entropy algorithm and the short-term combination forecasting model for PV power is established. The results show that this method can effectively improve the prediction accuracy of PV power and is of great significance to real-time economical dispatch.

1. Introduction

Photovoltaic (PV) power generation is widely used in human life and built around the world in recent years. However, due to the obvious impact of meteorological factors, PV power generation has some characteristics such as uncertainty, volatility, and intermittency. It is detrimental to do safe dispatch and energy management [1]. The operational risk will increase when PV power generation system is connected to the grid. Accurate prediction of PV output power can provide a basis for dispatching decision and it is of great significance to reduce system operating costs and ensure the safety and stability of the power system.

Until now, much research has been devoted to the forecasting of PV power generation. According to the difference of forecasting methods, it can be divided into two categories: direct method and indirect method [2]. The indirect method predicts solar irradiance by using meteorological data of PV power plants and then uses the relevant calculation formulas or algorithms to calculate the PV output power [3, 4]. The direct method combines the meteorological data with historical PV power data to predict output power [5–8]. Yuan et al. [5] used BP neural network method to predict the PV power and the weather types have been considered. Lan et al. [6] established an ARMA and Markov chain prediction model for short-term PV power forecast. Jing et al. [7] used extreme learning machine method for the short-term PV power forecasting. In [8], a forecast model for PV power was established by least square support vector machine, but the prediction accuracy is affected by the model parameters.

Single direct forecast methods mentioned above may have some limitations and the prediction accuracy can be further improved. Appropriate combination forecasting methods can effectively improve the prediction accuracy of PV power generation [9–12]. Wang et al. [9] proposed a combination forecast method based on improved grey BP neural network; fuzzy C means method was used to classify the historical data and then select similar days as the model training samples. Li et al. [10] adopted the grey relational analysis method to determine the meteorological factors of the highest impact on PV power generation. Then combined with the advantages of each individual prediction model, it proposed the combined forecasting model based on IOWA operators. Yang et al. [11] established the combination forecast method for PV power based on entropy method and obtained appropriate combination weights. Yang and Chen [12] proposed a combination method in PV power forecasting based on the correlation coefficient and the prediction accuracy was improved. Although these combination forecasting methods can effectively improve the prediction accuracy, the weights of the single forecast methods are fixed in these combination methods and cannot reflect the real-time changes of PV power.

Accordingly, in order to further improve PV power forecasting accuracy, this paper proposes a short-term PV power forecasting method based on similar day and cross entropy theory. First of all, the main influencing factors of PV power are analyzed. Distance entropy and grey relation entropy are introduced. The similarity degree selection index is proposed to select similar days. Then least square support vector machine (LSSVM), autoregressive and moving average (ARMA) method, and BP neural network are used to forecast PV power, respectively. Using the cross entropy algorithm to dynamically set the weights of three single forecast methods, the short-term PV power combination forecasting model based on the cross entropy theory is established. The correctness and superiority of this model are verified by cases analysis and comparison with the combination method based on the sum of the squared errors and the combination method based on the correlation coefficient.

2. Similar Day Selection

2.1. Entropy Theory. Entropy is a measure of the degree of chaos in a sequence. Entropy theory originated from the laws of thermodynamics and has been widely used in many fields such as systems science, information science, and management science.

Entropy can describe the degree of chaos in the sequence. The probability of occurrence of each variable in the sequence is $p(w_i)$ $(i = 1, 2, \ldots, l)$; the entropy value of the sequence is defined as

$$H = -C \sum_{i=1}^{l} p(w_i) \ln p(w_i), \quad (1)$$

where C is a constant and l is the number of variables.

2.2. Distance Entropy. Distance entropy is a combination of Euclidean distance and information entropy [13]. Euclidean distance is a method to effectively measure the similarity between two sequences, but it treats the differences between variables of different nature in the sequence as equivalent and sometimes cannot meet the actual requirements. Here, the combination of Euclidean distance and information entropy can overcome the shortcomings of Euclidean distance. The lower the distance entropy value, the more information is represented. That is to say, the less the difference between the comparison sequence and the reference sequence, the closer to the reference sequence.

The Euclidean distance of each meteorological characteristic variable between the historical day i and the forecast day is

$$d_i(k) = \left| x_i(k) - x'(k) \right|, \quad (2)$$

where $x'(k)$ and $x_i(k)$ are the meteorological characteristic variable k of the forecast day and the historical day i, respectively. Here $x'(k)$ and $x_i(k)$ need to be normalized first.

The ratio of the distance between each meteorological characteristic variable and the distance sum of all the characteristic variables of the historical day can be calculated. The ratio is the probability of occurrence of the meteorological characteristic variable for the historical day. Thereby, the distance entropy value of the historical day i can be calculated as

$$E_d(i) = -\frac{1}{\ln K} \sum_{k=1}^{K} \frac{d_i(k)}{\sum_{k=1}^{K} d_i(k)} \ln \frac{d_i(k)}{\sum_{k=1}^{K} d_i(k)}, \quad (3)$$

where $d_i(k)$ is the distance of the meteorological characteristic variable k between the historical day i and the forecast day; K is the total number of meteorological characteristic variables.

2.3. Grey Relation Entropy. Grey relation entropy is a combination of grey relational analysis method and information entropy. Based on the grey relational analysis method, the information entropy theory is used to quantitatively describe the degree of similarity between the influencing factors in the system, which can compensate for the local correlation tendency and personality information loss [14]. The greater the grey correlation entropy value is, the stronger the correlation between the comparison sequence and the reference sequence.

The correlation coefficient between the historical day i and the forecast day is

$$\varepsilon_i(k) = \frac{\min_i \min_k \left| x'(k) - x_i(k) \right| + \rho \max_i \min_k \left| x'(k) - x_i(k) \right|}{\left| x'(k) - x_i(k) \right| + \rho \max_i \max_k \left| x'(k) - x_i(k) \right|}, \quad (4)$$

where $x'(k)$ and $x_i(k)$ are the meteorological characteristic variable k of the forecast day and the historical day i. Here $x'(k)$ and $x_i(k)$ need to be normalized first.

The ratio of the correlation coefficient of each meteorological characteristic variable and the sum of the correlation coefficient of all characteristic variables for historical days can be calculated. The probability of occurrence of the meteorological characteristic variable k can be obtained as

$$p_i(k) = \frac{\varepsilon_i(k)}{\sum_{k=1}^{K} \varepsilon_i(k)}, \quad (5)$$

where $\varepsilon_i(k)$ is the correlation coefficient of the meteorological characteristic variable k of the historical day i and the forecast day; K is the total number of meteorological characteristic variables.

The grey relation entropy of the historical day i can be expressed as

$$E_\varepsilon(i) = -\frac{1}{\ln K}\sum_{k=1}^{K} p_i(k)\ln p_i(k). \tag{6}$$

2.4. Influence Factors of PV Power. The output power of PV power is related to many influencing factors, including solar irradiance, temperature, humidity, wind speed, and weather type. The output power of PV array is calculated as follows [15]:

$$P_s = \eta SI[1 - 0.005(t_0 + 25)], \tag{7}$$

where t_0 is the operating temperature (°C) of the photovoltaic cell, I is the solar irradiation intensity (kW/m^2), S is the area of the photovoltaic array (m^2), and η is the conversion efficiency of the photovoltaic cell (%).

It is assumed that the array area and conversion efficiency are constant for the short-term PV power prediction. It can be seen from equation (7) that when the two parameters are determined, the PV power is only related to two factors: solar radiance intensity and temperature. Due to the different seasons and weather types, the intensity of solar radiance is also quite different and the output power of PV arrays is also very different. Since the data of solar irradiance of the forecast day is generally difficult to be obtained, this paper uses the sunshine hours to replace it and selects the mean temperature, the maximum temperature, the minimum temperature, the relative humidity, the minimum humidity, and the sunshine hours as the main influencing factors to select similar days. These similar days are selected as training samples to predict PV power.

2.5. Similar Day Selection Based on Entropy Theory. The basic steps for similar day selection based on entropy theory are as follows:

Step 1. First, determine the type of season and weather of the forecast day, and then select the samples of historical days with the same season and weather type, and the number of samples is N.

Step 2. Select the mean temperature, maximum temperature, minimum temperature, relative humidity, minimum humidity, and sunshine hours of samples to form the meteorological characteristic vectors. The i_{th} $(i = 1, 2, \ldots, N)$ day meteorological characteristic vector is $C_i = [T_i\ T_{\max_i}\ T_{\min_i}\ H_i\ \text{H}_{\min_i}\ S_i]$. T_i, $T_{\max_i}$, and $T_{\min_i}$ indicate the mean temperature, maximum temperature, and minimum temperature of the i_{th} day, respectively; H_i and $\text{H}_{\min_i}$ indicate the relative humidity and minimum humidity of the i_{th} day; S_i indicates sunshine hours of the i_{th} day.

Step 3. Calculate the distance entropy $E_d(i)$ and the grey relation entropy $E_\varepsilon(i)$ of the i_{th} day.

Step 4. The distance entropy and grey relation entropy are combined into a similarity index R_i to characterize the similarity of historical day i. The calculation formula is

$$R_i = \frac{E_d(i)}{E_\varepsilon(i)}. \tag{8}$$

According to the comprehensive similarity index R_i, the six similar days are selected as the training samples for PV power prediction.

2.6. Cross Entropy Theory. Cross entropy (CE) is derived from the definition of entropy and is a measure method for the information difference between two random vectors. It is also known as the Kullback-Leibler (K-L) distance. K-L distance is not a physical distance in length but is used to describe the difference between two probability distribution functions. The lower the cross entropy value is, the more similar the two probability density distribution functions. The definition of cross entropy can be divided into two situations [16].

Discrete situation:

$$\begin{cases} D(f\|g) = \sum_{i=1}^{n} f_i \log_c \dfrac{f_i}{g_i}, \\ \sum_{i=1}^{n} f_i = 1, \\ \sum_{i=1}^{n} g_i = 1. \end{cases} \tag{9}$$

Continuous situation:

$$\begin{cases} D(f\|g) = \int_a^h f(x) \log_c \dfrac{f(x)}{g(x)} dx, \\ \int_a^b f(x)dx = 1, \\ \int_a^b g(x)dx = 1, \end{cases} \tag{10}$$

where f and g stand for probability vectors in equations (9) and (10).

The cross entropy stands for the distance between f and g. It is the description of the closeness degree between the two probabilities.

Theorem 1. *f and g are probability density functions; for $D(f = g) \geq 0$, it equals to zero only when $f = g$.*

Property 1. For $D(f = g) \geq 0$, it equals only when $f = g$ almost everywhere.

Cross entropy algorithm is as a random optimization method and can be used to simulate small probability event and solve the optimization problem. The cross entropy method has been applied to solve the practical problems such as combined optimization, multiobjective optimization, combined forecast, and machine learning.

2.7. Single Forecasting Methods. PV power is affected by solar irradiance, weather types, season types, temperature, humidity, etc. It is difficult to describe by a mathematical model. ARMA, BP neural network, and LSSVM method are often used to forecast PV power. ARMA is a linear model, which can predict the overall trend of the data. BP neural network and LSSVM model are nonlinear models with strong nonlinearity learning ability. Therefore, this paper selects these three single methods to forecast PV power.

2.8. ARMA. Auto regression moving average (ARMA) model is a random time sequence analysis model founded by Box and Jenkins; it is also called the B-J method. The ARMA model is a combination of autoregressive (AR) and moving average (MA) models [17]. The AR model uses past values, and the MA model uses past error values. The linear ARMA can be expressed as

$$x_t = \sum_{k=1}^{p} \phi_k x(t-kT) + \sum_{k=1}^{q} \theta_k x(t-kT) + \varepsilon(t), \tag{11}$$

where ϕ_k and θ_k are the AR and MA model coefficients. $x(t)$ is the output of the ARMA model and $\varepsilon(t)$ is the residual. T is the sampling time; p and q are the orders of AR and MA models, respectively.

In this paper, the input variables of ARMA model are historical PV power data of similar days. The forecast process mainly includes the test of sequence stability, model parameter estimation, and model ranking. Stationarity is tested by Augmented Dickey Fuller (ADF) unit root test. The model parameters are estimated by least square method and the model order is determined according to the Akaike info criterion (AIC).

2.9. BP. BP neural network is a multilayer feedforward neural network with error back propagation training, which has good self-organizing learning ability and can realize any nonlinear mapping from input to output. The prediction model mainly uses the input signal forward propagation and error signal back propagation to realize the training process, which can deal with parallel process of large-scale data and has the ability of robustness and fault tolerance [18]. The BP neural network basic structure is shown in Figure 1. w_1 is the connection weight between the input layer and the hidden layer node and w_2 is the connection weight between the hidden layer and the output layer node.

2.10. LSSVM. The least square support vector machine (LSSVM) was proposed by Suykens J. A. K in 1999, which is an improvement of the standard support vector machine (SVM) [19].

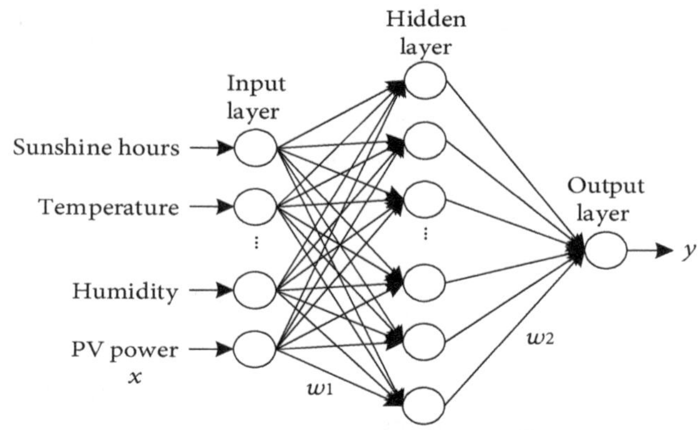

FIGURE 1: The structure of BP neural network.

LSSVM is mainly used to solve the problem of quadratic optimization of functions, which can be described as

$$\min_{w,b,e} \quad \varphi(w,b,e) = \frac{\|w\|^2}{2} + c\sum_{k=1}^{N} e_k^2 \qquad \text{s.t.} \quad y_k = w^T\varphi(x_k) + b + e_k, \tag{12}$$

where c is the penalty factor, e_k is the error variable, w is the weight vector, b is a constant, $\varphi(x_k)$ is the linear regression function, x_k is the input, y_k is the output, and n is the space vector dimension; N is the total number of samples.

The radial basis function is chosen as the kernel function.

$$\psi(x, x_k) = \exp\left(-\frac{\|x - x_k\|^2}{2\sigma^2}\right), \tag{13}$$

where σ^2 is the kernel parameter.

The estimation function of LSSVM can be expressed as

$$y(x) = \sum_{k=1}^{N} \alpha_k \psi(x, x_k) + b. \tag{14}$$

In the regression model of LSSVM, the penalty factor c and the kernel parameter σ^2 have the greatest influence on the model performance. This paper uses cross-validation to optimize parameters. The input variables of LSSVM are the PV power data of similar days, sunshine hours, mean temperature, maximum temperature, and minimum temperature of similar days and sunshine hours, mean temperature, maximum temperature, and minimum temperature of the forecast day.

2.11. Combination Forecasting Model Based on Cross Entropy. Before PV power generation forecast, the similar days need to be selected. It is very important for improving the efficiency and accuracy of the forecast model. The similar day selection method proposed in this paper is used to select samples.

The procedure of combination forecasting method based on the cross entropy theory is as follows:

Step 1. Predict the PV power at certain moment during forecasting period T.

Select similar days as forecasting samples. For the PV power during forecasting period T, using M kinds of algorithms to predict the PV power at the t point ($t = 1, 2, \ldots, N$), the m_{th} algorithm forecast results are P_{mt} ($m = 1, 2, \ldots, M$). Here, ARMA, BP neural network, and LSSVM methods are used to forecast PV power, respectively.

Step 2. Define the probability density function of PV power.

Here, the value of PV power at a certain point follows the normal distribution. Suppose PV power at the t point during forecasting period T follows normal distribution, $g_m(x)$ represents PV power distribution function at the t point by the m_{th} algorithm and it can be expressed as

$$g_m = \frac{1}{\sqrt{2\pi}\sigma_m} e^{-\left((x-\mu_m)^2/2\sigma_m^2\right)}, \tag{15}$$

where μ_m is the average; σ_m^2 is the variance.

Step 3. Calculate the digital characteristics of PV power distribution function.

PV power at the t point of similar day k ($k = 1, 2, \ldots, K$) is P_{kt}. The PV power P_{mt} predicted by the m_{th} algorithm can be regarded as the mean μ_m of probability distribution function and the variance can be calculated according to PV power of similar days.

$$\begin{cases} \mu_m = P_{mt}, \\ \sigma_m^2 = \dfrac{1}{K}\displaystyle\sum_{k=1}^{K} (P_{kt} - P_{mt})^2, \end{cases} \tag{16}$$

Suppose $g(x)$ is the probability density distribution function of PV power considering influence factors, $g_m(x)$ represents the PV power distribution function by the m_{th} algorithm, then

$$\begin{cases} \displaystyle\sum_{m=1}^{M} w_m = 1, \\ g(x) = \displaystyle\sum_{m=1}^{M} w_m g_m(x). \end{cases} \tag{17}$$

PV power at the t point obtained by equation (17) is the mathematical expectation of μ. The variance of the combination forecasting PV power is

$$\sigma^2 = \frac{1}{K}\sum_{k=1}^{K}\left(\sum_{m=1}^{M} w_m (P_{kt} - P_{mt})\right)^2. \tag{18}$$

Step 4. Establish support vector and obtain objective function of cross entropy.

Build support vector:

$$S = (S_1, S_2, \ldots, S_M). \tag{19}$$

According to the cross entropy definition, suppose

$$\begin{cases} S_m = D[g(x)\|g_m(x)] = \displaystyle\int g(x) \ln \frac{g(x)}{g_m(x)} dx \\ \text{s.t.} \quad \displaystyle\int g(x)dx = 1, \end{cases} \tag{20}$$

where S_m is used to indicate the mutual support degree between $g(x)$ and $g_m(x)$; S_m is smaller when the mutual support degree is greater. Define

$$A_m = \frac{S_m}{\sum_{m=1}^{M} S_m}, \tag{21}$$

in order to let the weights reflect the mutual support degree among different information sources better.

Define

$$w_m = \frac{1/A_m}{\sum_{m=1}^{M} 1/A_m} = \frac{1}{1 + \sum_{m=1, i\neq m}^{M} A_m/A_i}. \tag{22}$$

The objective function of the minimum cross entropy is

$$\begin{cases} F = \min\left(\displaystyle\sum_{m=1}^{M} D[g(x)\|g_m(x)]\right) \\ \text{s.t.} \quad \displaystyle\sum_{m=1}^{M} w_m = 1, 0 \le w_m \le 1, \quad m = 1, 2, \ldots, M. \end{cases} \tag{23}$$

Step 5. Solve the minimum cross entropy optimization problem.

It can be seen that the objective function is related to w_m from the above formula; the optimization problem is actually solving the minimum of F. In this paper, fmincon function in MATLAB is used to solve the minimum of F.

Step 6. Forecast PV power at other moment during forecast period.

For PV power at other moment of forecast period, it can be calculated by Step 5; then the PV power at each moment of forecast period T can be obtained.

3. Case Study

In order to verify the correctness and validity of the combination forecast method based on cross entropy theory proposed in this paper, the real data samples from a PV power station of Nanjing in Jiangsu Province from Oct. 1 to Dec. 31 are employed. The meteorological data are obtained from the local weather station. Here, ARMA, BP, and LSSVM method are used to forecast PV power, respectively. For specific analysis, three single forecast methods and three combination

Table 1: Meteorological parameters of forecasting day.

Weather type	Average temperature	Maximum temperature	Lowest temperature	Relative humidity	Minimum humidity	Sunshine time
Sunny	3.5°C	10.2°C	-1.6°C	56%	24%	8.9 h

Table 2: Selection results of similar days.

Number	1	2	3	4	5	6
Date	2013/11/30	2013/12/2	2013/11/29	2013/11/16	2013/11/28	2013/11/25
R_i	0.77	0.85	0.86	0.91	0.92	0.93

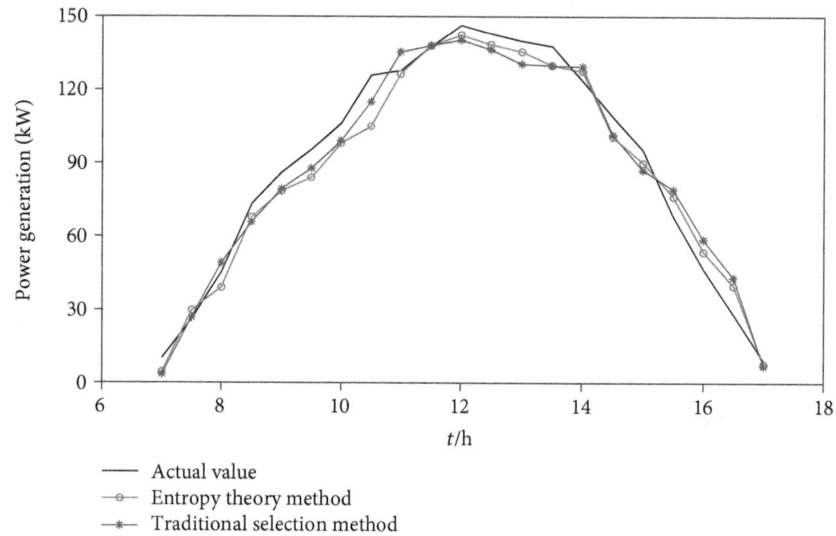

Figure 2: Prediction results of different similar day methods.

forecast methods are used to predict PV power, then the predicted results are compared and analyzed. Four typical forecast days with different weather types, sunny day (Dec. 10), cloudy day (Dec. 12), overcast day (Dec. 15), and rainy day (Dec. 17), are selected from data samples. The PV power data are recorded every 30 minutes. The prediction period is 7:00-17:00. The algorithms of all forecast models are implemented by MATLAB R2014a software.

3.1. Similar Day Selected Results. Taking the forecast date (Dec. 10) as an example, the meteorological parameters of the forecast day are shown in Table 1. According to the similar day selection method proposed in this paper, the selection results are shown in Table 2. All the selected results of similar days are taken as samples of three single prediction methods.

In order to further verify the rationality of the similar day selection method proposed in this paper, the traditional similar day selection method (traditional Euclidean distance selection method) is used for comparison and verification. The ARMA prediction model is used to predict results. The prediction results are shown as follows. Here, taking the forecast date (Dec. 10) as an example,

Table 3: Statistics of forecasting errors for different selection methods.

Method	Index	ARMA
Entropy theory method	MAPE	11.99
	TIC	0.0355
Traditional selection method	MAPE	13.50
	TIC	0.0402

Figure 2 and Table 3 show the prediction results and the statistics of forecasting errors by using different selection methods, respectively.

It can be seen from Figure 2 and Table 3 that the similarity day selection method proposed in this paper can better integrate the uncertainties of PV power generation because it considers distance entropy and grey correlation entropy synthetically. Therefore, the selected similar day samples are better and the prediction accuracy is higher.

3.2. Forecasting Results of Single Models. Figure 3 shows the forecasting results of PV power by using different single

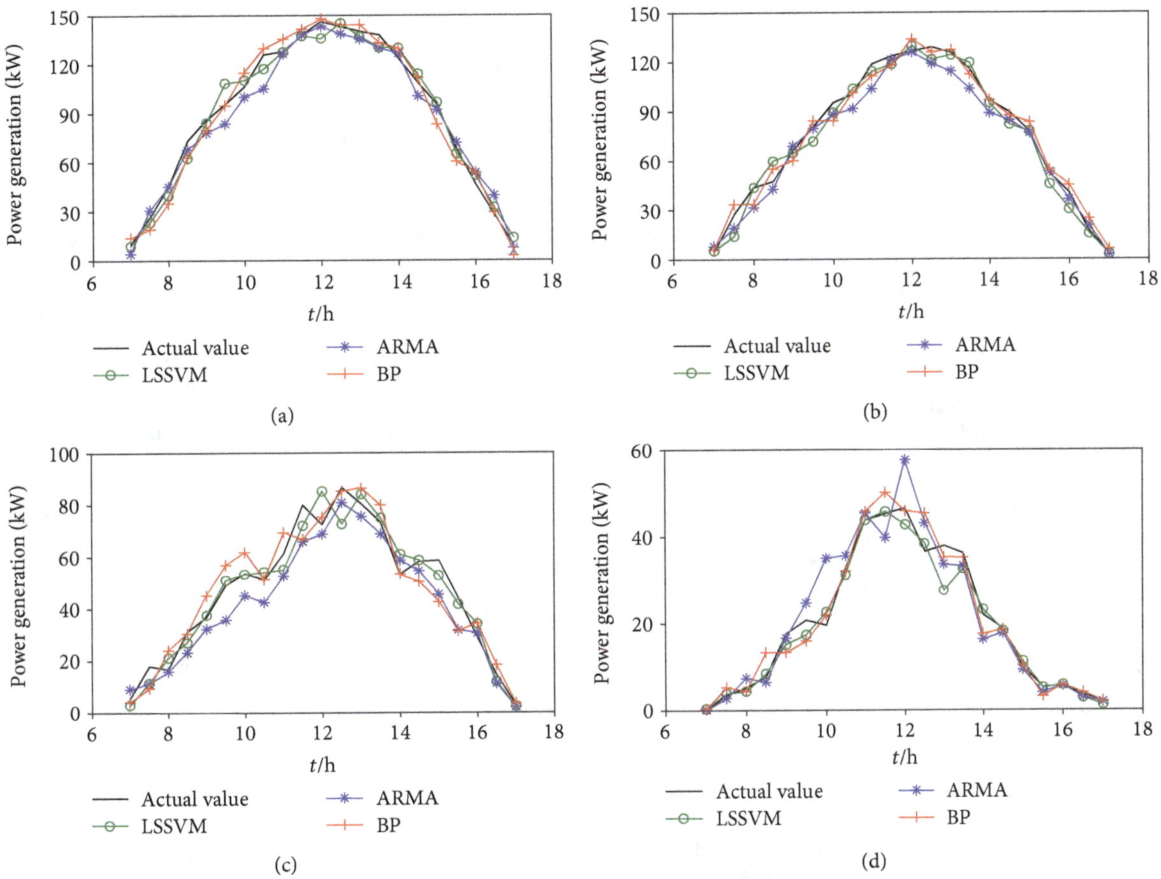

FIGURE 3: Forecasting results of single forecasting methods. (a) Sunny day. (b) Cloudy day. (c) Overcast day. (d) Rainy day.

TABLE 4: Statistics of forecasting errors for single models.

Weather	Index	LSSVM	ARMA	BP
Sunny	MAPE	8.63	11.99	10.89
	TIC	0.0298	0.0355	0.0312
Cloudy	MAPE	10.52	14.78	13.81
	TIC	0.0343	0.0460	0.0351
Overcast	MAPE	13.29	17.88	17.33
	TIC	0.0552	0.0749	0.0702
Rainy	MAPE	13.95	17.97	17.29
	TIC	0.0575	0.0945	0.0609

methods for different forecasting days. It is observed from Figure 3 that the prediction curves of three single prediction methods are close to the actual curve under the sunny day or cloudy day and the prediction results are ideal. But in overcast day or rainy day, the deviation between the predicted curves of three single prediction methods and the actual curve is more obvious.

In order to evaluate the validity of each forecast method, the mean absolute percent error (MAPE) and the Theil inequality coefficient (TIC) [20] are selected as evaluation indices. The smaller the TIC value, the better the prediction performance of the model is. MAPE and TIC are expressed as

TABLE 5: Weights of cross entropy model.

Weather	t (hour)	w_1	w_2	w_3
Sunny	8	0.001	0.860	0.139
	14	0.991	0.007	0.002
	17	0.371	0.052	0.577
Cloudy	7	0.157	0.092	0.751
	12	0.239	0.413	0.348
	15	0.106	0.506	0.388
Overcast	8.5	0.662	0.271	0.067
	12.5	0.000	0.590	0.410
	17	0.385	0.505	0.110
Rainy	7	0.359	0.191	0.450
	10	0.728	0.027	0.245
	16.5	0.052	0.751	0.197

$$\text{MAPE} = \frac{1}{n}\sum_{i=1}^{n}\left|\frac{y_i - y_{fi}}{y_i}\right| \times 100\%,$$
$$\text{TIC} = \frac{\sqrt{\sum_{i=1}^{n}\left(y_{fi} - y_i\right)^2 / n}}{\sqrt{\sum_{i=1}^{n} y_{fi}^2 / n} + \sqrt{\sum_{i=1}^{n} y_i^2 / n}}, \tag{24}$$

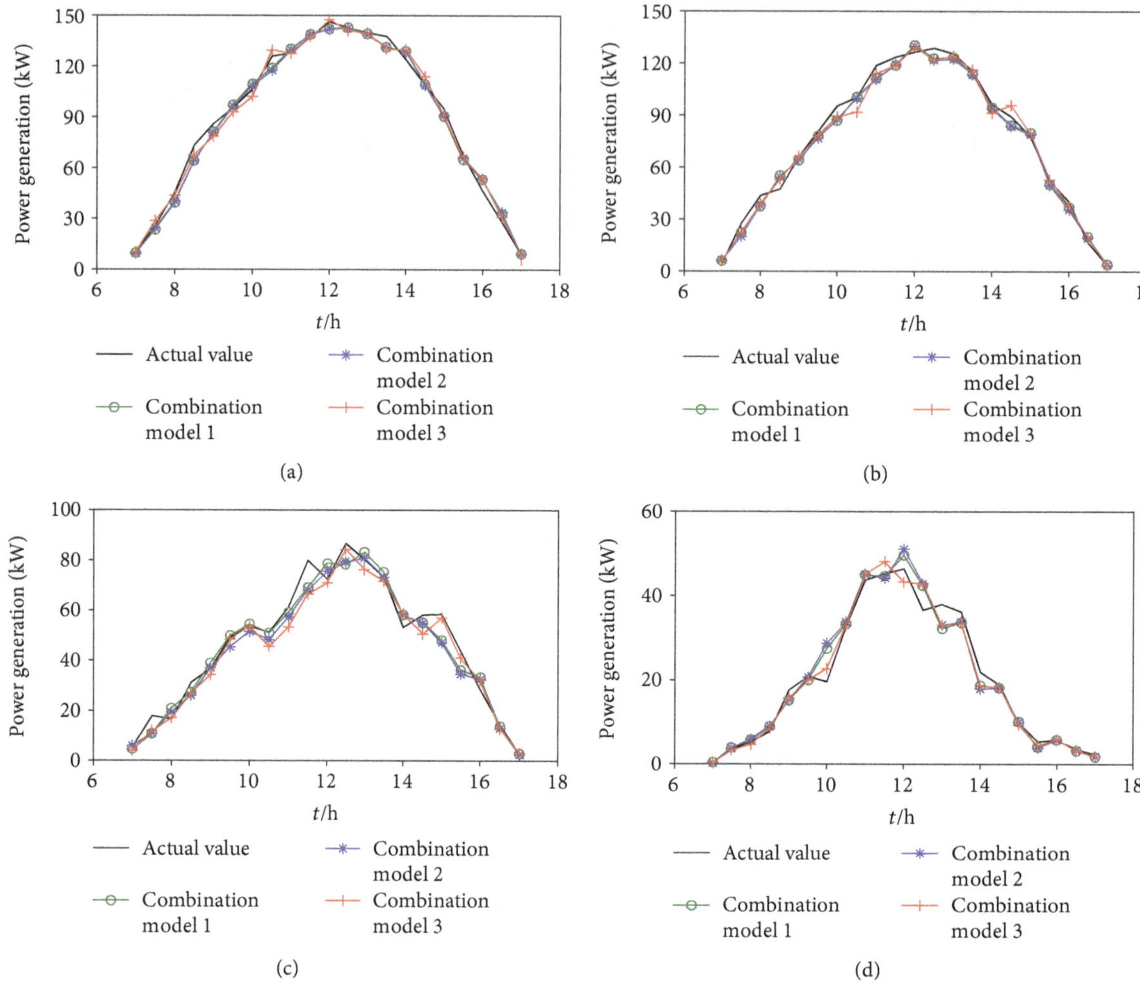

FIGURE 4: Forecasting results of combination forecasting models. (a) Sunny day. (b) Cloudy day. (c) Overcast day. (d) Rainy day.

where y_i is the actual power, y_{fi} is the forecasting power, and n is the number of forecasting points.

Table 4 shows the statistics of forecasting errors for single models for different weather types. LSSVM performs the best among the three single forecast methods. The MAPE index can be up to 8.63 in a sunny day. TIC is also smaller than the other two methods. For cloudy day, overcast day, or rainy day, the prediction error of the three single methods is increasing due to the fluctuation of PV power. In particular, during overcast and rainy days, MAPE of ARMA model can reach 17.88 and 17.97, respectively. Compared with sunny day, the prediction accuracy drops significantly. The input variables of BP such as sunshine hours, temperature, and humidity are taken into account, so its prediction error is lower than ARMA, but the prediction performance is not ideal compared with the LSSVM model.

3.3. Forecasting Results of Combination Model. Table 5 gives a part of weights of the cross entropy model; w_1, w_2, and w_3 stand for the weights of LSSVM, ARMA, and BP, respectively. It can be seen from Table 4 that the weights of the three single methods are different at different prediction points. At $t = 14$ h on sunny day, the prediction errors of the three methods are all larger. At this point, the mutual support degrees of ARMA and BP methods are less than those of the other single methods and the weights are all approximate to zero, which avoid the influence caused by the single prediction method error. At $t = 8.5$ h on overcast day and $t = 10$ h on rainy day, the prediction errors of the three methods are similar. Due to the greater mutual support degree of LSSVM, the weight of LSSVM needs to be increased. The weights are 0.662 and 0.728, respectively. Therefore, it can be seen that the weights obtained by the cross entropy theory can reflect the information fusion between different forecasting methods objectively.

Figure 4 shows results of PV power predicted by three combination forecasting methods for different forecasting days. Among them, the combination model 1 adopts the method based on the sum of the squared errors. The combination model 2 adopts the method based on the correlation coefficient and the combination model 3 is based on the cross entropy theory proposed in this paper. Compared with forecasting results which are shown in Figure 3, it can be clearly seen that the prediction PV power values obtained by combination models are closer to the actual power. Figures 4(a) and 4(b) show the results of sunny and cloudy days and the results are ideal. Figures 4(c) and 4(d) are results of overcast and rainy, respectively. Although the weather conditions in

Table 6: Statistics of forecasting errors for combination models.

Weather	Index	Model 1	Model 2	Model 3
Sunny	MAPE	5.29	5.45	4.39
	TIC	0.0215	0.0217	0.0198
Cloudy	MAPE	8.26	8.35	7.21
	TIC	0.0265	0.0276	0.0249
Overcast	MAPE	10.43	10.55	9.63
	TIC	0.0493	0.0518	0.0462
Rainy	MAPE	11.32	11.38	10.21
	TIC	0.0547	0.0605	0.0465

overcast and rainy days are more complex, the prediction accuracy of each combination forecasting models is still high, especially in combination model 3, which has the best prediction performance under different weather types.

Table 6 shows the statistics of forecasting errors for combination models under different weather types. It can be seen that combination methods can predict more accurately than single methods. In sunny day, MAPE of combination model 3 reduces 4.24 than the LSSVM model, which is still more accurate than combination models 1 and 2. TIC index shows that combined model 3 is also better than the other two combination models. In addition, TIC index values of combination model 3 under four weather types are 0.0198, 0.0249, 0.0462, and 0.0465, respectively, which are the lowest among the three combination prediction models, which can demonstrate the superiority of the combination model based on the cross entropy theory. Therefore, it can be seen that the cross entropy combination model has better performance.

4. Conclusions

This paper presents the short-term PV power generation combination forecasting method based on similar day and cross entropy. The selected samples of similar days by using the entropy theory are closer to the forecast days. The combination forecasting model can effectively reduce the prediction error under different weather types. Compared with the combination model based on the sum of the squared errors and the correlation coefficient, the combination model based on cross entropy has lower prediction error and better prediction performance. When the weather types are different, its prediction accuracy is also high. Therefore, the short-term PV power combination forecasting method based on similar day and cross entropy proposed in this paper is of great value and can be the reference for the research of operation and control of power system with PV power generation.

Data Availability

The photovoltaic generation power data is collected from a PV power plant in Nanjing, Jiangsu. The meteorological data are collected from the local weather station and can be obtained from the China National Meteorological Information Center website. In addition, the photovoltaic power data used to support the findings of this study is under license and so cannot be made freely available. If the reader needs data, he or she can contact the corresponding author: Qi Wang (email: wangqi@njnu.edu.cn).

Conflicts of Interest

The authors declare that there is no conflict of interest regarding the publication of this paper.

Acknowledgments

This work is supported by the project of State Key Laboratory of Smart Grid Protection and Control: Research on Collaborative Control Strategy of Multiple Reactive Source Field Network in Wind Farms (NARI-T-2-2017192-03).

References

[1] C. O. Inácio and C. L. T. Borges, "Stochastic model for generation of high-resolution irradiance data and estimation of power output of photovoltaic plants," *IEEE Transactions on Sustainable Energy*, vol. 9, no. 2, pp. 952–960, 2018.

[2] Y.-K. Wu, C.-R. Chen, and H. Abdul Rahman, "A novel hybrid model for short-term forecasting in PV power generation," *International Journal of Photoenergy*, vol. 2014, Article ID 569249, 9 pages, 2014.

[3] M. Diagne, M. David, P. Lauret, J. Boland, and N. Schmutz, "Review of solar irradiance forecasting methods and a proposition for small-scale insular grids," *Renewable & Sustainable Energy Reviews*, vol. 27, pp. 65–76, 2013.

[4] F. Wang, Z. Mi, S. Su, and H. Zhao, "Short-term solar irradiance forecasting model based on artificial neural network using statistical feature parameters," *Energies*, vol. 5, no. 5, pp. 1355–1370, 2012.

[5] X. L. Yuan, J. H. Shi, and J. Y. Xu, "Short-term power forecasting for photovoltaic generation considering weather type index," *Proceedings of the CSEE*, vol. 33, no. 34, pp. 57–64, 2013.

[6] H. Lan, Z. M. Liao, and Y. Zhao, "ARMA model of the solar power station based on output prediction," *Electrical Measurement & Instrumentation*, vol. 48, no. 54, pp. 31–34, 2011.

[7] B. Jing, Z. Qian, Y. Pei, and J. Wang, "Ultra short-term PV power forecasting based on ELM segmentation model," *The Journal of Engineering*, vol. 2017, no. 13, pp. 2564–2568, 2017.

[8] Q. Zhu and J. Tian, "Application of least square support vector machine in photovoltaic power forecasting," *Power System Technology*, vol. 35, no. 7, pp. 31–34, 2011.

[9] X. Wang, X. Zhou, J. Xing, and J. Yang, "A prediction method of PV output power based on the combination of improved grey back propagation neural network," *Power System Protection and Control*, vol. 44, no. 18, pp. 81–87, 2016.

[10] F. Li, J. Qian, Q. Yan et al., "A combination forecasting model based on IOWA operator for PV generation," in *International Conference on Renewable Power Generation*, pp. 1–4, Beijing, China, October 2015.

[11] X. Y. Yang, H. Liu, B. Zhang, and S. Chen, "A combination method for photovoltaic power forecasting based on entropy weight method," *Acta Energiae Solaris Sinica*, vol. 35, no. 5, pp. 744–749, 2014.

[12] X. Y. Yang and S. Chen, "A combination method in photovoltaic power forecasting based on the correlation coefficient," in *5th International Conference on Software Engineering and Service Science*, pp. 706–709, Beijing, China, June 2014.

[13] Q. Shi, Z. Chen, C. Fang, Y. Feng, and B. Xu, "Measuring the diversity of a test set with distance entropy," *IEEE Transactions on Reliability*, vol. 65, no. 1, pp. 19–27, 2016.

[14] T. A. Yongbo, G. Weihua, P. E. Tao, and O. Wei, "Transformer fault diagnosis method based on reconstruction-based contribution and grey relation entropy," *Chinese Journal of Scientific Instrument*, vol. 33, no. 1, pp. 133–138, 2012.

[15] A. Bracale, G. Carpinelli, and P. de Falco, "A probabilistic competitive ensemble method for short-term photovoltaic power forecasting," *IEEE Transactions on Sustainable Energy*, vol. 8, no. 2, pp. 551–560, 2017.

[16] Y. Tang, W. Qi, Q. Sha, N. Chen, and L. Zhu, "A combination forecast method based on cross entropy theory for wind power and application in power control," *Transactions of the Institute of Measurement and Control*, vol. 36, no. 7, pp. 891–897, 2014.

[17] M. Kim and J. Kim, "GA-ARMA model for predicting IGS RTS corrections," *International Journal of Aerospace Engineering*, vol. 2017, Article ID 6316590, 7 pages, 2017.

[18] S. Liu, T. Ren, H. Mu, and G. Yan, "Temperature prediction of the molten salt collector tube using BP neural network," *IET Renewable Power Generation*, vol. 10, no. 2, pp. 212–220, 2016.

[19] X. Lu, W. Liu, C. Zhou, and M. Huang, "Robust least-squares support vector machine with minimization of mean and variance of modeling error," *IEEE Transactions on Neural Networks and Learning Systems*, vol. 29, no. 7, pp. 1–12, 2017.

[20] B. Zhang, Z. Wei, J. Ren, Y. Cheng, and Z. Zheng, "An empirical study on predicting blood pressure using classification and regression trees," *IEEE Access*, vol. 6, pp. 21758–21768, 2018.

9

Research on Parameter Distribution Features of Photovoltaic Array under the Cover and Shadow Shading Conditions

Honglu Zhu,[1,2] Cao Yu,[2,3] Lingxing Lu,[2] Weiwei Lian,[2] Jianxi Yao,[1,2] and Yang Hu[1]

[1]*State Key Laboratory of Alternate Electrical Power System with Renewable Energy Sources, North China Electric Power University, Changping District, Beijing, China*
[2]*School of Renewable Energy, North China Electric Power University, Changping District, Beijing, China*
[3]*China Three Gorges New Energy Co., Ltd, Beijing, China*

Correspondence should be addressed to Yang Hu; hooyoung@ncepu.edu.cn

Academic Editor: Jegadesan Subbiah

The outdoor operating photovoltaic arrays have two different shading conditions, shadowing and covering. The shading causes a decrease in output power of photovoltaic system and may bring hot spots which causes physical damage to the array. This paper studies the electrical parameter distribution feature of photovoltaic array under different shading conditions by means of analog simulation and empirical testing. Through introducing theoretical computational method of the electrical parameters, it describes the distribution features of the electrical parameters of photovoltaic array. The results indicate that the influence of local shadowing on the current of array can be neglected. Shadowing decreases the optimal operating voltage while covering leads to a decrease in the optimal operating voltage and the open-circuit voltage. The drop magnitude of voltage is associated with the number of the shaded cell strings and the string voltage. The two shading types can be identified on the basis of distribution rules of open-circuit voltage and optimal operating voltage. Simulations and experiments verify the consistency of the rules. Relevant conclusions provide a reference for modeling, online fault diagnosis, and optimization design of the maximum power tracking algorithm of photovoltaic array under different shading conditions.

1. Introduction

As the most important method for solar energy utilization, photovoltaic power generation is developing rapidly. With the world cumulative photovoltaic (PV) installed capacity of up to 390 GWp in 2017, PV power generation plays an increasingly important role in the energy supply structure. In actual operation, PV arrays are usually distributed in a very large region due to their low power density. Therefore, shadowing from the front and rear rows of photovoltaic array is often inevitable due to the restriction of floor space. Meanwhile, photovoltaic arrays are mostly shaded by tall buildings, tall trees, fallen leaves, and birds' droppings. It is common that the photovoltaic arrays usually operate under different shading conditions [1, 2]. The temporary shadowing of PV arrays will lead to a decrease in array efficiency, increased power loss, and more complex output characteristics. The prolonged covering can cause excessively high local temperature, hot spots, and breakdown phenomenon, which adversely influence the output efficiency and service life of the PV modules [3, 4]. Therefore, the good understanding of electrical parameter transfer law of PV arrays under different shading conditions is significant for the performance assessment, fault diagnosis, and optimal operation of the photovoltaic power station.

At present, scholars both locally and abroad are conducting series of simulations and experimental studies on the output characteristics of the PV modules/arrays under shading conditions. In [5–7], MATLAB-Simulink is used to establish a simulation model of photovoltaic array under the shadowing condition and concluded that the output of the PV array exhibits a multipeak and stepped shape under the shadow condition. In [8], it is concluded that hot spots occur when the short-circuit current of the photovoltaic cells is smaller than the electric current flowing through the cells based on the simulations and experimental studies of the reverse

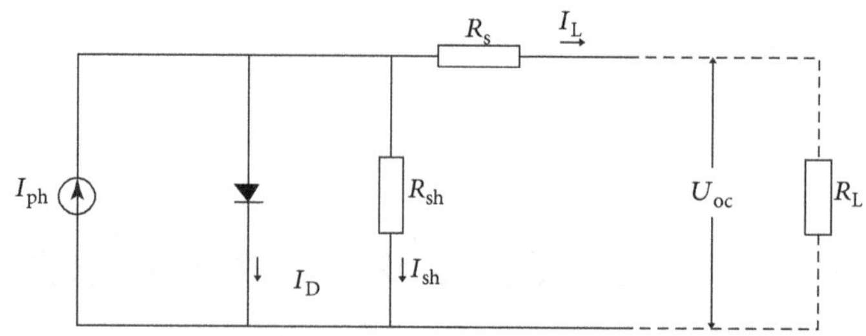

FIGURE 1: Equivalent circuit of the photovoltaic cell.

Bishop model of photovoltaic arrays. In [9, 10], it is concluded that the loss of the output power increases as the shading area increases and the more the cell strings are shaded, the greater the power loss under the same area is. In [11], it is pointed out that the conventional maximum power point tracking (MPPT) algorithms (such as the conductance increment method and the disturbance observation method) are susceptible to failure due to the local extreme point under the shadowing condition. In [12–14], some improved MPPT algorithms are proposed to raise the output efficiency of the photovoltaic array under the shadowing condition. In actual engineering, PV arrays may be subjected to two types of shading, that is, the shadow shading arising from tall buildings and the front and rear rows of array and the cover shading arising from objects such as fallen leaves and dust. In the above literature, however, only the output properties of photovoltaic array under the shadow condition by means of simulation or experiment are studied but no theoretical analyses or experimental studies are conducted on the cover shading. It is impossible to systematically describe the distribution rules of electrical parameters of photovoltaic array in actual operation. A research on the output characteristics and the rules of electrical parameter distribution of photovoltaic array under different shading conditions is therefore has significant theoretical and engineering application.

To illustrate the output characteristics and the electrical parameter distribution feature of photovoltaic array under the shadow and cover conditions, the paper first presents mathematical description of photovoltaic cells, modules, and arrays and establishes the appropriate Simulink model. Designing simulation experiments under the two shading conditions and discussing the theoretical maximum power point voltage and current as the contrasting parameters for empirical testing were considered in this study. Comparing the theoretical results with experimental results concludes the rules of electrical parameter distribution of the photovoltaic array under various shading conditions. The simulation results coincide with the empirical testing results.

2. Mathematical Models of Photovoltaic Cells, Modules, and Array

2.1. Equivalent Circuit Model of Photovoltaic Cells. Photovoltaic cell converts the luminous energy into electrical energy through the photovoltaic effect. The analysis process of this paper is based on the equivalent circuit of the photovoltaic cell of the monolithic diode model as shown in Figure 1, where I_{ph} is the photogenerated current, I_D is diode current, I_{sh} is parallel current, I_L is output current, R_{sh} is parallel resistance, R_s is series resistance, and R_L is external load.

The voltage-current equation can be expressed as follows:

$$I_L = I_{ph} - I_0\left\{\exp\left[\frac{q(U+R_sI)}{AKT}\right]-1\right\} - \frac{U+R_sI}{R_{sh}}, \tag{1}$$

where I_0 is reverse saturation current of diode, q is electron charge (1.602×10^{-19} C), A is ideal coefficient of diode, and K is Boltzmann constant (1.38×10^{-23} J/K).

The photogenerated current I_{ph} can be considered to approximate the short-circuit current of photovoltaic cell I_{sc}, and the tail term $(U+R_sI)/R_{sh}$ can be ignored; hence, it is much smaller than that of the short-circuit current. Equation (1) therefore can be rewritten as follows:

$$I_L = I_{sc}\left(1 - C_1\left(\exp\left(\frac{U}{C_2U_{oc}}\right)-1\right)\right), \tag{2}$$

where C_1 and C_2 are the definition coefficients. In case of maximum power point ($I=I_m$, $U=U_m$) and open-circuit voltage ($I=0$, $U=U_{oc}$) in (2), we can obtain

$$C_1 = \left[\frac{(I_{sc}-I_m)}{I_{sc}}\right] * \exp\left[\frac{-U_m}{(C_2 * U_{oc})}\right], \tag{3}$$

$$C_2 = \frac{(U_m/U_{oc}-1)}{\ln(1-I_m/I_{sc})}. \tag{4}$$

External operating environment is not considered in the external behavioral characteristics of the photovoltaic cells described in (2), (3), and (4). However, the output of the photovoltaic modules is primarily influenced by solar irradiance and temperature. Therefore, it is necessary to consider the changes in solar irradiance and temperature during the process of photovoltaic module simulation as below [15]:

$$I_{sc} = I_{scref}[1+a(T-T_{ref})]\cdot\frac{S}{S_{ref}}, \tag{5}$$

$$U_{oc} = U_{ocref}[1-c(T-T_{ref})]\ln\left[e+b\left(\frac{S}{S_{ref}}-1\right)\right], \tag{6}$$

$$I_m = I_{mref}[1+a(T-T_{ref})]\cdot\frac{S}{S_{ref}}, \tag{7}$$

$$U_m = U_{mref}[1-c(T-T_{ref})]\ln\left[e+b\left(\frac{S}{S_{ref}}-1\right)\right], \tag{8}$$

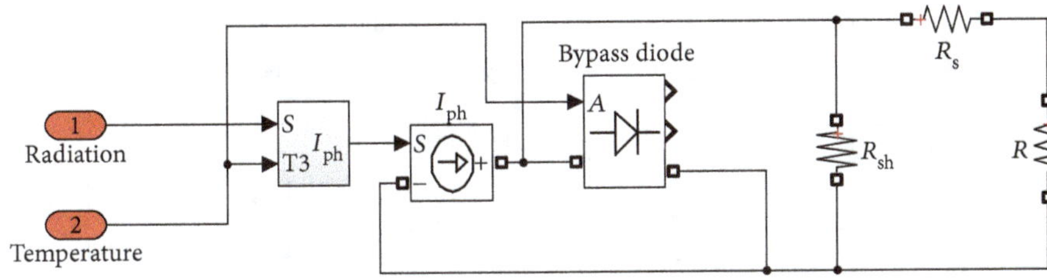

FIGURE 2: Simulink model of photovoltaic cells.

where S and S_{ref} are the solar irradiance received by photovoltaic cells and solar irradiance under standard conditions, respectively, T and T_{ref} are actual temperature and temperature of photovoltaic cells under standard conditions, respectively, and a, b, and c are correction coefficients, where $a = 0.0008/°C$, $b = 0.2$, and $c = 0.002/°C$.

Equations (5), (6), (7), and (8) describe the distribution features of theoretical short-circuit current, open-circuit voltage, and voltage and current at optimal operating point of photovoltaic cells under different solar irradiances and temperatures. The electrical parameters of photovoltaic array under different shading conditions are analyzed using (5), (6), (7), and (8) in the subsequent experimental process.

Based on the above mathematical model, the simulation model is established for photovoltaic cells as shown in Figure 2. There are two inputs, irradiance, and temperature. The photogenerated current I_{ph} does not change with the operative mode of photovoltaic cells when the irradiation intensity is constant. A constant current source I_{ph} is therefore used for simulation. The circuit also includes bypass diode, parallel resistance R_{sh}, and series resistance R_s, where R is the output load.

2.2. Model of Photovoltaic Modules and Array. The current and voltage generated by a single photovoltaic cell are not enough for the load. Generally, many photovoltaic cells are connected in series or in parallel to create a photovoltaic module [16]. In a photovoltaic power station, several modules are connected in series or in parallel. The modules and arrays might be connected in a way as shown in Figure 3.

It is assumed that the number of cells in series in the photovoltaic module is n and the number of cells in parallel is m. Then, the parameters of the photovoltaic module may be expressed as follows:

$$\begin{aligned} U_{\text{oc-module}} &= n \cdot U_{\text{oc}}, \\ U_{\text{m-module}} &= n \cdot U_{\text{m}}, \\ U_{\text{module}} &= n \cdot U, \\ I_{\text{sc-module}} &= m \cdot I_{\text{sc}}, \\ I_{\text{m-module}} &= m \cdot I_{\text{m}}, \\ I_{\text{module}} &= m \cdot I. \end{aligned} \tag{9}$$

It is assumed that the photovoltaic array comprises of N photovoltaic modules in series and M photovoltaic modules in parallel. Then, we can obtain

$$\begin{aligned} R_{\text{s-array}} &= \frac{N}{M} \cdot R_{\text{s-module}}, \\ R_{\text{sh-array}} &= \frac{N}{M} \cdot R_{\text{sh-module}}, \\ I_{\text{sc-array}} &= M \cdot I_{\text{sc}}, \\ I_{\text{array}} &= M \cdot I, \\ I_{\text{m-array}} &= M \cdot I_{\text{m}}, \\ U_{\text{oc-array}} &= N \cdot I_{\text{oc-module}}, \\ U_{\text{array}} &= N \cdot I_{\text{module}}, \\ U_{\text{m-array}} &= N \cdot U_{\text{m-module}}. \end{aligned} \tag{10}$$

The simulation model of the cells is encapsulated, and the appropriate parameters are set to obtain the simulation model of the modules. The modules are connected in series and parallel to create a photovoltaic array. Figure 4 presents the simulation model of a $N \times M$ photovoltaic array. A 1×3 photovoltaic array was used in the simulations and experiments for the next analysis simplicity.

2.3. Output Characteristics of Modules. A typical 245 W module is used for the analysis in the paper. The cell connection within the module and the module parameters are shown in Table 1 and Figure 5.

The simulation model is established for the modules according to (9). The output characteristics of the modules under typical temperature and light conditions are shown in Figure 6.

Figure 6 shows that photovoltaic module is a nonlinear direct-current power source instead of a constant voltage source or a constant current source. Its output current approximates a constant one under most operating voltages. The current declines sharply as it approaches open-circuit voltage point. The red line in the figure denotes the variation trend of the optimal operating point with change in solar irradiance and temperature. It can be seen from (a) and (b) that the current parameters of photovoltaic modules are directly proportional to irradiance at a fixed temperature. An increase in irradiance leads to an increase in short-circuit current and current at optimal operating point, but the effect of the irradiance on the

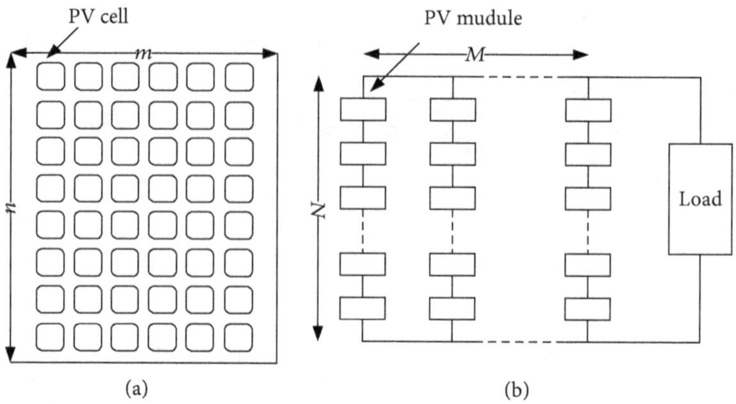

FIGURE 3: (a) Schematic diagram for photovoltaic module. (b) Schematic diagram for photovoltaic array.

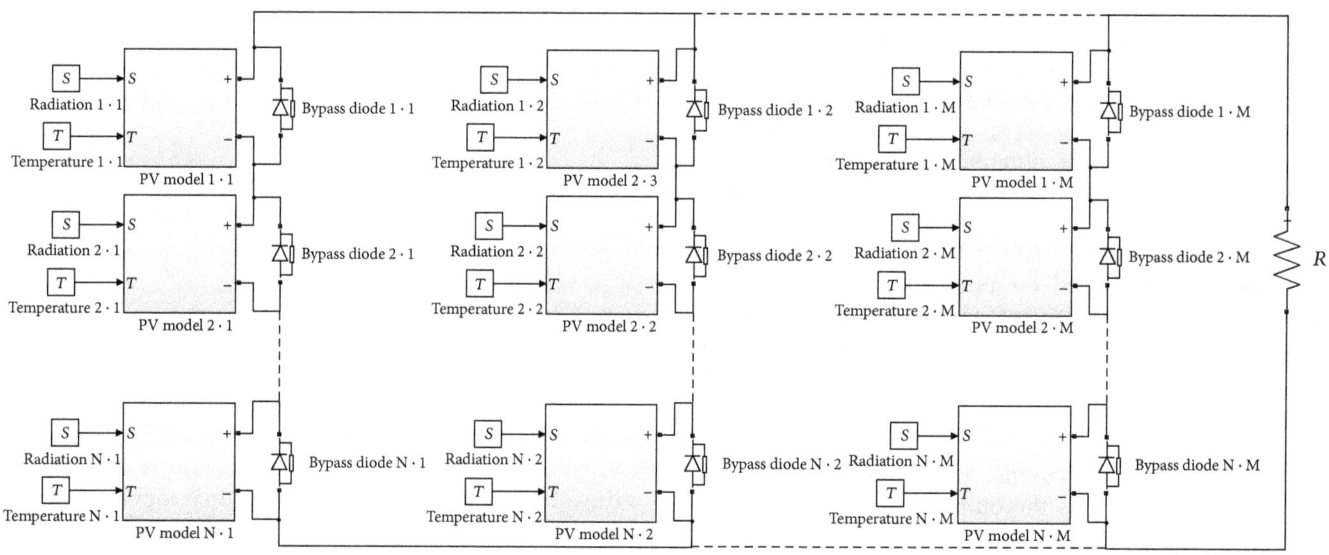

FIGURE 4: Simulation model of photovoltaic array. By entering different temperatures and irradiance, we can get the IV/PV curves of PV arrays under different operating conditions.

TABLE 1: Basic parameters for the module.

Parameter	Variable	Value
Maximum power	P_m	245 (W)
Optimum operating voltage	V_m	30.1 (V)
Optimum operating current	I_m	8.14 (A)
Open-circuit voltage	U_{oc}	37.5 (V)
Short-circuit current	I_{sc}	8.76 (A)
Temperature coefficient of I_{sc}	α	0.06 (%/°C)
Temperature coefficient of V_{oc}	β	−0.31 (%/°C)

Specific parameters of photovoltaic module.

open-circuit voltage and the optimal operating voltage is minimal. It can be seen from (c) and (d) that the voltage and temperature parameters of the photovoltaic module are inversely proportional. A lower temperature leads to a higher open-circuit voltage and a higher optimal operating voltage, but the effect of temperature on the short-circuit current and the current at the optimal operating point is little and insignificant. The relationship between various electrical parameters of photovoltaic modules, irradiance, and temperature is specifically described in (5), (6), (7), and (8).

3. Simulation of the Output Features of Photovoltaic Array under Various Shading Conditions

The shading of a photovoltaic array which working in outdoor can be classified into two, one is the shadow shading, arising from the surrounding buildings or the front and rear rows of the arrays. Such shading moves with sun position causing no physical damage to the photovoltaic modules or arrays, and the direct radiation received by the shadowed cells decreases substantially but small

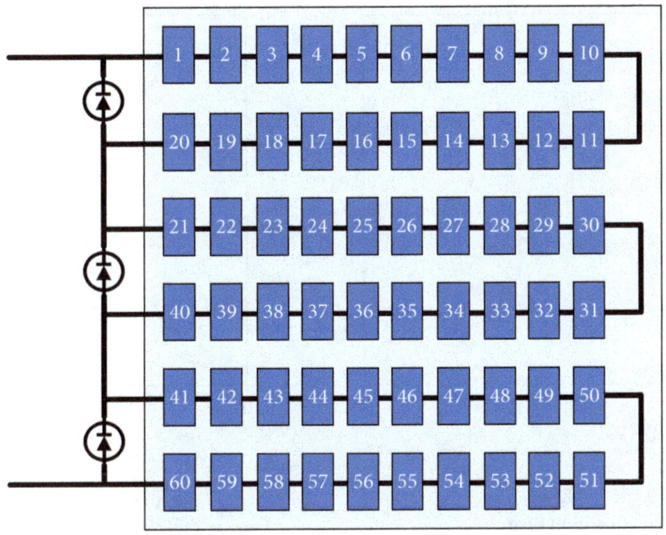

FIGURE 5: Internal wiring diagram for the module. The module consists of 60 cells in series.

amount of scattered radiation being absorbed. In simulation, the irradiation intensity of the shadowed photovoltaic modules is set at 15% of the irradiation intensity absorbed by the unshadowed modules [17]. Cover shading, on the other hand, refers to that arising from direct covering of the array surface by objects like leaves and falling products. In such a case, the irradiation intensity received by the covered cells approximates to zero. Such shading has often a fixed position and causes permanent hot spots which jeopardizes the safe operation of photovoltaic array if not identified in time. In simulation, the solar irradiation intensity of such shading is set at zero. Different shading types and the setting of shading conditions are shown in Figure 7 and Table 2.

3.1. Shadow Shading. The simulation results under the shadow conditions are shown in Figure 8. Shadowing leads to nonuniform irradiation received by the array, and the output presents multiple peaks and ladder-like features. Shadows 1, 2, and 4 have consistent output characteristics. Similarly, shadows 3 and 5 are consistent. The output characteristics decrease in a ladder manner from shadow 4 to shadow 6. Table 3 lists the distribution of the electrical parameters of the array under different shadowing conditions. The open-circuit voltage U_{oc} and short-circuit current I_{sc} change a little, but the electrical parameters of the optimal operating point vary regularly.

According to Table 3, from shadow 4 to shadow 6, shading leads to a drop in the optimum operating voltage U_m and the output power P while the open-circuit voltage U_{oc} and the short-circuit current I_{sc} basically remain unchanged. The drop magnitude of U_m is associated with the sum of the optimum operating voltages of the shadowed cell strings. Meanwhile, the shading positions of shadows 1, 2, and 4 are in the same cell string. The voltage at the optimal operating point U_m therefore decreases when the area of the shadow is larger than or equal to the area of a single cell. Based on the simulation results, the parameter computation formula of a PV array under shadowing conditions can be obtained as follows:

$$
\begin{aligned}
I'_{sc} &= I_{sc}, \\
U'_{oc} &= U_{oc}, \\
I'_{m} &= I_{m}, \\
U'_{m} &= U_{m} - u_1 \cdot n_1, \\
P'_{m} &= I'_{m} \cdot U'_{m},
\end{aligned}
\tag{11}
$$

where I'_{sc} and I'_{m} are the short-circuit current and the current at the optimal operating point of the array under the shadowing condition, respectively, U'_{oc} and U'_{m} are the open-circuit voltage and the voltage at the optimal operating point of the array under the shadowing condition, respectively, and I_{sc} and I_m are the short-circuited current and the current at the optimal operating point of the array in the absence of shading, respectively. From (5) and (7), we can obtain U_{oc} and U_m which are the open-circuit voltage and the voltage at the optimal operating point of the array in the absence of shading. We can obtain u_1 and n_1 from (6) and (8), where u_1 is the optimum operating voltage of the cell string and n_1 is the number of the shadowed cell strings.

Based on the above analysis, we can obtain the distribution rules of the electrical parameters of photovoltaic array under the following shadowing conditions:

(1) Local shadowing does not influence U_{oc} and I_{sc} of the photovoltaic array.

(2) U_m drops when the area of the shadow in a cell string is larger than or equal to the area of a single cell. The drop intensity is proportional to the number of the shadowed cell strings and the optimum operating voltage of the cell tandem. The output power therefore decreases.

(3) In actual application, shadowing in photovoltaic array arises from structures such as the front and rear

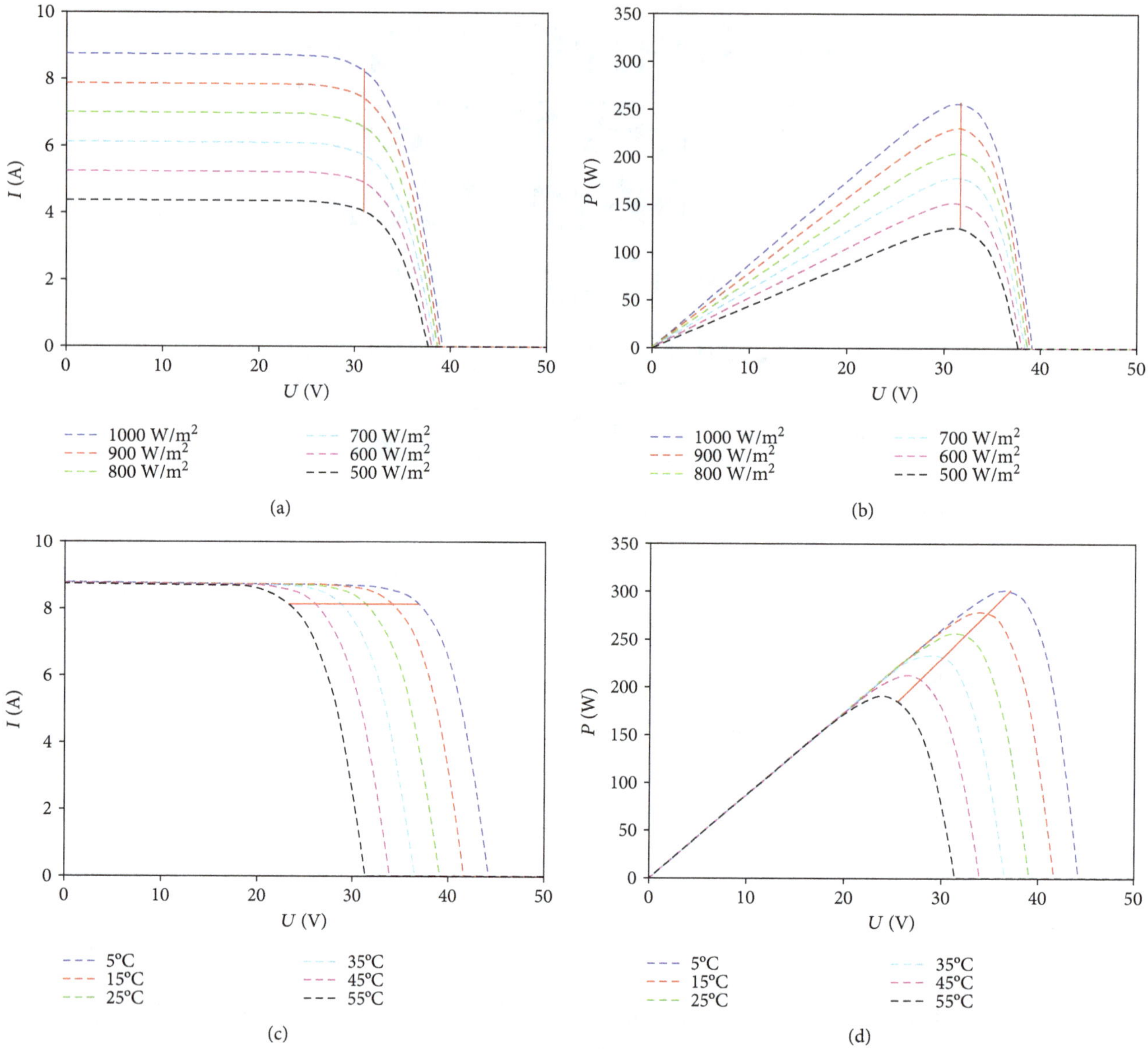

FIGURE 6: (a) IV curve under different irradiances, (b) PV curve under different irradiances, (c) IV curve under different temperatures, (d) PV curve under different temperatures. Describing the IV and PV curves under different operating conditions.

rows of the array and tall buildings. The area of the shadow is generally larger than that of a single photovoltaic cell. The voltage effect on the array arising from a shadow with an area less than that of a single photovoltaic cell can be ignored.

3.2. Cover Shading. The simulation results under the cover conditions are shown in Figure 9. The bypass diode is usually in a nonconducting state, and the module output is normally under uniform light in the absence of covering. The cells will have negative voltage and trigger the conduction of bypass diode when the negative voltage reaches a certain upper limit. In Figure 9, the output characteristics have no multiple peaks due to the fact that the covered cells with negative voltage trigger the conduction of the corresponding bypass diode. Covers 1, 2, and 4 have consistent output characteristics. Similarly, covers 3 and 5 are consistent. The output decreases in a ladder pattern from cover 4 to cover 6. In terms of the electrical parameters, I_{sc} and I_m remain unchanged while U_{oc}, U_m, and P_m decline regularly. The distribution of the electrical parameters of the array under cover shading is shown in Table 4.

Based on the above analysis, we can obtain the distribution rules of the electrical parameters of photovoltaic array under cover shading as follows:

(1) Cover shading does not influence I_{sc} and I_m of the photovoltaic array.

(2) U_{oc} and U_m drop when the area of the covered portion within the cell string is larger than or equal

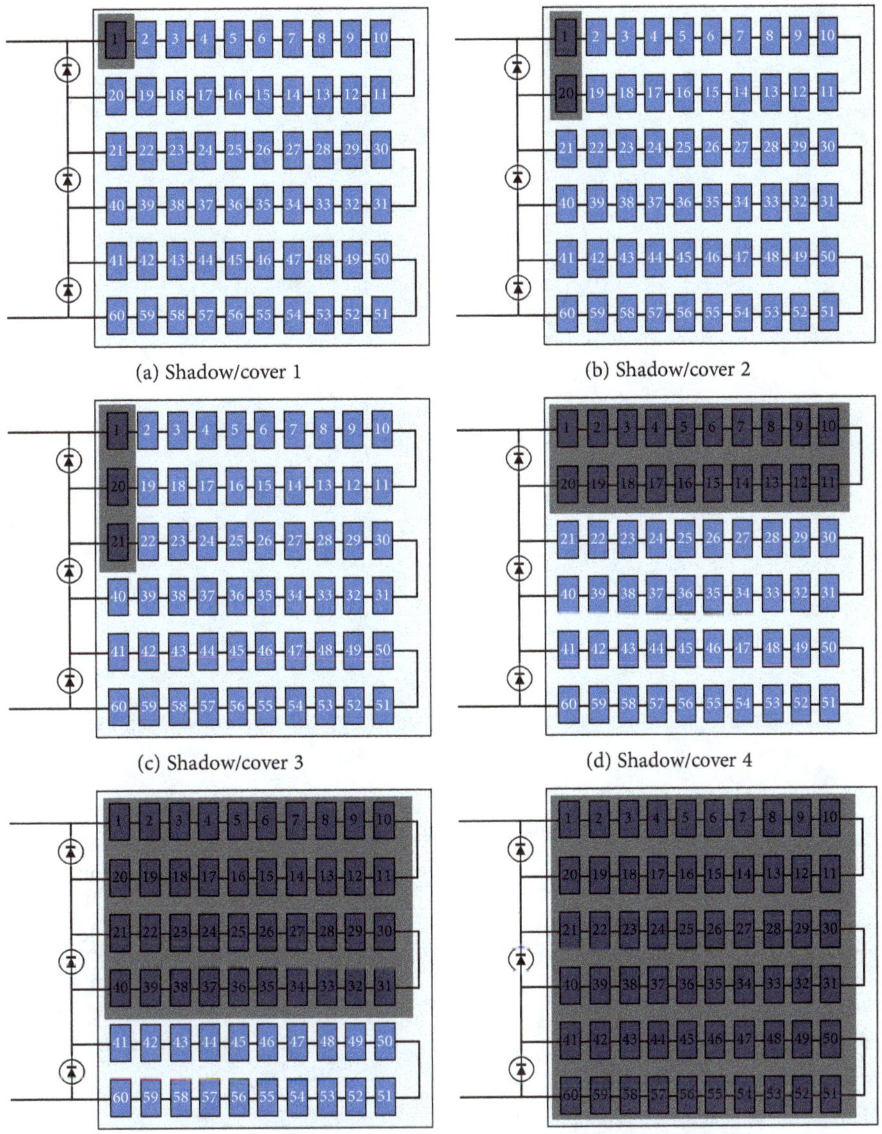

(a) Shadow/cover 1

(b) Shadow/cover 2

(c) Shadow/cover 3

(d) Shadow/cover 4

(e) Shadow/cover 5

(f) Shadow/cover 6

FIGURE 7: Setting of different shading conditions. Shading one cell/cell string, two cells/cell strings, three cells/cell strings, respectively.

to the area of a single cell. The drop intensity is proportional to the number of the covered cell strings and the cell string voltage. The output power therefore decreases.

(3) Cover shading with an area less than that of a single cell can be considered negligible.

The computation formula of the electrical parameters of the array under cover shading is as follows:

$$\begin{aligned}
I_{sc}' &= I_{sc}, \\
U_{oc}' &= U_{oc} - u_2 \cdot n_2, \\
I_m' &= I_m, \\
U_m' &= U_m - u_1 \cdot n_2, \\
P_m' &= I_m' \cdot U_m',
\end{aligned} \tag{12}$$

TABLE 2: Setting of simulation conditions.

Type of shading	Solar irradiation intensity of the shaded portion (W/m^2)	Solar irradiation intensity of the unshaded portion (W/m^2)	Module temperature (°C)
Shadow shading	150	1000	25
Cover shading	0	1000	25

The irradiation intensity of the shadowed photovoltaic cells is set at 15% of the irradiation intensity absorbed by the unshadowed modules. The solar irradiation intensity of cover is set at zero. Temperature is 25°C.

where n_2 is the number of the covered cell strings under the covering condition and u_2 is the open-circuit voltage of the cell string.

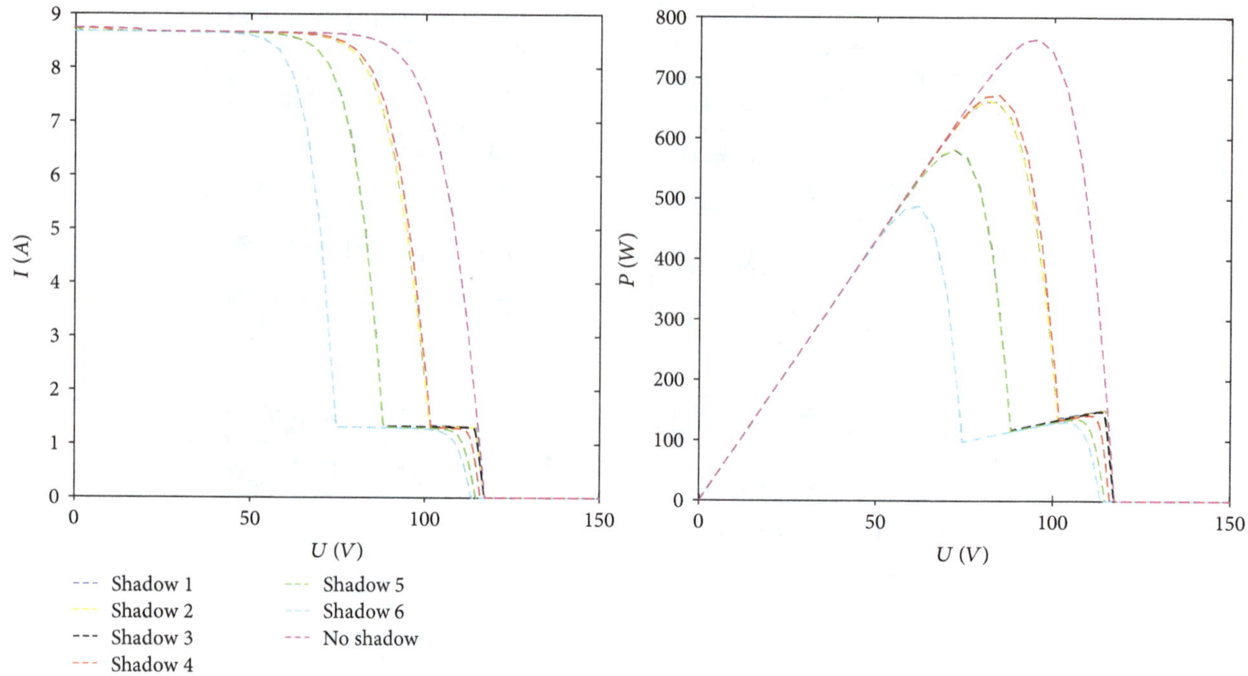

FIGURE 8: Simulation result for shadowing. Under the shadow conditions, the output curve presents a multipeak feature.

TABLE 3: Distribution of electrical parameters under shadowing condition.

Shadow conditions	U_{oc} (V)	I_{sc} (A)	P (W)	U_m (V)	I_m (A)
Normal	117.37	8.75	744.63	91.93	8.10
Shadow 1	117.30	8.74	661.47	80.51	8.22
Shadow 2	117.24	8.74	662.46	81.12	8.17
Shadow 3	117.10	8.73	581.47	71.95	8.08
Shadow 4	116.03	8.74	652.70	81.04	8.05
Shadow 5	114.68	8.73	580.40	70.77	8.20
Shadow 6	113.34	8.74	487.82	61.95	7.87

The distribution of electrical parameters under different shading conditions.

3.3. Summary. To illustrate the distribution features of the electrical parameters of the photovoltaic array under different shading conditions, Figure 10 presents the IV and PV curves of the array under the two types of shading. Shadow 4, shadow 5, cover 4, and cover 5 are selected for comparison. The output features and parameter distribution rules of the photovoltaic array under the shadow and cover conditions are as follows:

(1) The output of the array under shadowing is characterized by multiple peaks and ladder pattern while cover shading has no such characteristics.

(2) Shadowing and covering have no effect on I_{sc} of the array.

(3) When the shaded area is larger than or equal to a single cell, the shadow causes a drop in U_m while covering causes a drop in U_m and U_{oc}. The voltage drop extent is associated with the number of the shaded cell strings and the string voltage. Any shading with an area less than that of a single cell is considered negligible.

(4) The difference between shadowing and covering conditions is whether U_{oc} is falling or not. It can be judged from the distribution characteristics of U_m and U_{oc}.

4. Empirical Validation of the Electrical Properties of Photovoltaic Array under Different Shading Conditions

4.1. Method for Empirical Validation. Three modules were connected in series to create a photovoltaic array in the experiment. To investigate the distribution characteristics of the electrical parameters under different shading conditions, an opaque paperboard was used to set the shadow and covering conditions, respectively, as shown in Figure 7. The electrical parameters of the array under different shading conditions were recorded with an IV scanner. To explain the distribution rules of the electrical parameters, (8) was introduced to obtain the voltage at the maximum power point. Based on (11) and (12), we can obtain the voltage at the theoretical maximum power point under different shading conditions and compare with the experiment results. The experimental platform comprised of a photovoltaic array constituted by Jinko JKM245P modules, an irradiator for measuring the solar irradiance, a temperature sensor for measuring the

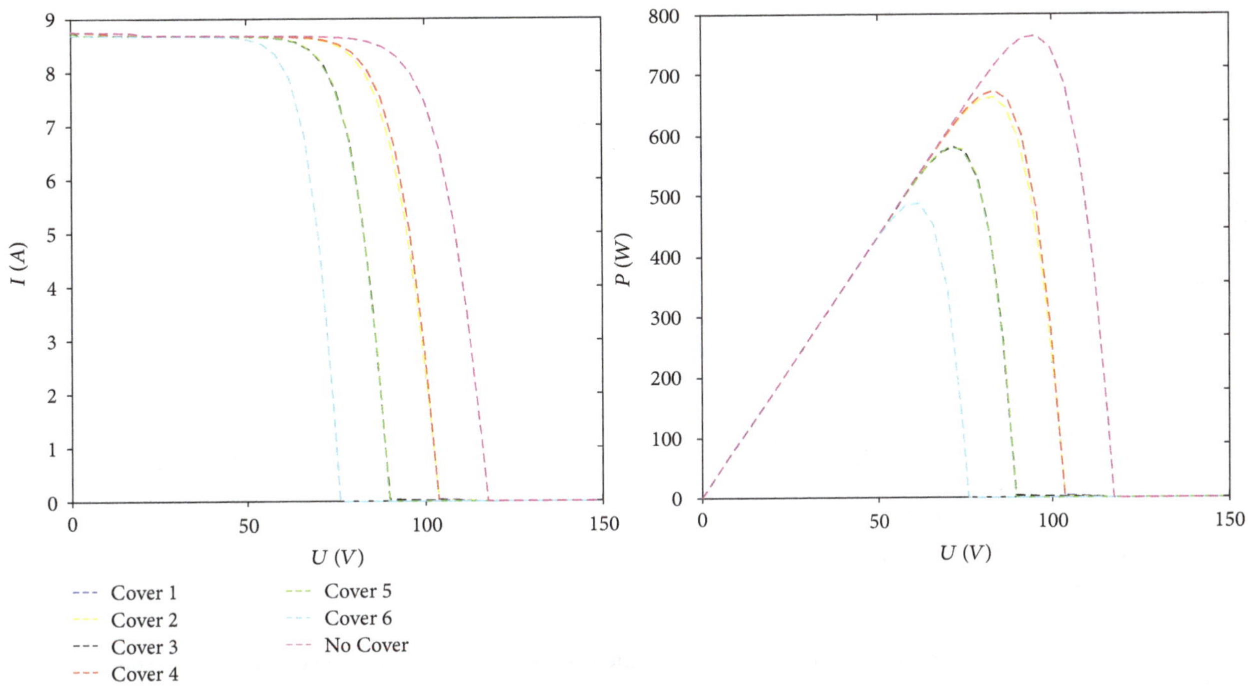

FIGURE 9: Simulation result for cover shading. The output curve without multipeak feature under cover shading.

TABLE 4: Distribution of electric parameters under cover shading.

Covering conditions	U_{oc} (V)	I_{sc} (A)	P (W)	U_m (V)	I_m (A)
Normal	117.37	8.75	743.97	91.11	8.17
Cover 1	103.17	8.74	660.92	80.26	8.23
Cover 2	103.50	8.74	662.99	82.39	8.05
Cover 3	89.64	8.73	578.84	70.08	8.26
Cover 4	103.44	8.74	652.07	81.90	7.96
Cover 5	89.68	8.73	574.56	70.24	8.18
Cover 6	75.83	8.69	489.36	60.99	8.02

The distribution of electrical parameters under different cover conditions.

backboard temperature, and an IV tester. Figure 11 presents the structural diagram.

4.2. Analysis of the Empirical Testing Results. As seen in Figure 12, the results indicate that the output of the array under shadow is characterized by multiple peaks. Shadowing only influences the optimal operating voltage. According to (a) and (b), shadows 1 and 2 nearly have consistent output characteristics. Compared to the normal operating condition (nonshaded), the voltage at the optimal operating point moves left while the open-circuit voltage remains unchanged. The voltage at the optimal operating point further moves left in the case of shadow 3 compared to that in the case of shadow 2. According to (c) and (d), with the increase of the number of shaded cell strings, the voltage at the optimal operating point of the arrays decreases in sequence.

As seen in Figure 13, the results indicate that the output of the array under cover have no multiple peaks. According to (a) and (b), cover 1 and cover 2 have nearly consistent output characteristics. The voltage at the optimal operating point and the open-circuit voltage move left compared to the normal operating condition (noncovered). The bypass diode is thus activated when the area of the covering is larger than the area of a single cell. The voltage at the optimal operating point and the open-circuit voltage in the case of cover 3 move left compared with that in the case of cover 2. At this point, two bypass diodes are activated. According to (c) and (d), the voltage at the optimal operating point of the array and the open-circuit voltage decrease in sequence with the increase of the covered cell strings.

As seen in Figure 14, cover shading influences both the voltage at the maximum operating point U_m and the open-circuit voltage U_{oc} while shadow shading only influences the voltage at the maximum operating point U_m, which coincides with the simulation results.

Equations (11) and (12) present the computational formulae for the voltage at the optimal operating point and the open-circuit voltage U_{oc} under the shadow and covering conditions, which will be analyzed in the following paragraphs. Under shading conditions, the relative error between the voltage at the theoretical optimal operating point and the voltage at the measured optimal operating point is expressed as e_1 and the relative error between the theoretical open-circuit voltage and the measured open-circuit voltage is expressed as e_2, as shown in (13).

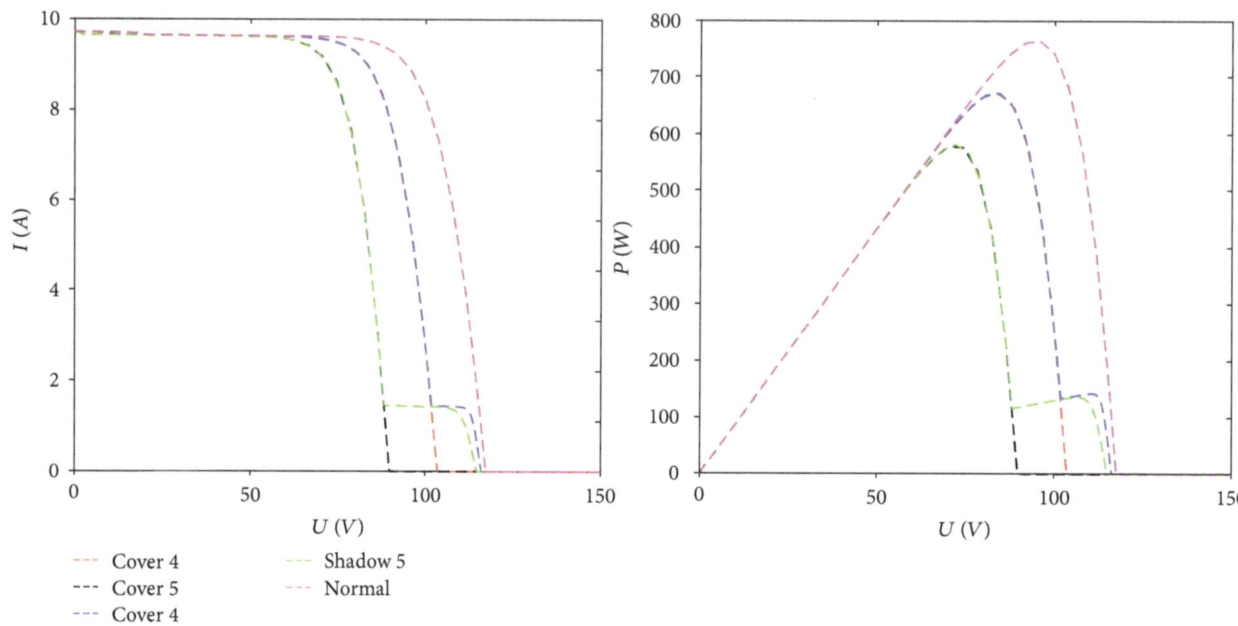

FIGURE 10: Comparison between shadow and cover conditions. The difference between shadow and cover is whether the open-circuit voltage drops or not.

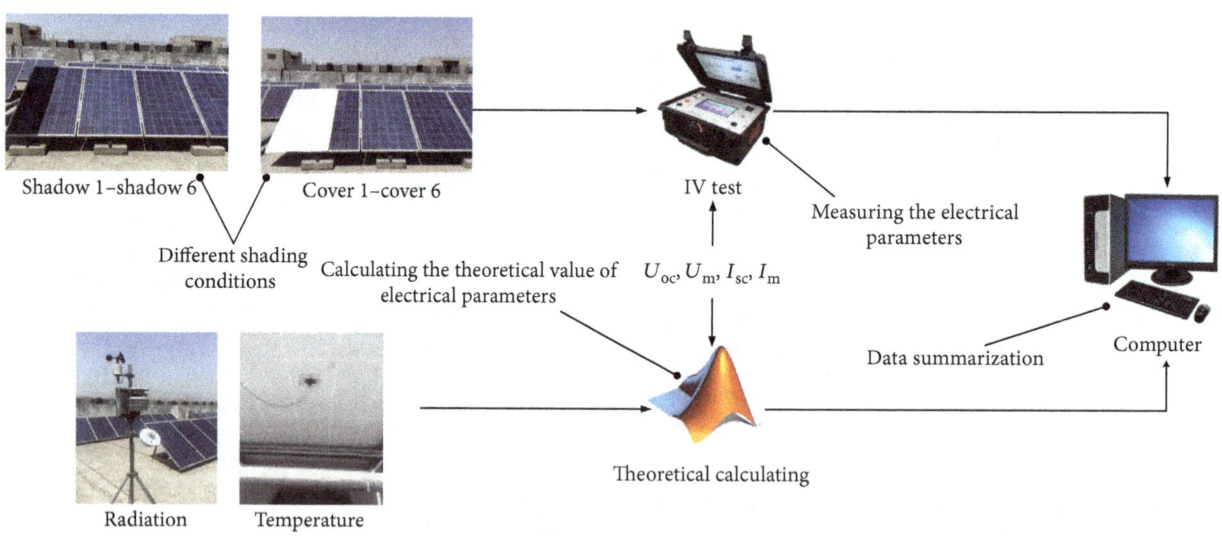

FIGURE 11: Block diagram for experimental testing. By comparing the theoretical value and the measured value, summarize the rules.

$$e_1 = \frac{\left|U'_{\text{m-test}} - U'_{\text{m-calculate}}\right|}{U'_{\text{m-calculate}}} \times 100\%,$$
$$e_2 = \frac{\left|U'_{\text{oc-test}} - U'_{\text{oc-calculate}}\right|}{U'_{\text{oc-calculate}}} \times 100\%, \tag{13}$$

where $U_{\text{m-test}}'$ and $U_{\text{oc-test}}'$ are the measured voltage at the optimal operating point and the open-circuit voltage under shading conditions, respectively, and $U_{\text{m-calculate}}'$ and $U_{\text{oc-calculate}}'$ are the theoretical voltage at the optimal operating point and the open-circuit voltage under the shading conditions.

Table 5 lists the distribution of the voltage at the optimal operating point and the open-circuit voltage under different shading conditions and the relative error between the measured shaded voltage and the computed shaded voltage. As seen in the table, the maximum error between the measured voltage and the computed voltage is less than 2% under shading conditions. The empirical results completely coincide with the simulation results.

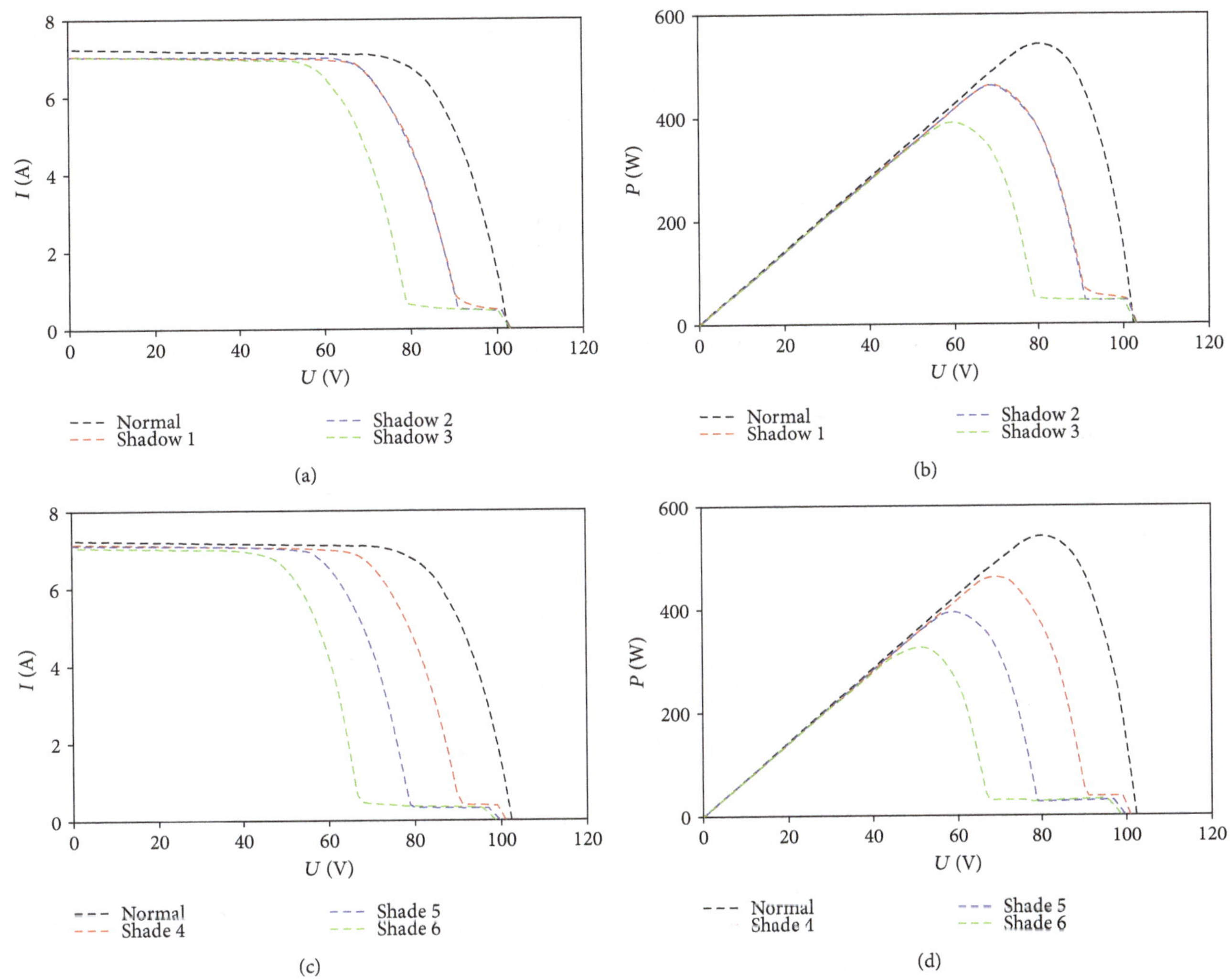

FIGURE 12: Results for the shadowing experiment. The curves are coincide with the simulation results.

Based on Table 5, it can be concluded that

(1) the computation results of the electrical parameters of the theoretical optimal operating point under the shading conditions are accurate and applicable,

(2) the number of the shaded modules can be known from the voltage at the theoretical optimal operating point and the voltage at the measured optimal operating point. The computation formula is as follows:

$$n_3 = \left| \frac{U'_{\text{m-calculate}} - U'_{\text{m-test}}}{u} \right|, \quad (14)$$

where u is the optimum operating voltage of a cell string and n_3 is the number of the shaded modules.

5. Conclusion

The paper illustrates the distribution rules of the electrical parameters of photovoltaic array under two types of shading conditions, shadow and cover shadings, both by simulation and empirical testing drawing several conclusions as below:

(1) Shadow and cover shadings have different effects on photovoltaic array in actual engineering. This model accurately reflects the output properties of the photovoltaic array under different shading conditions.

(2) Shadowing only influences the voltage at the optimal operating point of the array while covering influences both the open-circuit voltage and the voltage at the optimal operating point. The type of shading, that is, the shadow and cover can be judged from the distribution of the open-circuit voltage and the voltage at the optimal operating point. Shading does not influence the current of the array.

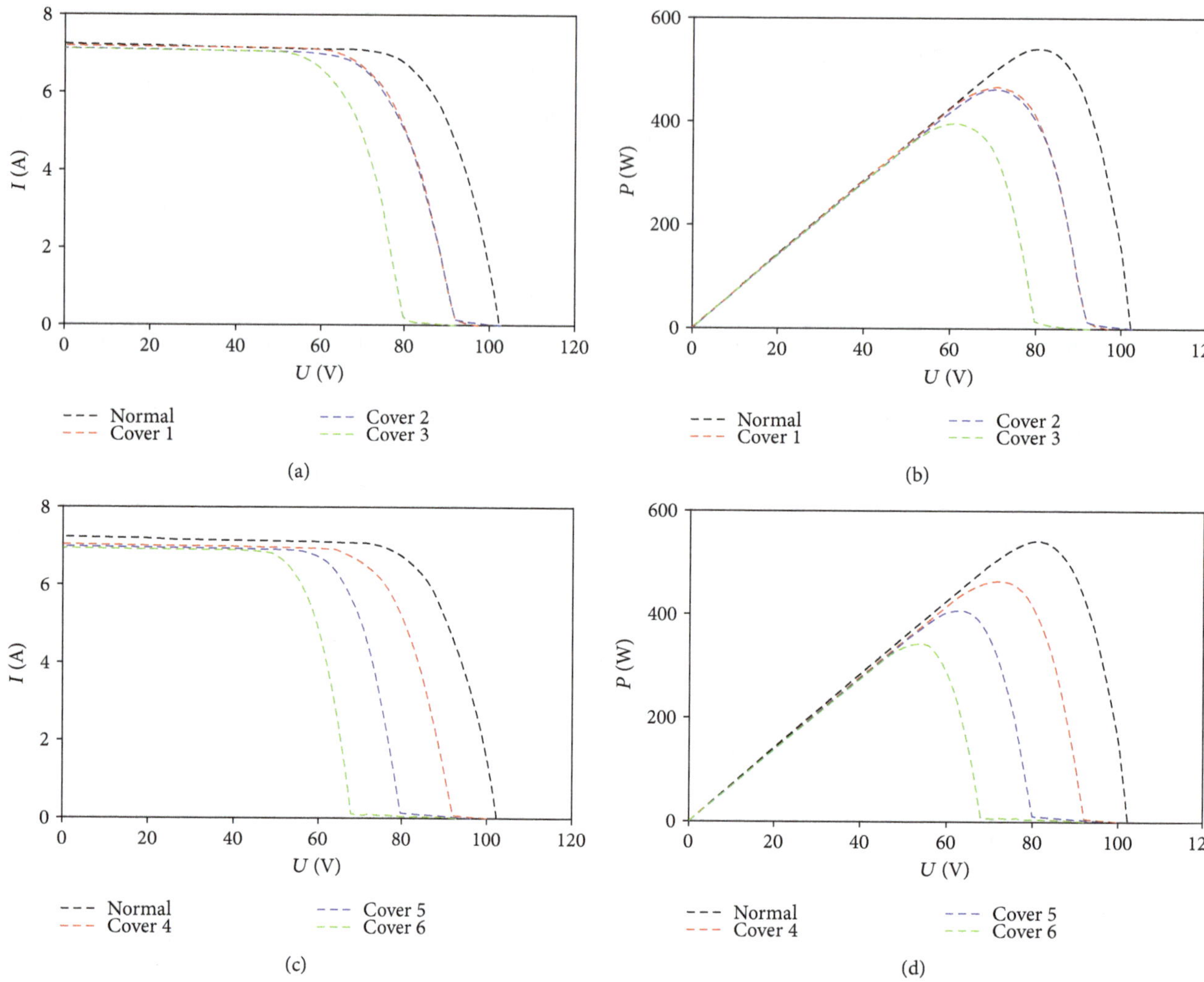

FIGURE 13: Results for the cover shading experiment. The curves are coincide with the simulation results.

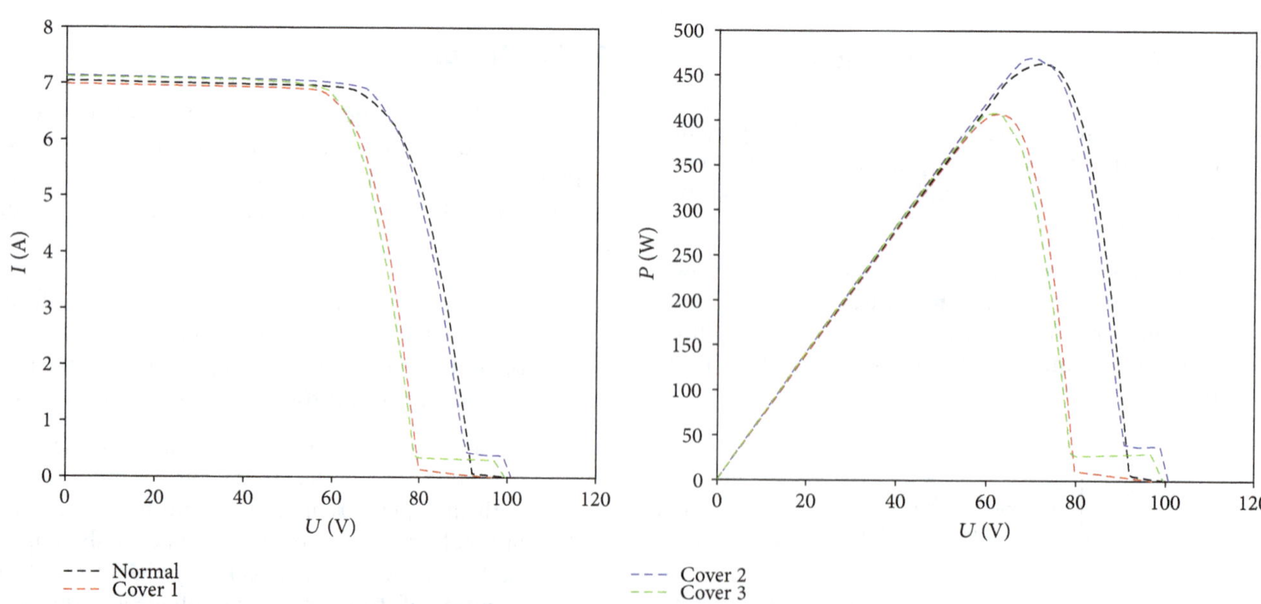

FIGURE 14: Comparison of the results of the shadow and the cover shading experiments. The difference between shadow and cover is whether the open-circuit voltage drops or not.

TABLE 5: Error analysis for computation result.

Shading type	$U'_{\text{m-test}}$ (V)	$U'_{\text{oc-test}}$ (V)	$U'_{\text{m-calculate}}$ (V)	$U'_{\text{oc-calculate}}$ (V)	e_1 (%)	e_2 (%)	S (W/m^2)	T (°C)
Normal	87.0	108.4	88.2	109.8	1.3%	1.3%	982	36
Shadow 1	77.4	108.6	78.1	109.0	0.8%	1.0%	932	35
Shadow 2	77.2	108.5	78.1	109.3	1.1%	1.1%	936	35
Shadow 3	69.1	108.8	68.3	110.3	1.1%	1.1%	945	34
Shadow 4	76.5	108.4	77.9	109.6	1.7%	1.0%	931	36
Shadow 5	66.7	108.3	67.9	109.3	1.8%	1.0%	939	36
Shadow 6	57.4	108.3	57.9	109.1	0.8%	1.1%	936	36
Cover 1	77.1	97.4	78.1	97.81	1.3%	0.4%	973	36
Cover 2	77.6	96.1	78.3	98.03	0.9%	1.9%	973	35
Cover 3	68.1	87.1	68.4	86.08	0.3%	1.1%	952	34
Cover 4	77.8	98.3	78.5	98.23	0.8%	0.1%	942	33
Cover 5	67.9	83.8	68.3	85.25	0.5%	1.7%	963	35
Cover 6	58.6	74.2	58.1	73.69	0.9%	0.6%	959	36

The data of shading voltage under different conditions and the relative error between the measured occlusion voltage and the theoretical occlusion voltage.

(3) A drop occurs in voltage when the area of shading in a cell string is larger than or equal to that of a single cell. The magnitude of the drop is proportional to the number of the shaded strings and the string voltage.

(4) The number of the shaded modules can be effectively judged from the computed results of the theoretical maximum power point and the measured results.

Conflicts of Interest

The authors declare that they have no conflicts of interest.

Acknowledgments

This work was supported by the Fundamental Research Funds for the Central Universities (2016MS52, 2016MS31) and China Three Gorges New Energy Co. Ltd.

References

[1] P. Guerriero, F. Di Napoli, F. Cominale, V. d'Alessandro, and S. Daliento, "Accurate analysis of small shadows effects on photovoltaic systems yield," in *2014 International Symposium on Power Electronics, Electrical Drives, Automation and Motion*, pp. 987–992, Ischia, Italy, June 2014.

[2] J. Qi, X. Zhang, Y. Zhang, and W. Zhou, "Study on simulation algorithm of PV array considering shade effect," *Proceedings of the CSEE*, vol. 32, pp. 131–138, 2012.

[3] C. H. Wu, D. Q. Zhou, and Z. H. Li, "Hot spot detection and fuzzy optimization control method of PV module," *Proceedings of the CSEE*, vol. 33, pp. 50–61, 2013.

[4] Y. Haoyuan, Y. Shuo, S.-C. Tan, and S. Y. R. Hui, "Dynamic modeling of partial shading on photovoltaic arrays," in *2015 IEEE Energy Conversion Congress and Exposition (ECCE)*, pp. 6616–6621, Montreal, QC, Canada, September 2015.

[5] K. Ding, X. G. Bian, and H. H. Liu, "Matlab-Simulink based modeling to study the influence of nonuniform insolation photovoltaic array," in *2011 Asia-Pacific Power and Energy Engineering Conference*, pp. 1–4, Wuhan, China, March 2011.

[6] P. Burns and N. Anani, "Modelling and simulation of photovoltaic arrays under varying conditions," in *2014 9th International Symposium on Communication Systems, Networks & Digital Sign (CSNDSP)*, pp. 831–834, Manchester, UK, July 2014.

[7] G. Celsa and G. M. Tina, "Matlab/Simulink model of photovoltaic modules/strings under uneven distribution of irradiance and temperature," in *IREC'2015. The Sixth International Renewable Energy Congress*, pp. 1–6, Sousse, Tunisia, March 2015.

[8] Q. Tian, Z. Zhao, Y. Deng, L. Yuan, and F. He, "Simulation and experimental study about reverse model of photovoltaic cells," *Proceedings of the CSEE*, vol. 31, no. 23, pp. 121–128, 2011.

[9] A. Kumar, R. K. Pachauri, and Y. K. Chauhan, "Experimental analysis of SP/TCT PV array configurations under partial shading conditions," in *2016 IEEE 1st International Conference on Power Electronics, Intelligent Control and Energy Systems (ICPEICES)*, pp. 1–6, Delhi, India, July 2016.

[10] F. Zhicheng, W. Yahui, and W. Lulu, "Experimental study on characteristics of PV module under partially shaded conditions," *Acta Energiae Solaris Sinica*, vol. 36, no. 2, pp. 392–398, 2015.

[11] T. P. Zhou and W. Sun, "Maximum power point tracking of photovoltaic array under nonuniform shadow conditions," *Automation of Electric Power Systems*, vol. 39, no. 10, pp. 42–49, 2015.

[12] Y. P. Wang, X. B. Ruan, and Y. Li, "A rapid tracking method of maximum power point for solar units in series under uneven solar irradiance," *Proceedings of the Chinese Society for Electrical Engineering*, vol. 35, pp. 4870–4878, 2015.

[13] Y. W. Zhu, X. C. Shi, Y. Q. Dan et al., "Application of PSO algorithm in global MPPT for PV array," *Proceedings of the CSEE*, vol. 32, pp. 42–48, 2012.

[14] X. Yuan, D. Yang, and H. Liu, "MPPT of PV system under partial shading condition based on adaptive inertia weight particle

swarm optimization algorithm," in *2015 IEEE International Conference on Cyber Technology in Automation, Control, and Intelligent Systems (CYBER)*, pp. 729–733, Shenyang, China, June 2015.

[15] X. Liu, F. Zhuo, Y. Chen, and L. Xiong, "Development of fast simulation models for photovoltaic generation system based on Simulink," in *2015 IEEE Energy Conversion Congress and Exposition (ECCE)*, pp. 3265–3270, Montreal, QC, Canada, September 2015.

[16] Y. J. Wang and S. S. Lin, "Analysis of a partially shaded PV array considering different module connection schemes and effects of bypass diodes," in *2011 International Conference & Utility Exhibition on Power and Energy Systems: Issues and Prospects for Asia (ICUE)*, pp. 1–7, Pattaya City, Thailand, September 2011.

[17] D. Q. Zhou, C. H. Wu, Z. H. Li, L. Fu, and Y. Z. Wang, "Simulation and experimental study of the photovoltaic model under partial shading," *Acta Energiae Solaris Sinica*, vol. 35, pp. 2098–2105, 2014.

10

On the Link between the Operating Point and the Temperature Distribution in PV Arrays Working under Mismatching Conditions

Luigi Costanzo and Massimo Vitelli

Department of Engineering, Università degli Studi della Campania "Luigi Vanvitelli", Aversa, 81031 Caserta, Italy

Correspondence should be addressed to Luigi Costanzo; luigi.costanzo@unicampania.it

Academic Editor: K. R. Justin Thomas

Mismatching operating conditions negatively affect the extracted energy in photovoltaic (PV) systems. They may also lead to dangerous localized heating phenomena (hot spots) that can cause, in turn, accelerated ageing and reduced reliability. Since the adoption of bypass diodes or smart active switches does not prevent the occurrence of hotspots, it is necessary to investigate alternative strategies. A promising solution is represented by the proper regulation of the operating point of the PV cells in the current vs. voltage (I-V) or power vs. voltage (P-V) planes when mismatching conditions occur. In particular, in this paper, the existence of operating points allowing a suitable compromise between maximization of the extracted power and minimization of thermal stresses, due to hot spots, is experimentally evidenced. Experimental results highlighting the link existing between the operating point in the I-V plane and the PV cell temperature distribution under uniform and mismatching operating conditions are presented and discussed. On the basis of the obtained experimental results, it is possible to state that, when mismatching conditions occur, it is mandatory to properly choose the operating point: the global maximum power point may not be the best operating point. Hence, it is crucial to gain information about the eventual occurrence of mismatching conditions in order to be able to properly choose the best operating point. Therefore, another crucial aspect that is evidenced in this paper is represented by the fact that the detection of the occurrence of mismatching conditions, based on the analysis of the shape of the I-V and/or P-V characteristics, is effective only if the analysis is carried out for both positive and negative voltages.

1. Introduction

It is well known that, in photovoltaic (PV) systems, mismatching due to partial shading, shadows of neighboring objects, dirtiness, clouds, different orientation angles of modules of the PV field, soiling, manufacturing tolerances, or ageing can cause reduced reliability and significant losses in the energy yield [1–6]. Even with the commonly used bypass diodes, mismatched cells may become reverse biased [7] and dissipate power, leading to dangerous localized heating phenomena called hot spots that in turn may cause accelerated ageing or even the damage of the cells. The hot spot in a reverse-biased PV cell occurs when its temperature increases, above the temperature of the cells surrounding it, due to power dissipation which can occur in the entire cell or a portion of it.

This increase in the cell temperature will gradually degrade the output power generated by the PV module [8, 9]. As discussed in [4], the operating temperature of the PV modules has a huge impact on their aging. In fact, the higher the operating temperature, the higher the probability of discoloration of the encapsulant material, module delamination, creation of bubbles, and corrosion. Moreover, higher working temperatures also accelerate the degradation of the ribbon and solder bonds and the PID effect [10], as well as cause cell cracks and failures of junction boxes and bypass diodes. Many long-term studies made on field-aged PV generators have shown that, despite the presence of bypass diodes, localized heating phenomena represent one of the main causes of PV module failures [11–13]. On the basis of the above considerations, it is evident that it is advisable to limit as much as possible the operating temperature of PV modules. Many distributed power electronic architectures have been introduced in the literature in order to mitigate the mismatching effects (hot spots) and increase the energy yield of PV systems [14–19]. Also, the dynamical reconfiguration of PV arrays has been proposed in the literature in order to face the drawbacks that are associated to mismatching phenomena [20–25]. PV array reconfiguration can be obtained by means of a reconfigurable switch matrix made of MOSFETs or of electromechanical relays. In [26–30], it has been shown that distributed converters can improve system reliability by reducing the occurrence of hot spots. In fact, reliability represents a crucial aspect to take into account in PV applications. In [31], the following innovative concept has been introduced as concern reliability: the objective of the maximization of the energy production of a PV array during its lifetime may be in contrast with the objective of the maximization of the power production in any operating condition. That is, it may be in contrast with maximum power point tracking (MPPT) objectives, if reliability is given due consideration. Even if [31] is focused in particular on PV reconfiguration, the above concept is of course valid also when distributed power architectures are adopted. At first sight, the link existing between the reliability of a PV array and the efforts to extract its maximum possible power (MPPT) may seem to be weak. However, as shown in this paper, different PV array operating points lead to different temperature distributions, which in turn lead to different durations of life or at least to different degradation rates of the array itself [4]. Therefore, it is clear that the aging process, which is related to the PV module temperature distribution, can be accelerated or delayed by properly regulating the operating point. At this stage, the following question arises: is the MPP the best operating point in any working condition of the PV array? In this paper, it will be shown that the answer to the previous question is negative. In particular, in Section 2 the considered experimental set-up will be briefly described. In Section 3 (Section 4), experimental results highlighting the link existing between the operating point in the current vs. voltage (I-V) plane and the PV cell temperature distribution under uniform (mismatching) operating conditions are presented and discussed. On the basis of these experimental results, it is possible to state that, if mismatching conditions occur, it is mandatory to properly choose the operating point: the global MPP may not be the best operating point. Therefore, it is crucial to gain information about the eventual occurrence of mismatching conditions in order to be able to properly choose the best operating point. Hence, in Section 5, it will be experimentally shown that the detection of mismatching conditions can be efficiently supported by the analysis of the shape of the I-V and/or power vs. voltage (P-V) characteristics not only for positive but also for negative voltages. Conclusions end the paper.

TABLE 1: Main electrical characteristics of the adopted Photonsolar PM0020 [39] Modules.

Description	Symbol	Value
MPP voltage	V_{MPP}	16.5 V
MPP current	I_{MPP}	1.21 A
MPP power	P_{MPP}	20 W
Open-circuit voltage	V_{OC}	20 V
Short-circuit current	I_{SC}	1.38 A
Nominal operating cell temperature	NOTC	45°C
Number of series cells	N_s	36

2. The Experimental Set-up

The main prevention method for hot spotting of PV modules is represented by passive bypass diodes that are placed in parallel with strings of PV cells. Bypass diodes turn on to provide an alternative current path and attempt to prevent extreme reverse bias voltages, and hence hot spots on PV strings. A number of long-term field studies have found that, nonetheless, hot spots still occur on systems employing bypass diodes [32–36]. Active bypass switches are an improvement over bypass diodes but do not resolve the problem [7]. There are two further methods to prevent hot spots: method 1—regulation of the operating point (make the PV string work in open circuit in extreme cases) [26, 37, 38] and method 2—ensure that the cell can fully dissipate the worst-case power scenario without damages [7]. For example, cells with low reverse-breakdown voltages limit the power dissipated during hot spots and may be an effective prevention method. Further studies are needed to determine the susceptibility of cells with low reverse-breakdown voltages to hot spot damages [7]. The focus of this paper is on method 1. In particular, it will be shown that once the occurrence of mismatching is detected, it is possible to reduce system degradation, increase longevity, and improve lifetime energy harvesting by properly regulating the PV operating point and the associated temperature distribution.

In the following, without any loss of generality, the results of targeted experimental activities carried out on Photonsolar PM0020 Modules [39], with the main characteristics that are reported in Table 1, will be presented and discussed.

Since Photonsolar PM0020 Modules lack bypass diodes, such modules will be called PV units (PVUs) in order to underline the fact that they are different from traditional PV modules that are equipped with bypass diodes. As a consequence, the validity of the experimental findings contained in Sections 3, 4, and 5 refers to PV modules that are not equipped with bypass diodes.

Initially, two PVUs have been simultaneously analysed. In particular, the two considered PVUs have been connected in series during the first part of the experimental campaign. Two different experimental set-ups have been considered. In Figure 1(a), the experimental set-up that has been used for tracing the I-V and the P-V curves of the considered PVUs is shown. In particular, such curves are obtained by means of sweeps carried out by using the IV Tracker HT

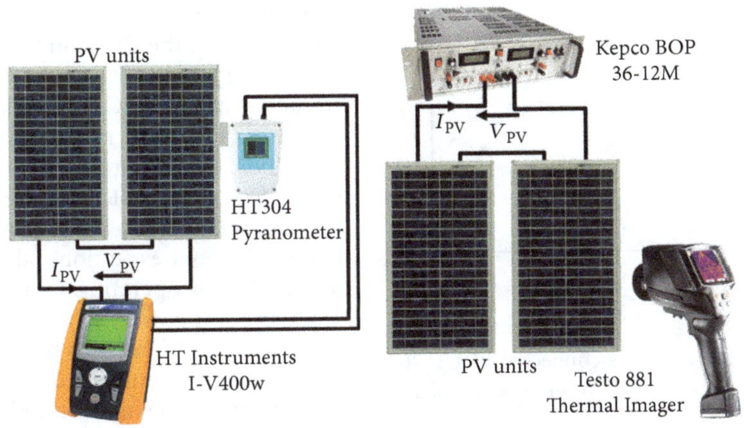

FIGURE 1: Experimental set-up adopted for (a) tracing I-V and P-V characteristics and (b) measuring PVU temperature as a function of the operating point.

FIGURE 2: Two series-connected PVUs equipped with the IV Tracker. 26 September 2017, Aversa (CE), Italy. $S = 600\ \mathrm{W/m^2}$ and ambient temperature $T_{amb} = 25°\mathrm{C}$.

Instruments I-V400w [40, 41] accompanied by the HT304 Pyranometer [40]. This is a system that is able to trace the I-V and P-V characteristics of PV modules or fields up to a maximum of 1000 V (with a voltage resolution of 0.1 V) and 15 A (with a current resolution of 0.01 A) [40]. A second experimental set-up has been adopted in order to get the temperature distribution of the considered PVUs as a function of the operating point. Such experimental set-up is shown in Figure 1(b). In particular, the desired operating point is set by means of a power supply (that fixes V_{PV}) and a thermographic image of the PVUs is obtained by means of a thermal camera [42]. The adopted power supply is a Kepco BOP 36-12M [43], a device that is able to operate in all four quadrants of the I-V plane. It is a linear power supply with two bipolar selectable control channels (voltage or current mode) that are individually controllable either by front panel controls (front panel voltage or front panel current control mode) or by remote control signals (remote voltage or remote current control mode) [43]. In the considered application, the Kepco works in remote voltage control mode. The adopted thermal camera is a Testo 881 Thermal Imager whose main characteristics are the following [44]: 3.5″ LCD display with 320 × 240 pixel resolution, temperature measuring range 0–350°C (32–662°F), accuracy ± 2°C (±3.6°F) or ±2% of the reading (whichever is greater), reproducibility ± 1°C (±1.8°F) or ±1% (whichever is greater), and on-time (time to image) 30 s.

3. The Link between the Operating Point in the I-V Plane and the PV Cell Temperature Distribution under Uniform Operating Conditions

The first set of experimental results are characterized by the absence of mismatching conditions. In particular, a picture of the two series-connected PVUs equipped with the IV Tracker is reported in Figure 2. The tests have been carried out on 26 September 2017 in Aversa (CE), Italy (40°57′56.2″N, 14°12′43.2″E), under a clean-sky day with an irradiance of $S = 600\ \mathrm{W/m^2}$ and an ambient temperature of $T_{amb} = 25°\mathrm{C}$.

The I-V and P-V characteristics of the two series-connected PVUs have been obtained by means of the experimental set-up shown in Figure 1(a). Such curves are shown in Figures 3 and 4.

Indeed, another curve is also shown in Figure 4. It is the average temperature of the two PVUs as a function of the operating voltage. Such a curve has been obtained by means of thermographic images carried out by means of the

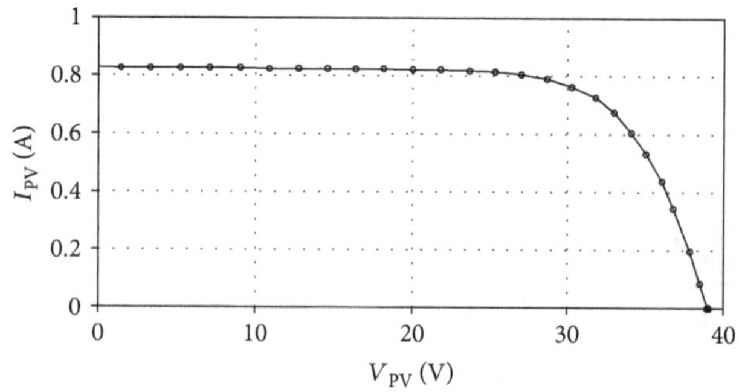

FIGURE 3: I-V characteristic of the two series-connected PVUs. 26 September 2017, Aversa (CE), Italy. $S = 600\ W/m^2$ and ambient temperature $T_{amb} = 25°C$.

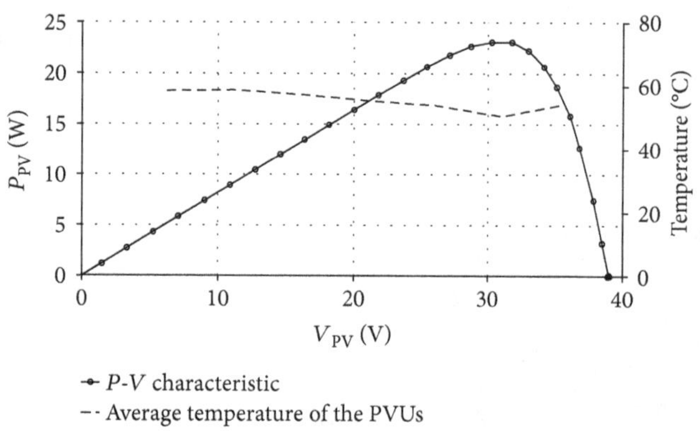

FIGURE 4: P-V characteristic of the two series-connected PVUs and average temperature of the two PVUs as a function of the operating voltage. 26 September 2017, Aversa (CE), Italy. $S = 600\ W/m^2$ and ambient temperature $T_{amb} = 25°C$.

experimental set-up shown in Figure 1(b). In particular, an example of thermographic image is shown in Figure 5. It has been obtained when the two PVUs operate in the MPP. Of course, in the absence of whatever mismatching condition, the distribution of the temperature on the two PVUs is nearly uniform. The most interesting, even if somewhat predictable, result is clearly shown by the dashed curve in Figure 4. In the absence of whatever mismatching condition, not only is the temperature distribution uniform whichever is the operating point in the I-V plane, but, in the MPP it attains its minimum value (50.5°C). Such a result is predictable since, when the total sun energy hitting a PV unit is more or less constant (as in the case of Figure 2 which refers to a clean-sky day), the operating point which allows extracting the maximum electrical power (that is the MPP) must be characterized by the lowest temperature. In fact, a simple analysis concerning the energetic balance of the system composed by the PVUs allows stating that, in the MPP, the energy coming from the sunlight and which remains available for the heating of the PVUs attains its minimum value just because the extracted electric energy attains its maximum value. On the other hand, in short-circuit (SC) and open-circuit (OC) conditions, the energy which is available for the heating of the PVUs attains its maximum value just because the extracted electric energy is equal to zero. Indeed, as shown in Figure 4, the difference among the value of the temperature of the PVUs in the MPP and the corresponding ones taking place in SC and/or in OC conditions is quite limited and is around 8°C.

This is a general result. In the absence of mismatching conditions, the working of the PV system in the MPP is preferable for a two-fold reason: the maximization of the extracted power and the minimization of the aging effects that are, without exceptions, all intensified with the increase of the PV cell operating temperature [1–6]. A number of additional experimental characterizations confirming the above statement have also been carried out in different days and in different ambient operating conditions as concerns the irradiance level and the ambient temperature. They have not been reported here for the sake of brevity.

4. The Link between the Operating Point in the I-V Plane and the PV Cell Temperature Distribution under Mismatching Operating Conditions

A similar analysis has been carried out by artificially creating a mismatching condition as shown in Figure 6. It is well known that in PV systems, mismatching can be due to partial shading, shadows of neighboring objects, dirtiness, bird droppings, leaves, snow, clouds, different orientation angles of modules of the PV field, soiling, manufacturing tolerances, or ageing [1–6]. In particular, mismatching due to moving clouds and also to other possible moving bodies is characterized by both relatively fast dynamics (short duration) and also a relatively low degree of nonuniformity as concerns cell operating conditions. Therefore, moving clouds very rarely lead to dangerous hot spot phenomena. On the other hand, dirtiness, bird droppings, leaves, snow, etc. are characterized by both a relatively long duration (indeed they require clean-up operations in order to be removed) and also a very high degree of nonuniformity as concerns cell operating conditions. Therefore, they nearly always lead to hot spot phenomena. This is the main motivation why, in the following experimental tests, mismatching conditions have been artificially induced by covering one or more cells. Firstly, a whole cell of one of the two PVUs (PVU A) has been coated with an adhesive plastic in order to partially shade such a cell. In this way, two different irradiance levels S_1 and S_2 can be identified. $S_1 = 616\ W/m^2$ characterizes the unshaded cells, and $S_2 = 145\ W/m^2$ characterizes the shaded cell. The I-V and P-V characteristics of the two series-connected PVUs have been obtained by means of the experimental set-up shown in Figure 1(a). Such curves are shown in Figures 7 and 8.

The presence of the two PVUs allows the simultaneous comparison, under the same weather conditions, of the temperature distribution of the PVU in mismatching operating conditions (one shaded cell) and of the identical PVU in uniform operating conditions. Such a comparison is possible by analysing Figures 9–12. Each of such figures shows the thermographic image that has been obtained in a specific operating point in the I-V plane by means of the experimental arrangement depicted in Figure 1(b). Of course, every

FIGURE 5: Thermographic image obtained when the two PVUs operate in the MPP. 26 September 2017, Aversa (CE), Italy. $S = 600\ \mathrm{W/m^2}$ and ambient temperature $T_{amb} = 25°\mathrm{C}$.

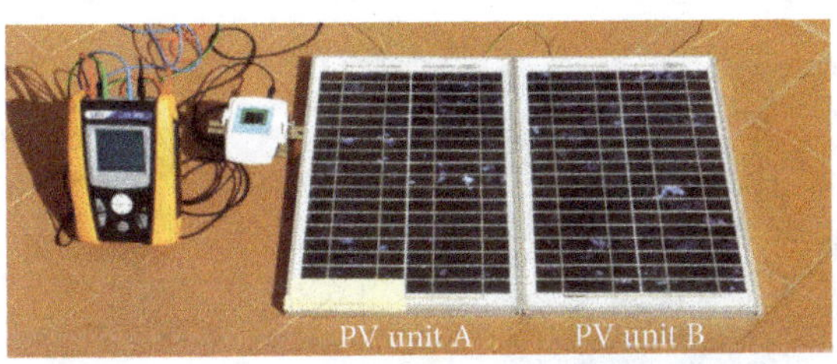

FIGURE 6: PVUs in mismatching conditions. 26 September 2017, Aversa (CE), Italy. $S_1 = 616\ \mathrm{W/m^2}$ characterizes the unshaded cells, and $S_2 = 145\ \mathrm{W/m^2}$ characterizes the shaded cell; $T_{amb} = 25°\mathrm{C}$.

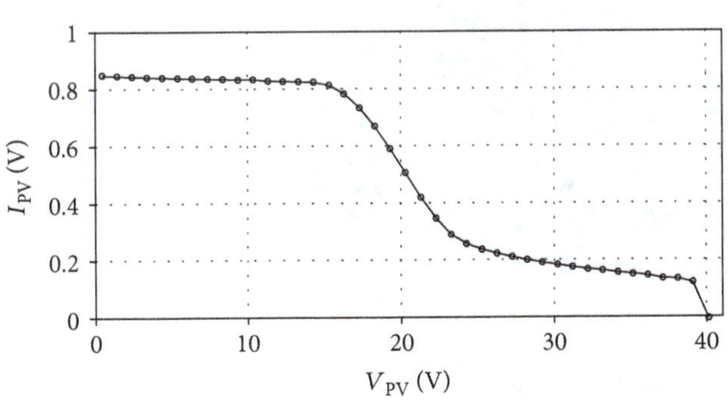

FIGURE 7: I-V characteristic of the two series-connected PVUs. 26 September 2017, Aversa (CE), Italy. $S_1 = 616\ \mathrm{W/m^2}$ characterizes the unshaded cells, and $S_2 = 145\ \mathrm{W/m^2}$ characterizes the shaded cell; $T_{amb} = 25°\mathrm{C}$.

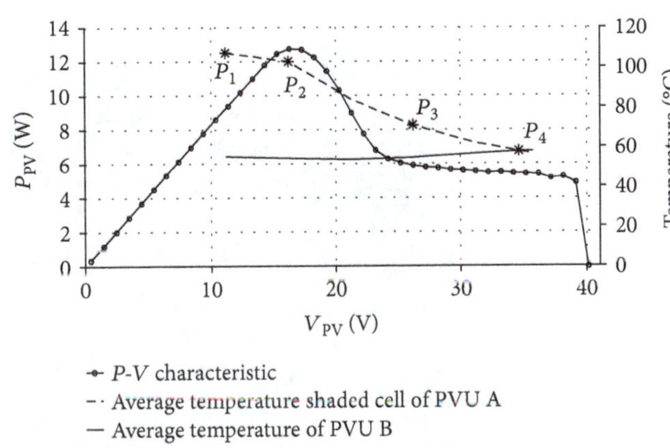

FIGURE 8: P-V characteristic of the two series-connected PVUs: average temperature of the shaded cell of the PVU A and average temperature of the PVU B as a function of the operating voltage. 26 September 2017, Aversa (CE), Italy. $S_1 = 616\ \mathrm{W/m^2}$ characterizes the unshaded cells, and $S_2 = 145\ \mathrm{W/m^2}$ characterizes the shaded cell; $T_{amb} = 25°\mathrm{C}$.

thermographic image has been obtained after temporarily removing the shading plastic. The chosen operating points are the following ones: P_1 ($V_{PV} = 11\ \mathrm{V}$), P_2 ($V_{PV} = 16\ \mathrm{V}$, MPP), P_3 ($V_{PV} = 26\ \mathrm{V}$), and P_4 ($V_{PV} = 34\ \mathrm{V}$). In particular, P_1 has been chosen as an example of an operating point located in the high-current region where the shaded cell is strongly reverse biased. P_2 is located at the best operating voltage from the MPPT point of view, but it is also characterized by a relatively high temperature of the shaded cell. P_3 represents an example of an operating point which allows getting a possible compromise between the extracted power and temperature level of the shaded cell. P_4 has been chosen as an example of an operating point located in the high-voltage region where the shaded cell is not reverse biased at all.

The analysis of Figures 9–12 allows drawing the following considerations. First of all, in mismatching conditions, the temperature distribution is far from being uniform. In particular, as it is well known, the PV cell region characterized by the highest temperature T_H is just the shaded one. Many papers in the literature have put in evidence the above aspect [2, 4, 31, 45]. Much less investigated instead is the strong role

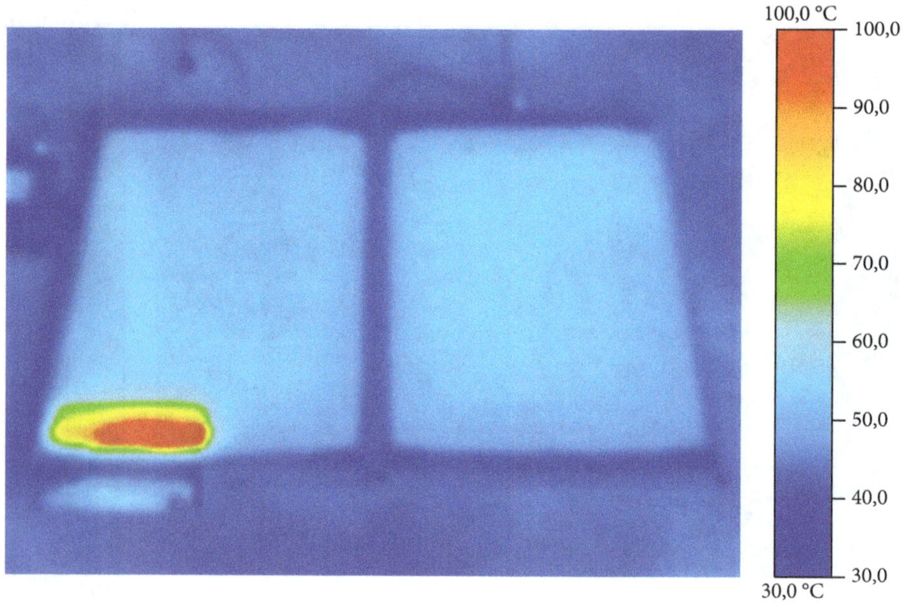

FIGURE 9: Thermographic image obtained when the two PVUs operate in P_1 ($V_{PV} = 11$ V). 26 September 2017, Aversa (CE), Italy. $S_1 = 616\,W/m^2$ characterizes the unshaded cells, and $S_2 = 145\,W/m^2$ characterizes the shaded cell; $T_{amb} = 25°C$.

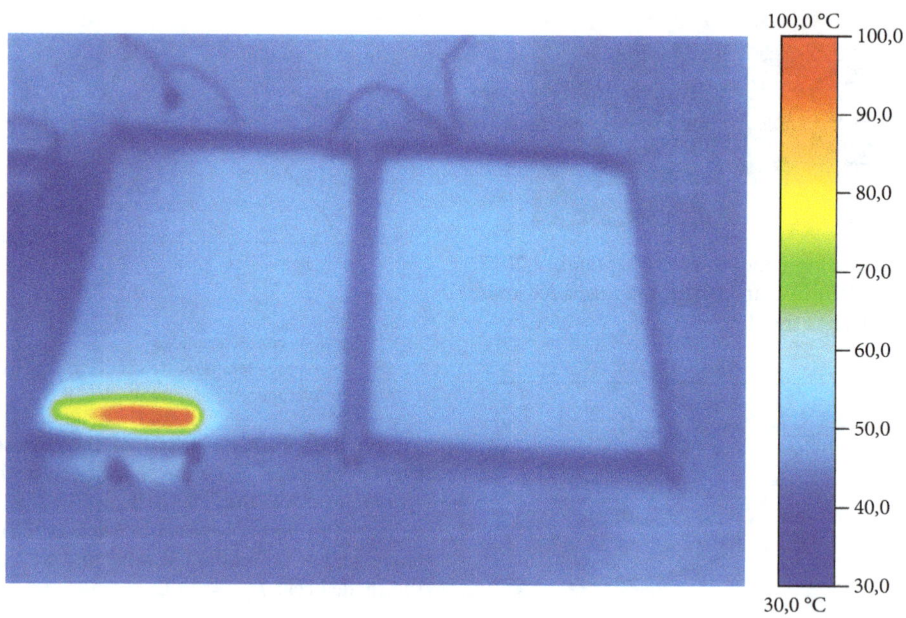

FIGURE 10: Thermographic image obtained when the two PVUs operate in P_2 ($V_{PV} = 16$ V, MPP). 26 September 2017, Aversa (CE), Italy. $S_1 = 616\,W/m^2$ characterizes the unshaded cells, and $S_2 = 145\,W/m^2$ characterizes the shaded cell; $T_{amb} = 25°C$.

played by the operating point in the P-V plane. In particular, it is evident that, differently from the uniform case that has been analysed in Section 3, in the presence of mismatching the global MPP is characterized by potentially dangerous temperatures. In the considered case for example, the temperature T_H in P_2 is equal to 100°C.

In any case, it is possible to state that, in mismatching conditions, if reliability considerations are taken in due consideration, it may happen that the working in the MPP is to be avoided (as in the case of Figures 9–12). As an example, in the considered case, it is reasonable to operate in P_4 if any overheating is to be avoided, even if around 40% of the available electric power is wasted. Instead, a compromise (between maximization of the extracted power and minimization of localized heating phenomena) operating point is represented for example by P_3.

5. Detection of Mismatching Operating Conditions

As discussed in Section 4, if mismatching conditions occur, it is mandatory to properly choose the operating point since it

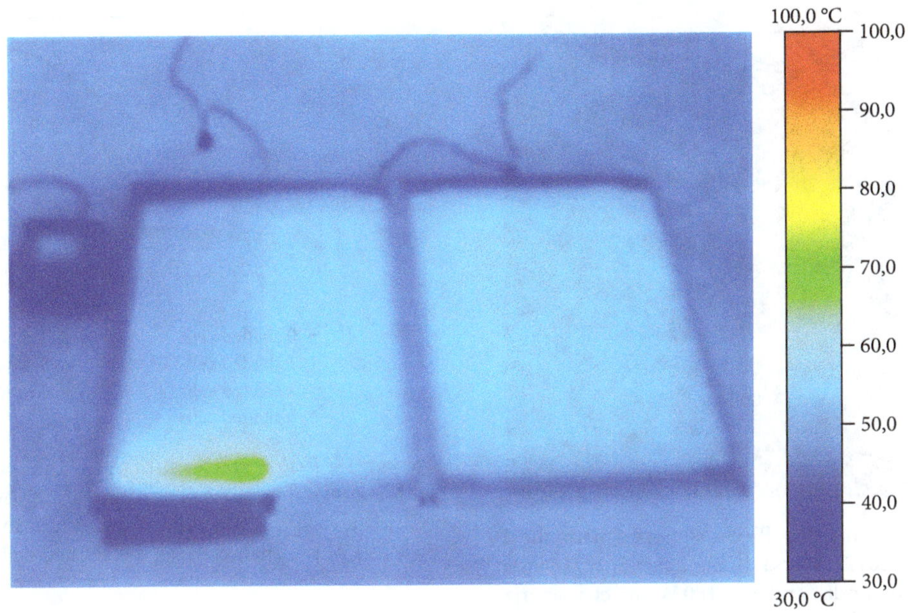

FIGURE 11: Thermographic image obtained when the two PVUs operate in P_3 ($V_{PV} = 26\,V$). 26 September 2017, Aversa (CE), Italy. $S_1 = 616\,W/m^2$ characterizes the unshaded cells, and $S_2 = 145\,W/m^2$ characterizes the shaded cell; $T_{amb} = 25°C$.

FIGURE 12: Thermographic image obtained when the two PVUs operate in P_4 ($V_{PV} = 34\,V$). 26 September 2017, Aversa (CE), Italy. $S_1 = 616\,W/m^2$ characterizes the unshaded cells, and $S_2 = 145\,W/m^2$ characterizes the shaded cell; $T_{amb} = 25°C$.

frequently happens that the global MPP is not the best operating point. Hence, it is crucial to develop robust techniques able to provide accurate information on the eventual occurrence of mismatching conditions (mismatching detection). Such techniques cannot do without the analysis of the shape of the *I*-*V* and/or *P*-*V* characteristics. Many papers in the literature have been devoted to such an analysis (that is carried out by analysing the shape of the *I*-*V* and/or *P*-*V* characteristics for positive voltages). The analysis is nearly always coupled with the adoption of irradiance and/or temperature sensors [46–50]. The accuracy of the mismatching detection is strongly related to the capillarity of the distribution of the adopted sensors. It is worth noting that, for economic and practical reasons, of course only a limited number of sensors can be used in practical PV applications. Therefore, the possibility of mismatching detection errors is real. In this section, it will be shown that the mismatching detection can be efficiently supported by the analysis of the shape of the *I*-*V* and/or *P*-*V* characteristics that has to be carried out not only for positive but also for negative voltages. In other words, the probability of mismatching detection errors cannot be strongly decreased unless the shape of the *I*-*V* and/or *P*-*V*

FIGURE 13: Three series-connected PVUs equipped with the IV Tracker. 27 September 2017, Aversa (CE), Italy. $S_1 = 775\,\text{W/m}^2$ characterizes the unshaded cells, and $S_2 = 180\,\text{W/m}^2$ characterizes the shaded cells; $T_{amb} = 26°\text{C}$.

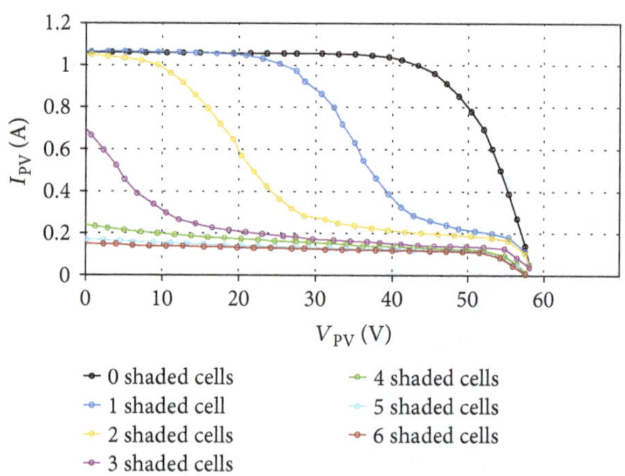

FIGURE 14: I-V characteristics of the three series-connected PVUs with a growing number of shaded cells. 27 September 2017, Aversa (CE), Italy. $S_1 = 775\,\text{W/m}^2$ characterizes the unshaded cells, and $S_2 = 180\,\text{W/m}^2$ characterizes the shaded cells; $T_{amb} = 26°\text{C}$.

characteristics for negative voltages is properly taken into account. The following results have been obtained by adopting the experimental set-up of Figure 1(a) but with three instead of two PVUs, as shown in Figure 13. In particular, a growing number of cells have been coated with an adhesive plastic in order to partially shade them. In this way, two different irradiance levels S_1 and S_2 have been identified. $S_1 = 775\,\text{W/m}^2$ characterizes the unshaded cells, and $S_2 = 180\,\text{W/m}^2$ characterizes the shaded cells. The corresponding I-V and P-V characteristics are shown in Figures 14 and 15.

From Figures 14 and 15, it is evident that the knee point in the I-V plane progressively moves towards the second quadrant as long as the number of shaded cells increases. In the cases of 1 and 2 shaded cells, two maxima appear in the P-V curve, while in the remaining cases only one MPP is

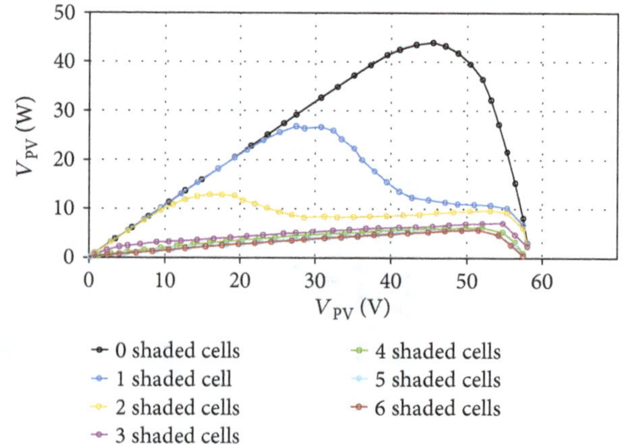

FIGURE 15: P-V characteristics of the three series-connected PVUs with a growing number of shaded cells. 27 September 2017, Aversa (CE), Italy. $S_1 = 775\,\text{W/m}^2$ characterizes the unshaded cells, and $S_2 = 180\,\text{W/m}^2$ characterizes the shaded cells; $T_{amb} = 26°\text{C}$.

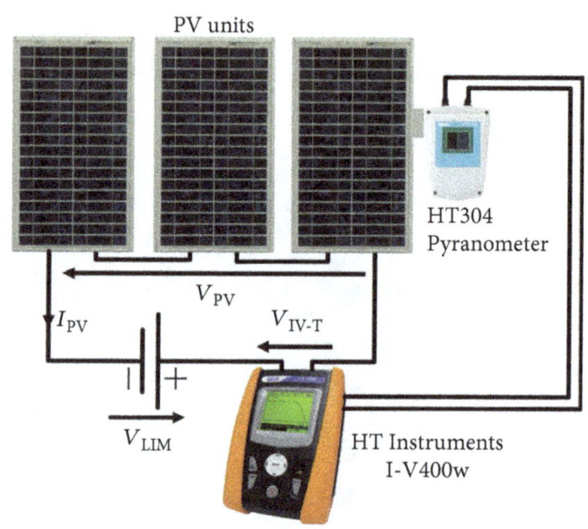

FIGURE 16: Experimental set-up adopted for tracing I-V and P-V characteristics both for positive and negative voltages.

present. In particular, the global MPP is located at low voltages for the P-V curves corresponding to 0, 1, and 2 shaded cells and at high voltages for the remaining cases. An important aspect to underline is that, the mere analysis of the I-V and/or of the P-V characteristics for positive voltages only, does not allow mismatching detection in the case where the number of shaded cells is greater than 2. Hence, in such cases, it is necessary to extend the analysis by including also the second quadrant of the I-V plane (negative voltages). In order to get the sweep of the characteristics not only for positive but also for negative voltages, it has been necessary to adopt the experimental set-up that is schematically shown in Figure 16.

It has been necessary to employ the battery pack (V_{LIM}) because the adopted IV Tracker (HT Instruments I-V400w) is able to trace I-V and P-V characteristics only for positive voltages [40]. In practice, the following equality holds:

$$V_{PV} = V_{IV\text{-}T} - V_{LIM}, \tag{1}$$

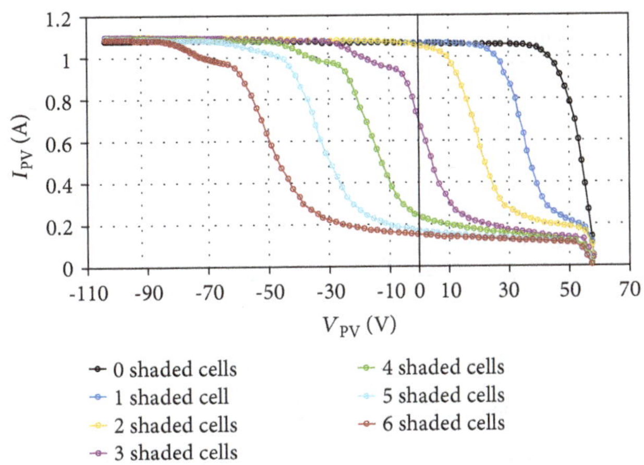

FIGURE 17: Full version (for both positive and negative voltages) of the I-V characteristics of the three series-connected PVUs with a growing number of shaded cells. 27 September 2017, Aversa (CE), Italy. $S_1 = 775\ \mathrm{W/m^2}$ characterizes the unshaded cells, and $S_2 = 180\ \mathrm{W/m^2}$ characterizes the shaded cells; $T_{amb} = 26°\mathrm{C}$.

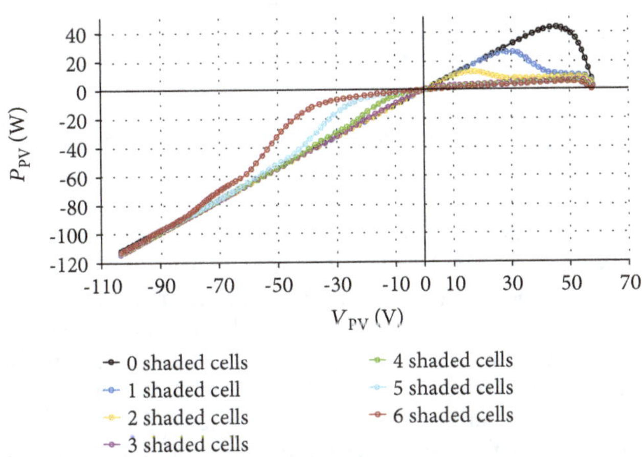

FIGURE 18: Full version (for both positive and negative voltages) of the P-V characteristics of the three series-connected PVUs with a growing number of shaded cells. 27 September 2017, Aversa (CE), Italy. $S_1 = 775\ \mathrm{W/m^2}$ characterizes the unshaded cells, and $S_2 = 180\ \mathrm{W/m^2}$ characterizes the shaded cells; $T_{amb} = 26°\mathrm{C}$.

where $V_{LIM} \approx 106\ \mathrm{V}$. Hence, it is possible to keep the output voltage V_{IV-T} of the IV Tracker for positive values (in fact it is able to trace the I-V and P-V characteristics only from 0 V up to a maximum of 1000 V [40]) while allowing V_{PV} to assume negative values. In particular, the minimum allowed negative value that can be assumed by V_{PV} is equal to $-V_{LIM}$.

Figures 17 and 18, respectively, represent the full versions (for both positive and negative voltages) of the I-V and the P-V characteristics of Figures 14 and 15.

The analysis of Figures 14, 15, 17, and 18 clearly reveals that the inclusion of negative voltages allows the mismatching detection (based on the detection of knees or on the analysis of the shape of the characteristics) also for the cases corresponding to a number of shaded cells that are greater than 2.

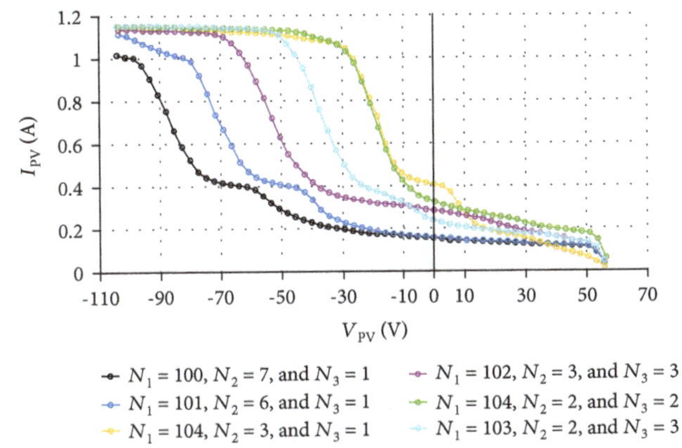

FIGURE 19: Full version (for both positive and negative voltages) of the I-V characteristics of the three series-connected PVUs with different numbers of shaded cells. 27 September 2017, Aversa (CE), Italy. $S_1 = 841\ \mathrm{W/m^2}$, $S_2 = 198\ \mathrm{W/m^2}$, and $S_3 = 330\ \mathrm{W/m^2}$; $T_{amb} = 27°\mathrm{C}$.

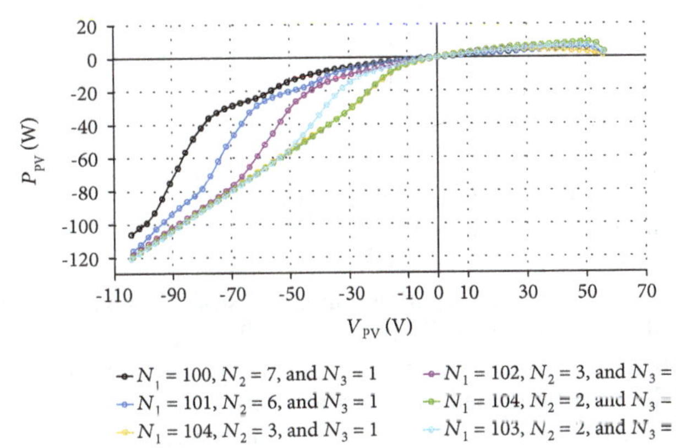

FIGURE 20: Full version (for both positive and negative voltages) of the P-V characteristics of the three series-connected PVUs with different numbers of shaded cells. 27 September 2017, Aversa (CE), Italy. $S_1 = 841\ \mathrm{W/m^2}$, $S_2 = 198\ \mathrm{W/m^2}$, and $S_3 = 330\ \mathrm{W/m^2}$; $T_{amb} = 27°\mathrm{C}$.

Obviously, the mismatching detection based on the analysis of the shape of the characteristics also for negative voltages is affected by the following limitations. Bringing cell operating points in the second quadrant (negative voltages) may cause their breakdown [51]. Therefore, in order to avoid permanent damages to the cells, the value of V_{LIM} must be limited on the basis of the physical characteristics of the adopted cells. However, it is worth noting that, obviously, the analysis in the second quadrant is possible only for $V_{PV} \geq -V_{LIM}$. Hence, should the knees be located at voltages $V_{PV} \ll -V_{LIM}$ mismatching detection can be carried out with the help of a number of irradiance sensors.

Moreover, the mismatching detection based on the monitoring of the I-V module characteristic also for negative voltages, unavoidably leads to the necessity of the substitution of the usual bypass diodes with suitable "smart" bypass diodes to be implemented with active switches. The use of "smart" bypass diodes has multiple functions. In fact, "smart" bypass diodes allow not only the characterization

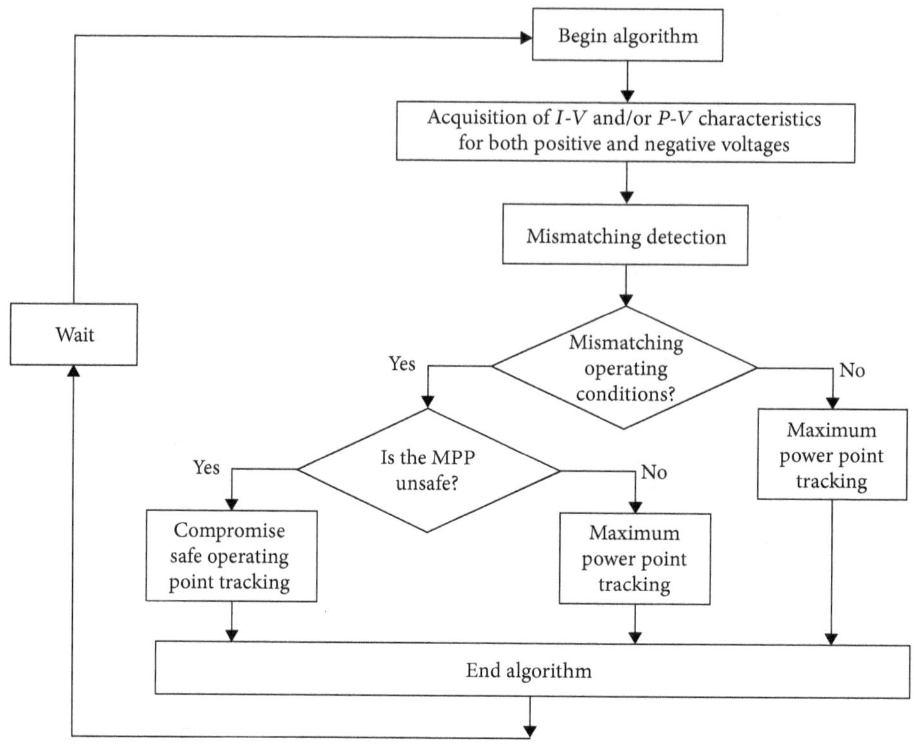

FIGURE 21: Flow chart example of an algorithm aimed at identifying a proper operating point allowing a suitable compromise between maximization of extracted power and minimization of thermal stresses.

of PV modules for negative voltages ("smart" bypass diode *off*) but also the possibility of controlling the value of the voltage in correspondence to which the "smart" bypass diode must be turned *on*. Such a function can be beneficial in terms of controlling the severity of reverse bias voltage values on the shaded cells, and therefore in terms of controlling losses (and hence of temperatures). The control of the above losses means the control of the associated increase of the cell temperature that can reach, in extreme cases, 500°C [13, 52, 53].

Furthermore, it is worth noting that, of course, the voltage sweeps of the characteristics of the PV modules not only for positive but also for negative voltages must be as fast as possible (depending on the value assumed by the cell's capacitance). The repetition frequency of the sweep must be customized on the basis of the considered application. As an example, if it is known that surrounding objects may project a shadow during a given interval of time of the day, the sweep could be carried out much more frequently just during such an interval of time. Of course, the adopted frequency must be a compromise between contrasting requirements. The higher the repetition frequency, the higher the robustness of the diagnostic. Nevertheless, at the same time, the higher the repetition frequency, the higher the energy losses due to the lack of production during sweep operations.

It is worth noting that Figures 14, 15, 17, and 18 refer to a scenario in which only two different irradiance levels S_1 and S_2 can be identified. $S_1 = 775\ \mathrm{W/m^2}$ characterizes the unshaded cells, and $S_2 = 180\ \mathrm{W/m^2}$ characterizes the shaded cells. Further tests have been carried out by using the experimental set-up of Figure 16 but with three different irradiance levels $S_1 = 841\ \mathrm{W/m^2}$, $S_2 = 198\ \mathrm{W/m^2}$, and $S_3 = 330\ \mathrm{W/m^2}$, obtained with different types of coating plastics. The corresponding I-V and P-V characteristics, obtained both for positive and negative voltages, are shown in Figures 19 and 20. In such figures, N_1 indicates the number of cells characterized by an irradiance level S_1, N_2 the number of cells characterized by an irradiance level S_2, and N_3 the number of cells characterized by an irradiance level S_3. Also in this scenario, it is evident that in some cases the mismatching detection based on the analysis of the shape of the I-V and/or P-V characteristics is possible only by including negative voltages.

The above considerations can be profitably exploited as explained in the sequel. In particular, in Figure 21, a flow chart example of an algorithm aimed at identifying a proper operating point that allows getting a suitable compromise between the maximization of extracted power and minimization of thermal stresses is reported. In such a figure, the main steps that should be followed are evidenced. First of all, the algorithm should perform the acquisition of the I-V and/or P-V characteristics of PV modules for both positive and negative voltages. On the basis of such curves (and, if necessary, with the help of additional irradiance and/or temperature sensors) the mismatching detection can be carried out. Hence, if mismatching conditions are not present, a classical MPPT algorithm can be carried out. If, instead, mismatching conditions are detected it means that, depending on the considered case, the MPP could be an unsafe operating point from the thermal stresses' point of view. In such a case, the tracking of a compromise safe operating point different from the MPP must be carried out.

The detailed design of a solution for the identification of the optimal compromise safe operating point is beyond the objectives of this paper. Further work is in progress on such an aspect. In particular, the identification algorithm that has been proposed in [31] can be considered as a first attempt of identification of a compromise safe operating point. In fact, in [31], with specific reference to the series parallel PV array architecture, it is shown that, by means of the reconfiguration of the PV module connections and with the help of a proper objective function, it is possible to identify, in any operating condition, both the optimal array configuration and the associated optimal operating point that are able to lead to the maximization of the extracted energy during the whole PV array lifetime.

6. Conclusions

Since the adoption of bypass diodes or smart active switches does not prevent the occurrence of hotspots in PV applications, it is necessary to investigate alternative possibilities. A promising solution is represented by the proper regulation of the operating point of the PV cells in the I-V plane when mismatching conditions occur. In this paper, the existence of operating points allowing a suitable compromise between the maximization of the extracted power and minimization of thermal stresses has been experimentally evidenced. The main conclusion that can be drawn is that, during the life of a PV system, it is certainly preferable to give up a part of the available energy if it is possible to gain greater energy in the future. Therefore, situations exist in which MPPT is to be avoided, especially when it is obtained at the price of too high and dangerous thermal stresses that will certainly lead to the accelerated derating of the PV array due to accelerated aging or, in extreme cases, to its premature failure. Of course, the proper regulation of the operating point is strictly related to the robust detection of the occurrence of mismatching conditions. Therefore, another crucial aspect that has been evidenced, in this paper, is represented by the fact that the mismatching detection based on the analysis of the shape of the I-V and/or P-V curves is effective only if the sweep of the characteristics is carried out both for positive and negative voltages. The authors feel that the best strategy in order to avoid or limit the occurrence of hotspots and of their associated drawbacks is represented by the combined use of a careful regulation of the operating point (which does not always mean MPPT) coupled with the adoption of PV cells with low reverse-breakdown voltages able to intrinsically limit the power dissipated during hot spots.

Conflicts of Interest

The authors declare that there is no conflict of interest regarding the publication of this paper.

References

[1] G. Makrides, B. Zinsser, M. Schubert, and G. E. Georghiou, "Performance loss rate of twelve photovoltaic technologies under field conditions using statistical techniques," *Solar Energy*, vol. 103, pp. 28–42, 2014.

[2] W. B. Xiao, F. Y. Hu, H. M. Zhang, and H. M. Wu, "Experimental investigation of the effects of partial shading on photovoltaic cells' electrical parameters," *International Journal of Photoenergy*, vol. 2015, Article ID 191603, 7 pages, 2015.

[3] J. Guerrero, Y. Muñoz, F. Ibáñez, and A. Ospino, "Analysis of mismatch and shading effects in a photovoltaic array using different technologies," *IOP Conference Series: Materials Science and Engineering*, vol. 59, no. 1, article 012007, 2014.

[4] P. Manganiello, M. Balato, and M. Vitelli, "A survey on mismatching and aging of PV modules: the closed loop," *IEEE Transactions on Industrial Electronics*, vol. 62, no. 11, pp. 7276–7286, 2015.

[5] M. A. Chaaban, L. El Chaar, and M. Alahmad, "An adaptive photovoltaic topology to overcome shading effect in PV systems," *International Journal of Photoenergy*, vol. 2015, Article ID 294872, 9 pages, 2015.

[6] W. He, F. Liu, J. Ji, S. Zhang, and H. Chen, "Safety analysis of solar module under partial shading," *International Journal of Photoenergy*, vol. 2015, Article ID 907282, 8 pages, 2015.

[7] K. A. Kim and P. T. Krein, "Reexamination of photovoltaic hot spotting to show inadequacy of the bypass diode," *IEEE Journal of Photovoltaics*, vol. 5, no. 5, pp. 1435–1441, 2015.

[8] M. Dhimish, V. Holmes, B. Mehrdadi, M. Dales, and P. Mather, "Output-power enhancement for hot spotted polycrystalline photovoltaic solar cells," *IEEE Transactions on Device and Materials Reliability*, vol. 18, no. 1, pp. 37–45, 2018.

[9] M. Simon and E. L. Meyer, "Detection and analysis of hot-spot formation in solar cells," *Solar Energy Materials and Solar Cells*, vol. 94, no. 2, pp. 106–113, 2010.

[10] V. Naumann, D. Lausch, A. Hähnel et al., "Explanation of potential induced degradation of the shunting type by Na decoration of stacking faults in Si solar cells," *Solar Energy Materials and Solar Cells*, vol. 120, pp. 383–389, 2014.

[11] S. Kaplanis and E. Kaplani, "Energy performance and degradation over 20 years performance of BP c-Si PV modules," *Simulation Modelling Practice and Theory*, vol. 19, no. 4, pp. 1201–1211, 2011.

[12] H. Ziar, S. Mansourpour, E. Afjei, and M. Kazemi, "Bypass diode characteristic effect on the behavior of solar PV array at shadow condition," in *2012 3rd Power Electronics and Drive Systems Technology (PEDSTC)*, pp. 229–233, Tehran, Iran, February 2012.

[13] M. Köntges, S. Kurtz, C. Packard et al., *Performance and Reliability of Photovoltaic Systems, Subtask 3.2: Review of Failures of Photovoltaic Modules, IEA PVPS Task 13, External final report IEA-PVPS*, International Energy Agency, Photovoltaic Power Systems Programme, 2014.

[14] N. Femia, G. Lisi, G. Petrone, G. Spagnuolo, and M. Vitelli, "Distributed maximum power point tracking of photovoltaic arrays: novel approach and system analysis," *IEEE Transactions on Industrial Electronics*, vol. 55, no. 7, pp. 2610–2621, 2008.

[15] S. M. MacAlpine, R. W. Erickson, and M. J. Brandemuehl, "Characterization of power optimizer potential to increase energy capture in photovoltaic systems operating under nonuniform conditions," *IEEE Transactions on Power Electronics*, vol. 28, no. 6, pp. 2936–2945, 2013.

[16] M. Balato, L. Costanzo, and M. Vitelli, "Chapter 5. DMPPT PV system: modeling and control techniques," in *Advances in*

Renewable Energies and Power Technologies, I. Yahyaoui, Ed., pp. 163–205, Elsevier, 2018.

[17] C. Olalla, C. Deline, D. Clement, Y. Levron, M. Rodriguez, and D. Maksimovic, "Performance of power-limited differential power processing architectures in mismatched PV systems," *IEEE Transactions on Power Electronics*, vol. 30, no. 2, pp. 618–631, 2015.

[18] M. Vitelli, "On the necessity of joint adoption of both distributed maximum power point tracking and central maximum power point tracking in PV systems," *Progress in Photovoltaics: Research and Applications*, vol. 22, no. 3, pp. 283–299, 2012.

[19] R. C. N. Pilawa-Podgurski and D. J. Perreault, "Submodule integrated distributed maximum power point tracking for solar photovoltaic applications," *IEEE Transactions on Power Electronics*, vol. 28, no. 6, pp. 2957–2967, 2013.

[20] D. La Manna, V. Li Vigni, E. R. Sanseverino, V. Di Dio, and P. Romano, "Reconfigurable electrical interconnection strategies for photovoltaic arrays: a review," *Renewable and Sustainable Energy Reviews*, vol. 33, pp. 412–426, 2014.

[21] M. Balato, L. Costanzo, and M. Vitelli, "Optimization of both the energetic efficiency and the duration of life of PV arrays by means of the dynamical reconfiguration of PV modules connections," in *2015 Tenth International Conference on Ecological Vehicles and Renewable Energies (EVER 2015)*, pp. 1–6, Monte Carlo, Monaco, March–April 2015.

[22] D. Nguyen and B. Lehman, "An adaptive solar photovoltaic array using model-based reconfiguration algorithm," *IEEE Transactions on Industrial Electronics*, vol. 55, no. 7, pp. 2644–2654, 2008.

[23] M. Balato, L. Costanzo, and M. Vitelli, "Series-parallel PV array re-configuration: maximization of the extraction of energy and much more," *Applied Energy*, vol. 159, pp. 145–160, 2015.

[24] H. Obane, K. Okajima, T. Oozeki, and T. Ishii, "PV system with reconnection to improve output under nonuniform illumination," *IEEE Journal of Photovoltaics*, vol. 2, no. 3, pp. 341–347, 2012.

[25] M. Balato, L. Costanzo, and M. Vitelli, "Multi-objective optimization of PV arrays performances by means of the dynamical reconfiguration of PV modules connections," in *2015 International Conference on Renewable Energy Research and Applications (ICRERA)*, pp. 1646–1650, Palermo, Italy, November 2015.

[26] C. Olalla, D. Maksimovic, C. Deline, and L. Martinez-Salamero, "Impact of distributed power electronics on the lifetime and reliability of PV systems," *Progress in Photovoltaics: Research and Applications*, vol. 25, no. 10, pp. 821–835, 2017.

[27] P. S. Shenoy, K. A. Kim, B. B. Johnson, and P. T. Krein, "Differential power processing for increased energy production and reliability of photovoltaic systems," *IEEE Transactions on Power Electronics*, vol. 28, no. 6, pp. 2968–2979, 2013.

[28] C. Olalla, M. Hasan, C. Deline, and D. Maksimović, "Mitigation of hot-spots in photovoltaic systems using distributed power electronics," *Energies*, vol. 11, no. 4, p. 726, 2018.

[29] S. Roy, M. K. Alam, F. Khan, J. Johnson, and J. Flicker, "An irradiance-independent, robust ground-fault detection scheme for PV arrays based on spread spectrum time-domain reflectometry (SSTDR)," *IEEE Transactions on Power Electronics*, vol. 33, no. 8, pp. 7046–7057, 2018.

[30] M. Dhimish, V. Holmes, B. Mehrdadi, and M. Dales, "Comparing Mamdani Sugeno fuzzy logic and RBF ANN network for PV fault detection," *Renewable Energy*, vol. 117, pp. 257–274, 2018.

[31] M. Balato, L. Costanzo, and M. Vitelli, "Reconfiguration of PV modules: a tool to get the best compromise between maximization of the extracted power and minimization of localized heating phenomena," *Solar Energy*, vol. 138, pp. 105–118, 2016.

[32] C. Chamberlin, M. A. Rocheleau, M. W. Marshall, A. M. Reis, N. T. Coleman, and P. A. Lehman, "Comparison of PV module performance before and after 11 and 20 years of field exposure," in *2011 37th IEEE Photovoltaic Specialists Conference*, pp. 101–105, Seattle, WA, USA, June 2011.

[33] P. Sánchez-Friera, M. Piliougine, J. Peláez, J. Carretero, and M. Sidrach de Cardona, "Analysis of degradation mechanisms of crystalline silicon PV modules after 12 years of operation in southern Europe," *Progress in Photovoltaics: Research and Applications*, vol. 19, no. 6, pp. 658–666, 2011.

[34] K. A. Kim and P. T. Krein, "Photovoltaic hot spot analysis for cells with various reverse-bias characteristics through electrical and thermal simulation," in *2013 IEEE 14th Workshop on Control and Modeling for Power Electronics (COMPEL)*, pp. 1–8, Salt Lake City, UT, USA, June 2013.

[35] F. Fertig, S. Rein, M. Schubert, and W. Warta, "Impact of junction breakdown in multi-crystalline silicon solar cells on hot-spot formation and module performance," in *26th European Photovoltaic Solar Energy Conference and Exhibition*, pp. 1168–1178, Hamburg, Germany, September 2011.

[36] Q. Zhang and Q. Li, "Temperature and reverse voltage across a partially shaded Si PV cell under hot spot test condition," in *2012 38th IEEE Photovoltaic Specialists Conference*, pp. 1344–1347, Austin, TX, USA, June 2012.

[37] M. Dhimish, V. Holmes, P. Mather, and M. Sibley, "Novel hot spot mitigation technique to enhance photovoltaic solar panels output power performance," *Solar Energy Materials and Solar Cells*, vol. 179, pp. 72–79, 2018.

[38] P. Guerriero, L. Piegari, R. Rizzo, and S. Daliento, "Mismatch based diagnosis of PV fields relying on monitored string currents," *International Journal of Photoenergy*, vol. 2017, Article ID 2834685, 10 pages, 2017.

[39] "Datasheet," http://photonsolar.in/pdfdocs/pm0020.pdf.

[40] "Datasheet," https://www.ht-instruments.it/it-it/prodotti/fotovoltaico/misuratori-curva-i-v/i-v400w/.

[41] M. Herman, M. Jankovec, and M. Topič, "Optimal *I-V* curve scan time of solar cells and modules in light of irradiance level," *International Journal of Photoenergy*, vol. 2012, Article ID 151452, 11 pages, 2012.

[42] N. Samano, J. A. Padilla-Medina, and N. Vázquez, "Characterization of photovoltaic panels by means of thermograph analysis," *International Journal of Photoenergy*, vol. 2016, Article ID 8560728, 7 pages, 2016.

[43] "Datasheet," http://www.kepcopower.com/support/opmanls.htm#bop.

[44] "Datasheet," http://www.testo350.com/downloads/881/testo%20881%20Instruction%20Manual.pdf.

[45] E. Kaplani, "Detection of degradation effects in field-aged c-Si solar cells through IR thermography and digital image

processing," *International Journal of Photoenergy*, vol. 2012, Article ID 396792, 11 pages, 2012.

[46] M. Dhimish and V. Holmes, "Fault detection algorithm for grid-connected photovoltaic plants," *Solar Energy*, vol. 137, pp. 236–245, 2016.

[47] A. Triki-Lahiani, A. Bennani-Ben Abdelghani, and I. Slama-Belkhodja, "Fault detection and monitoring systems for photovoltaic installations: a review," *Renewable and Sustainable Energy Reviews*, vol. 82, pp. 2680–2692, 2018.

[48] S. R. Madeti and S. N. Singh, "Monitoring system for photovoltaic plants: a review," *Renewable and Sustainable Energy Reviews*, vol. 67, pp. 1180–1207, 2017.

[49] S. Daliento, A. Chouder, P. Guerriero et al., "Monitoring, diagnosis, and power forecasting for photovoltaic fields: a review," *International Journal of Photoenergy*, vol. 2017, Article ID 1356851, 13 pages, 2017.

[50] Y. Hu, W. Cao, J. Ma, S. J. Finney, and D. Li, "Identifying PV module mismatch faults by a thermography-based temperature distribution analysis," *IEEE Transactions on Device and Materials Reliability*, vol. 14, no. 4, pp. 951–960, 2014.

[51] B. Senitzky and P. D. Radin, "Effect of internal heating on the breakdown characteristics of silicon $p - n$ junctions," *Journal of Applied Physics*, vol. 30, no. 12, pp. 1945–1950, 1959.

[52] I. Geisemeyer, F. Fertig, W. Warta, S. Rein, and M. C. Schubert, "Prediction of silicon PV module temperature for hot spots and worst case partial shading situations using spatially resolved lock-in thermography," *Solar Energy Materials and Solar Cells*, vol. 120, Part A, pp. 259–269, 2014.

[53] S. Armstrong and W. G. Hurley, "A thermal model for photovoltaic panels under varying atmospheric conditions," *Applied Thermal Engineering*, vol. 30, no. 11–12, pp. 1488–1495, 2010.

11

Grid-Connected Photovoltaic System with Active Power Filtering Functionality

Joaquín Vaquero,[1] Nimrod Vázquez,[2] Ivan Soriano,[1] and Jeziel Vázquez[3]

[1]*Electronics Technology Area, Rey Juan Carlos University, 28933 Mostoles, Madrid, Spain*
[2]*Electronics Engineering Department, Technological Institute of Celaya, 38010 Celaya, GTO, Mexico*
[3]*Computer Engineering Department, Superior Technological Institute of South of Guanajuato, 38980 Uriangato, GTO, Mexico*

Correspondence should be addressed to Nimrod Vázquez; n.vazquez@ieee.org

Academic Editor: Jegadesan Subbiah

Solar panels are an attractive and growing source of renewable energy in commercial and residential applications. Its use connected to the grid by means of a power converter results in a grid-connected photovoltaic system. In order to optimize this system, it is interesting to integrate several functionalities into the power converter, such as active power filtering and power factor correction. Nonlinear loads connected to the grid generate current harmonics, which deteriorates the mains power quality. Active power filters can compensate these current harmonics. A photovoltaic system with added harmonic compensation and power factor correction capabilities is proposed in this paper. A sliding mode controller is employed to control the power converter, implemented on the CompactRIO digital platform from National Instruments Corporation, allowing user friendly operation and easy tuning. The power system consists of two stages, a DC/DC boost converter and a single-phase inverter, and it is able to inject active power into the grid while compensating the current harmonics generated by nonlinear loads at the point of common coupling. The operation, design, simulation, and experimental results for the proposed system are discussed.

1. Introduction

Nowadays, with the increasing energy demand and its price increment, along with the scarcity of nonrenewable resources such as oil, natural gas, and coal, researchers are looking for new energy sources to meet the current energy needs. This has led to innovative solutions with desirable characteristics, such as greater efficiency, more power, and less pollution when generating energy [1].

An important issue is power quality, which is influenced by the growing use of nonlinear loads by residential, commercial, and industrial consumers. This type of load generates high harmonic currents that interact with the grid impedance, causing harmonic voltages which affect all users connected to the same point of common coupling (PCC).

Among the problems caused by the presence of harmonics are distortion of the AC mains voltage within the facilities, high currents flowing through the neutral conductor, overheating of transformers and conductors, poor operation of switches and fuses, erroneous operation of electronic equipment, life reduction in incandescent lamps, resonance risk in fluorescent lamps, and overheating of rotating machines [2]. Active power filters have been proposed to compensate the current harmonics [3–5]. Although they provide a good compensation system, their high cost is a considerable disadvantage.

The growth of distributed generation (DG) has been favored by the use of renewable energies. The main power conversion stage in a DG system is the inverter, which is very flexible from a control point of view. This flexibility allows exploration of the possibility of injecting active power to the grid from a photovoltaic (PV) system while compensating current harmonics.

Some papers have proposed the use of multifunctional inverters [6–9]. Researchers in [6] proposed the use of resonant tanks which are capable of effectively removing selected harmonics, but many resonant tanks are needed to remove the number of harmonics. Instantaneous reactive power

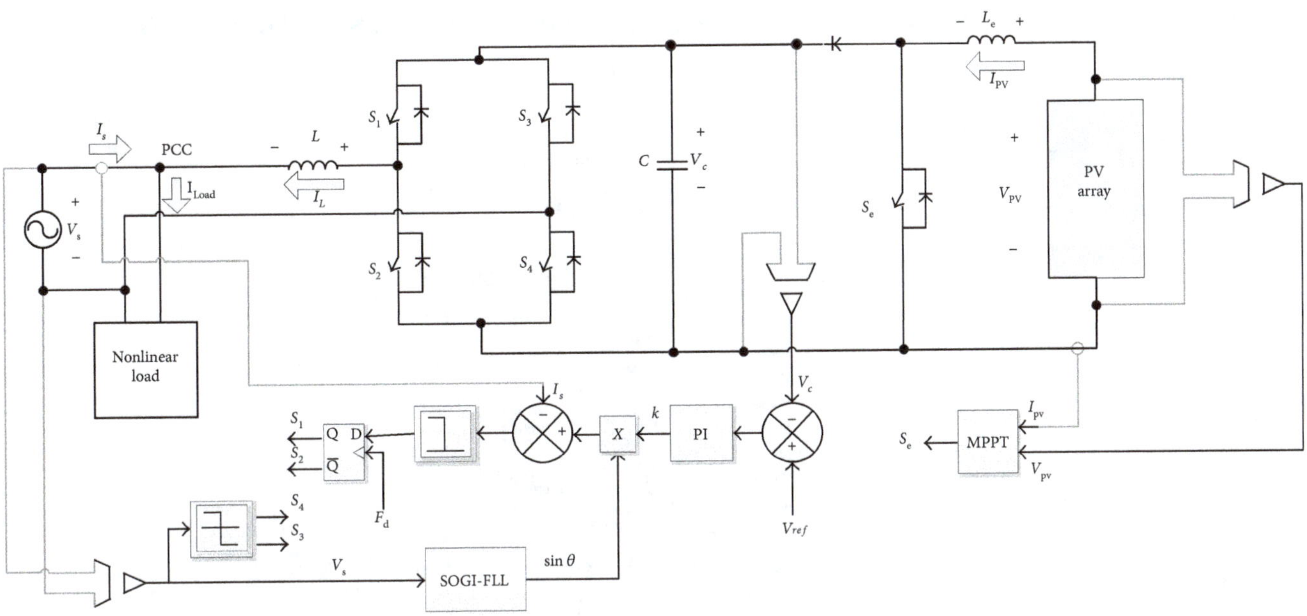

FIGURE 1: Block diagram of the power stage with a sliding mode controller.

theory has also been used in [7–9]. However, this is only applicable to three-phase systems. In [9], the synchronous reference frame theory is used, usually applied to three-phase systems, but it requires virtual signals when used in single-phase systems.

In this paper, a PV system with active power filtering functionality and a controller based on sliding mode is analyzed, implemented, and tested. The authors have previously presented the simulation results of this proposal in [10], adding now experimental results, discussion on implementation issues, and a complete design and analysis of the system, which corroborates the conclusions of the previous work. The system consists of a single-phase two-stage converter (Figure 1). This converter is able to inject both active and reactive power and, additionally, to compensate the harmonic currents at the PCC. The sliding mode control brings the advantage of easy implementation with few computational resources compared to other techniques discussed above. In addition, it does not need an independent control loop as in [6] to generate the reference signal.

The paper is organized as follows: the power stage, the controller, and implementation are described, the simulation and experimental results are then illustrated, and finally, the conclusions are given.

2. Work Bench and Multifunctional System

The power stage consists of a DC/DC boost converter, an energy storage capacitor, a full-bridge single-phase inverter, and an output inductor connected to the grid (Figure 1). The switching devices used are a new generation of silicon carbide (SiC) metal-oxide-semiconductor field-effect transistor (MOSFET). The digital platform for the controller implementation is the CompactRIO from National Instruments Corporation (NI), together with LabVIEW software, which allows a flexible visual programming environment.

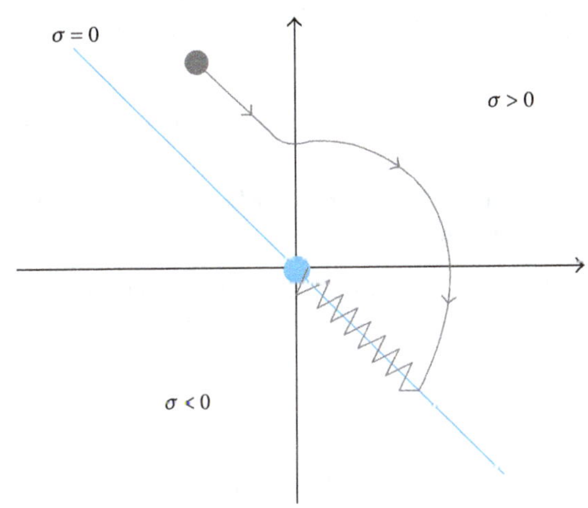

FIGURE 2: Evolution of the system with a sliding mode controller.

The DC/DC boost converter demands the available power from the PV panels by means of a maximum power point tracking algorithm (MPPT). The inverter injects active power generated by the PV panels to the grid and compensates the current harmonic content at the PCC. The entire control tasks for both converters have been implemented in the NI platform.

2.1. The Power Stage. In order to simplify the analysis of the power stage, the PV panels and the DC/DC boost converter can be considered to behave as a current source, since changes in the energy provided by the PV panels take longer than the AC mains period.

The converter is operated to produce a unipolar output voltage. Each branch of the inverter is switched independently, with S_3 and S_4 activated every half-period of the AC mains and S_1 and S_2 activated following a pulse width

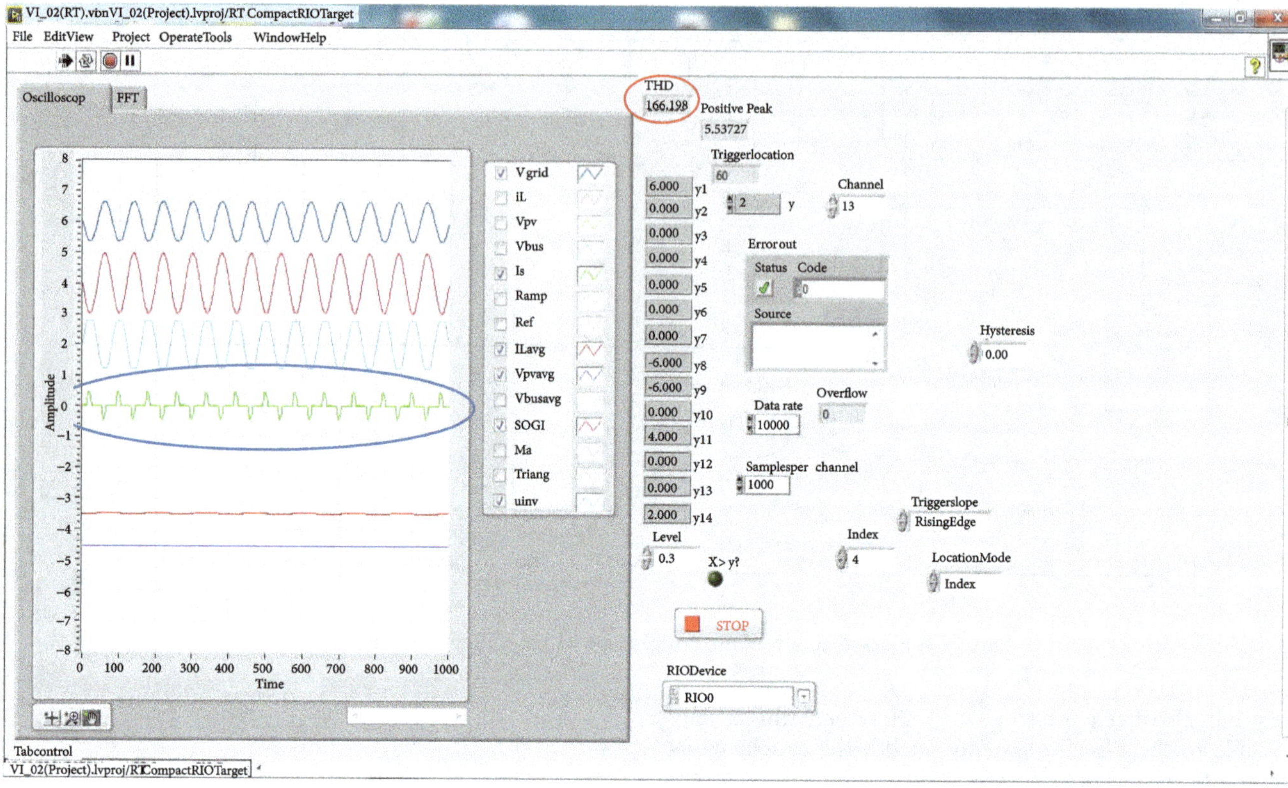

FIGURE 3: Implemented LabVIEW control panel.

modulation (PWM) pattern, in order to control the injected current to the grid.

2.1.1. Modelling the Power Stage. Considering the above simplification, the model of the system is

$$\begin{aligned}\frac{di_L}{dt} &= \frac{v_c}{L}(u_1 - u_3) - \frac{v_s}{L},\\ \frac{dv_c}{dt} &= -\frac{i_L}{C}(u_1 - u_3) + \frac{i_{PV}}{C},\end{aligned} \tag{1}$$

where i_L is the inductor current, v_c is the capacitor voltage, i_{PV} represents the energy provided by the PV panels and the boost converter, u_x is the control signal for switch S_x (1 = on, 0 = off), and v_s is the AC mains voltage.

The switches of the same branch are complementary; hence,

$$\begin{aligned}u_1 + u_2 &= 1,\\ u_3 + u_4 &= 1.\end{aligned} \tag{2}$$

For a model of the PV panel, see that proposed in [11].

2.2. Design of the Power Stage. The capacitor is calculated considering the stored energy and the low-frequency component of the capacitor voltage ripple. The following equation is first used:

$$\Delta E = \frac{1}{2}C\left(v_{cf}^2 - v_{co}^2\right), \tag{3}$$

where ΔE is the energy stored in the capacitor and v_{cf} and v_{co} are the capacitor voltages due to the ripple.

The low frequency of the capacitor voltage ripple is twice the grid frequency. Considering that the stored energy handled by the system in each half of the AC mains period is 10% of the nominal output energy and that the capacitor voltages due to the ripple are $v_{cf} = (v_c + \Delta v_c)/2$ and $v_{co} = (v_c - \Delta v_c)/2$, (4) is obtained:

$$C = \frac{0.1\,P}{2v_c \Delta v_c f_o}, \tag{4}$$

where Δv_c is the desired ripple at the capacitor voltage, f_o is the AC mains frequency, and P is the nominal output power.

If $v_c = 250$ V, $f_o = 60$ Hz, $P = 200$ W, and $\Delta v_c = 1\%$, a value of $C = 266\,\mu$F is obtained. It is approximated to the next highest commercially available value of 330 μF. For practical purposes, an electrolytic type capacitor was selected.

The output inductor can be calculated with

$$L = \frac{V_L D}{\Delta I_L f_s}, \tag{5}$$

where ΔI_L is the desired current ripple at the inductor, f_s is the switching frequency of the inverter, V_L is the voltage at the inductor, and D is the equivalent duty cycle of the inverter.

Considering a worst-case design for the duty cycle, the zero crossing the grid voltage, a duty cycle of $D = 0.5$ was

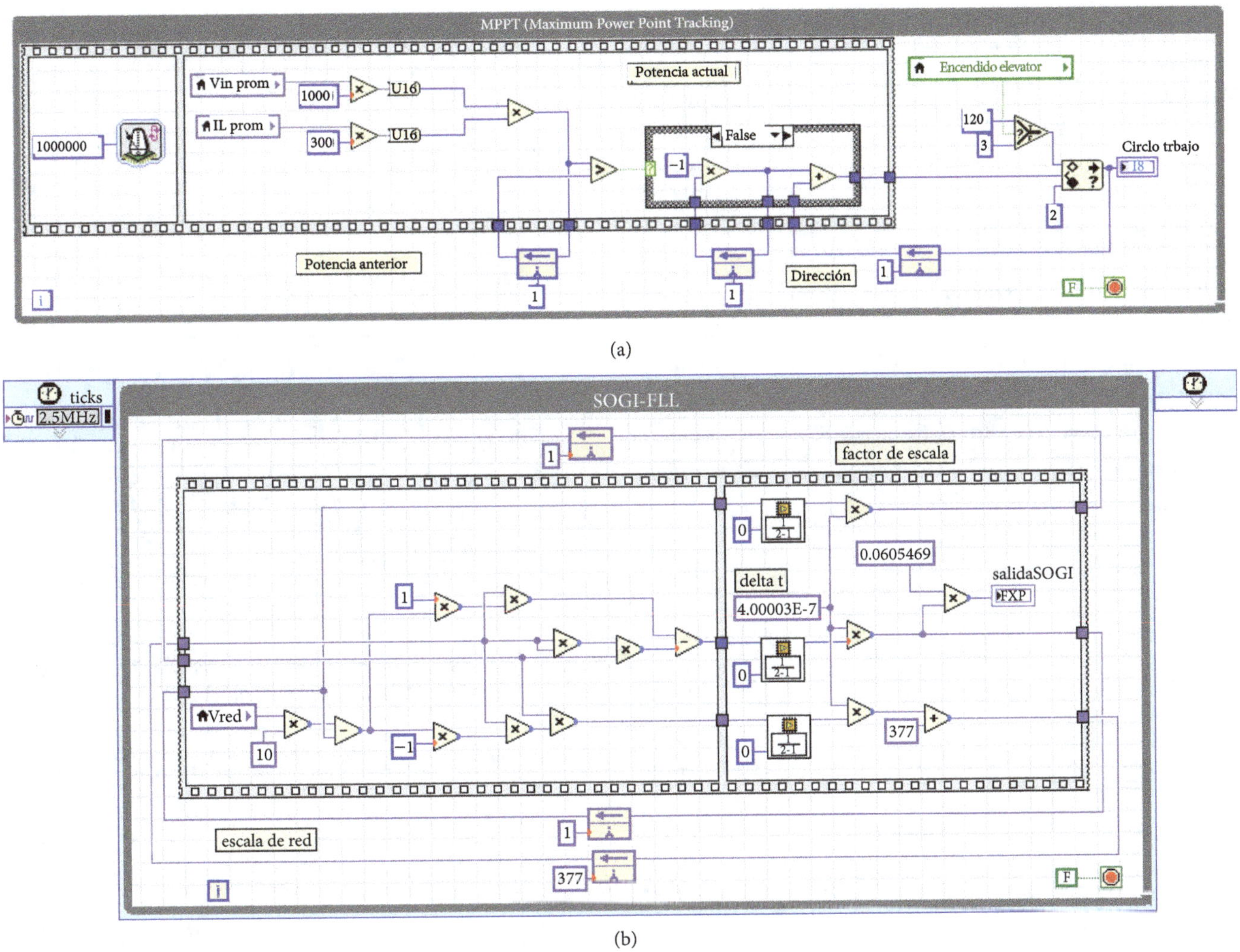

Figure 4: LabVIEW control blocks implemented on the CompactRIO platform. (a) MPPT algorithm. (b) Synchronization controller.

selected. If $f_s = 90\,\text{kHz}$, $V_L = 250\,\text{V}$, and an allowed current ripple $\Delta I_L = 10\%$, a value of $L = 3.96\,\text{mH}$ is obtained.

The design of the DC/DC boost converter can be calculated as a classical one, so it is not discussed in this paper.

2.2.1. Controller Stage. There are several MPPT algorithms [12–15] reported in the literature that can be used for the DC/DC boost converter. A classic perturb and observe (P&O) method has been used in this work.

For the inverter, a sliding mode control has been considered. It provides advantages like stability at large variations in load and voltage, robustness, good dynamic response, and easy implementation [16].

The sliding mode controller is based on the theory of variable structures [16, 17], so power converters are good candidates for this type of control, since they naturally operate in this way.

The controller design begins with the definition of a sliding surface, toward which the system must be attracted and must remain in it (Figure 2). The sliding surface can be a line or a plane in the generalized form. The conditions of existence and stability of the sliding mode control assure its proper operation [16].

Since the converter is intended to operate as an active power filter in addition to the power injection, the grid current i_s must be sinusoidal and in phase with the voltage source v_s, regardless of the existing nonlinear loads connected at the PCC at the grid. Then, the AC current must be

$$i_s = k v_s, \tag{6}$$

where k determines the actual power demanded by the load plus the energy available at the PV panels.

The proposed sliding surface is

$$\sigma = i_s - k v_s, \tag{7}$$

and the control laws considered are

$$u_1 = \begin{cases} 1, & \text{for } \sigma > 0, \\ 0, & \text{for } \sigma < 0, \end{cases} \qquad u_3 = \begin{cases} 1, & \text{for } v_s < 0, \\ 0, & \text{for } v_s > 0. \end{cases} \tag{8}$$

2.3. Existence of the Sliding Mode. The existence condition ensures that the equation system will be maintained in

TABLE 1: Simulation and experimental parameters.

Photovoltaic array	
Maximum power	200 W
Open-circuit voltage	92 V
Short-circuit current	4.76 A
Nominal voltage (MPP)	67 V
Nominal current (MPP)	3.6 A
Topology	
DC/DC boost converter inductor L_e	3.5 mH
DC link voltage	250 V
Energy storage capacitor C	330 μF
Output inductor L	4 mH
DC/DC boost converter switching frequency F_s	39 kHz
Inverter switching frequency F_d	90 kHz
PI controller	
Gain	0.05
Time constant	0.05
Nonlinear load	
Power factor	68%
Apparent power	105 VA
Grid	
Frequency	60 Hz
Grid voltage	120 V rms

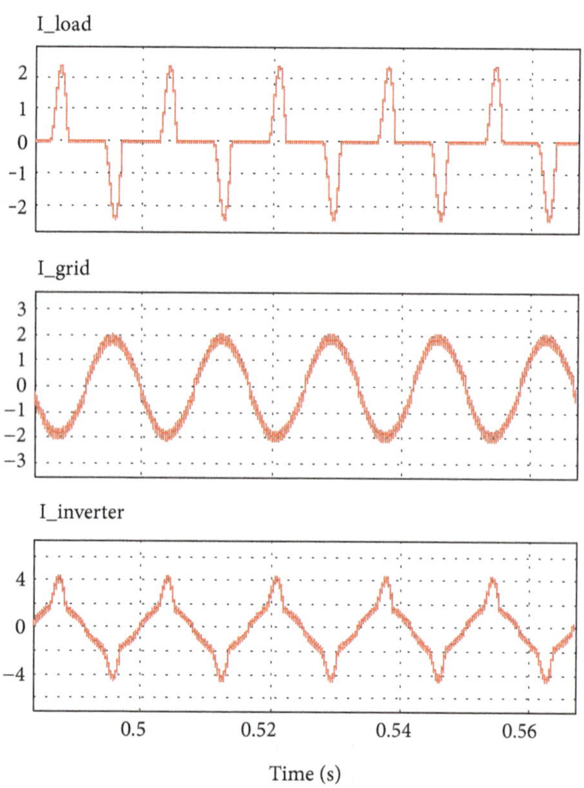

FIGURE 5: System injecting and alleviating the harmonic distortion. From (a) to (c): load current (1 A/div), grid current (1 A/div), and inverter current (2 A/div).

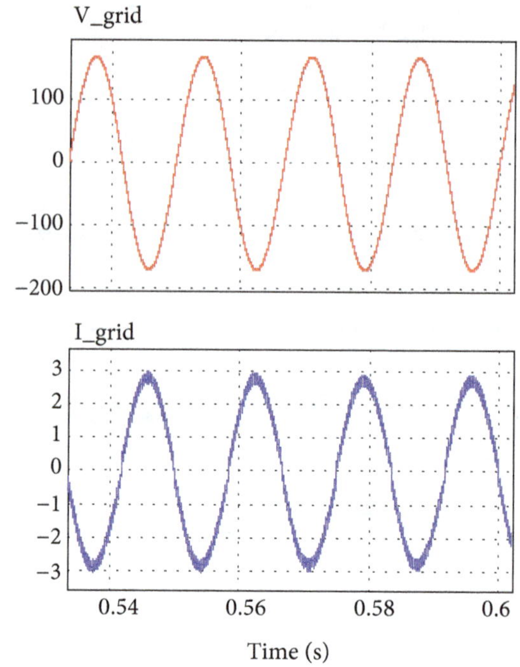

FIGURE 6: Topology injecting only active power. From (a) to (b): grid voltage (100 V/div) and grid current (1 A/div).

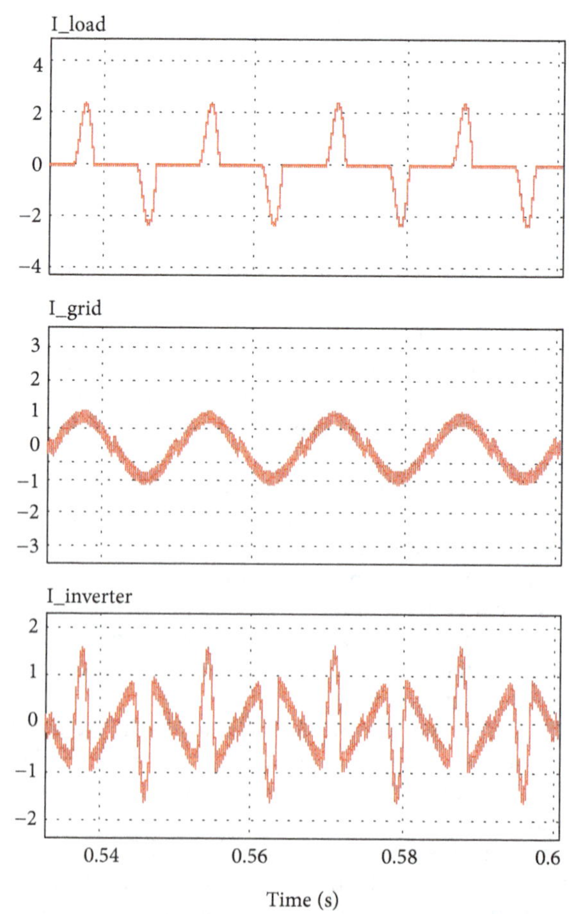

FIGURE 7: Operation as an active power filter. From (a) to (c): load current (2 A/div), grid current (1 A/div), and inverter current (1 A/div).

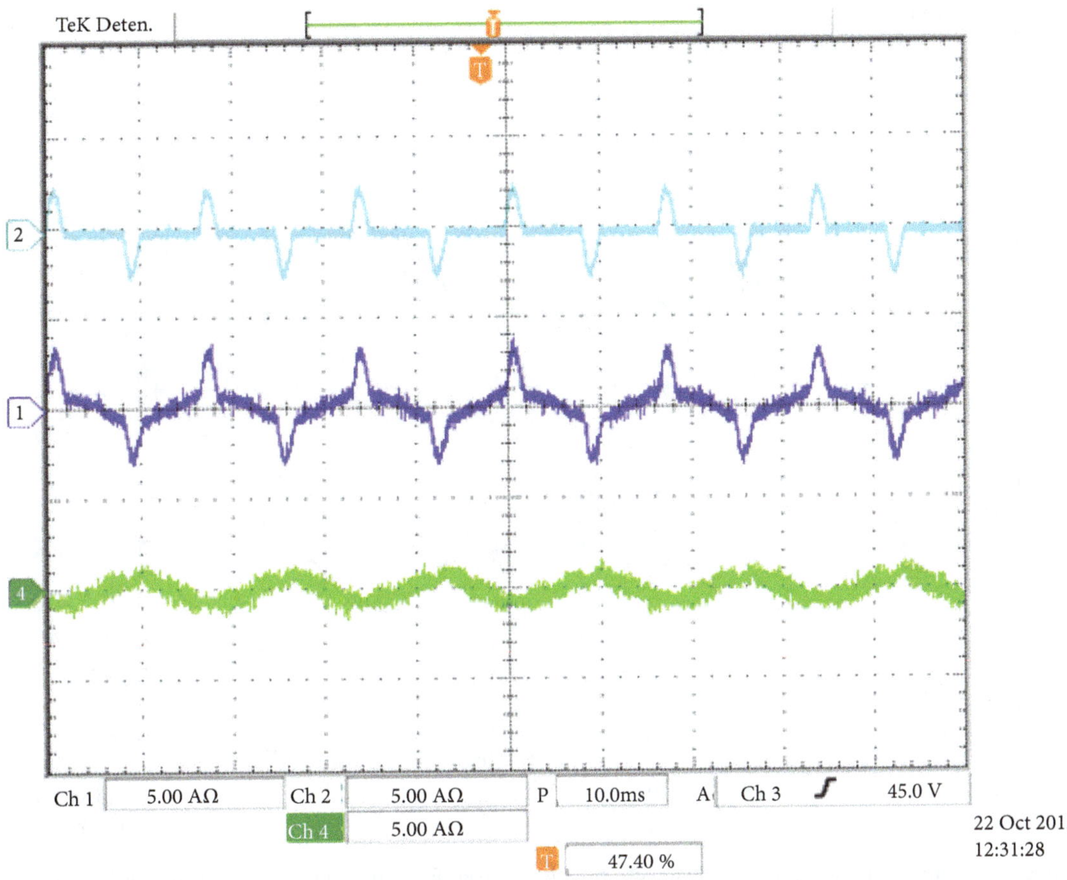

Figure 8: Experimental results for the system injecting and alleviating the harmonic distortion. From top to bottom: load current (5 A/div), inverter current (5 A/div), and grid current (5 A/div).

the sliding surface. It implies satisfying the following inequality [16]:

$$\sigma\sigma' < 0. \tag{9}$$

By deriving (7), (10) is obtained:

$$\sigma' = i_s' - kv_s'. \tag{10}$$

Considering Figure 1 and by applying the first Kirchhoff law at the PCC, (11) is obtained:

$$i_s = i_{\text{load}} - i_L. \tag{11}$$

With the derivative of (11) and by substituting it in (10) together with (1), (12) can be obtained:

$$\sigma' = i_{\text{load}}' - \frac{v_c}{L}(u_1 - u_3) + \frac{v_s}{L} - kv_s'. \tag{12}$$

Considering $v_s = V_P \sin(\omega t)$, using (8) and (9), the existence conditions of the sliding surface are obtained.

If σ and v_s are both positive then, due to the control laws, $u_1 = 1$ and $u_3 = 0$. Therefore,

$$-\frac{v_c}{L} + \frac{V_P \sin(\omega t)}{L} - k\omega V_P \cos(\omega t) < -i_{\text{load}}'. \tag{13}$$

If σ and v_s are both negative then, due to the control laws, $u_1 = 0$ and $u_3 = 1$. Therefore,

$$\frac{v_c}{L} + \frac{V_P \sin(\omega t)}{L} - k\omega V_P \cos(\omega t) > -i_{\text{load}}'. \tag{14}$$

If $\sigma < 0$ and $v_s > 0$, then, due to the control laws, $u_1 = 0$ and $u_3 = 0$. Then,

$$\frac{V_P \sin(\omega t)}{L} - k\omega V_P \sin(\omega t) > -i_{\text{load}}'. \tag{15}$$

If $\sigma < 0$ and $v_s < 0$, then, due to the control laws, $u_1 = 1$ and $u_3 = 1$.
Then,

$$\frac{V_P \sin(\omega t)}{L} - k\omega V_P \sin(\omega t) < -i_{\text{load}}'. \tag{16}$$

To satisfy (13) and (14) inequalities, the selected capacitor voltage v_c should be much higher than the peak of the AC mains V_p. To satisfy (15) and (16) inequalities, the load should be smooth.

The stability analysis is beyond the purpose of this paper, but previous analysis allows assurance of the proper operation of the system in the sliding surface. To determine the stability, an equivalent control method can be employed [17]. An equivalent control has to be found once the system is in

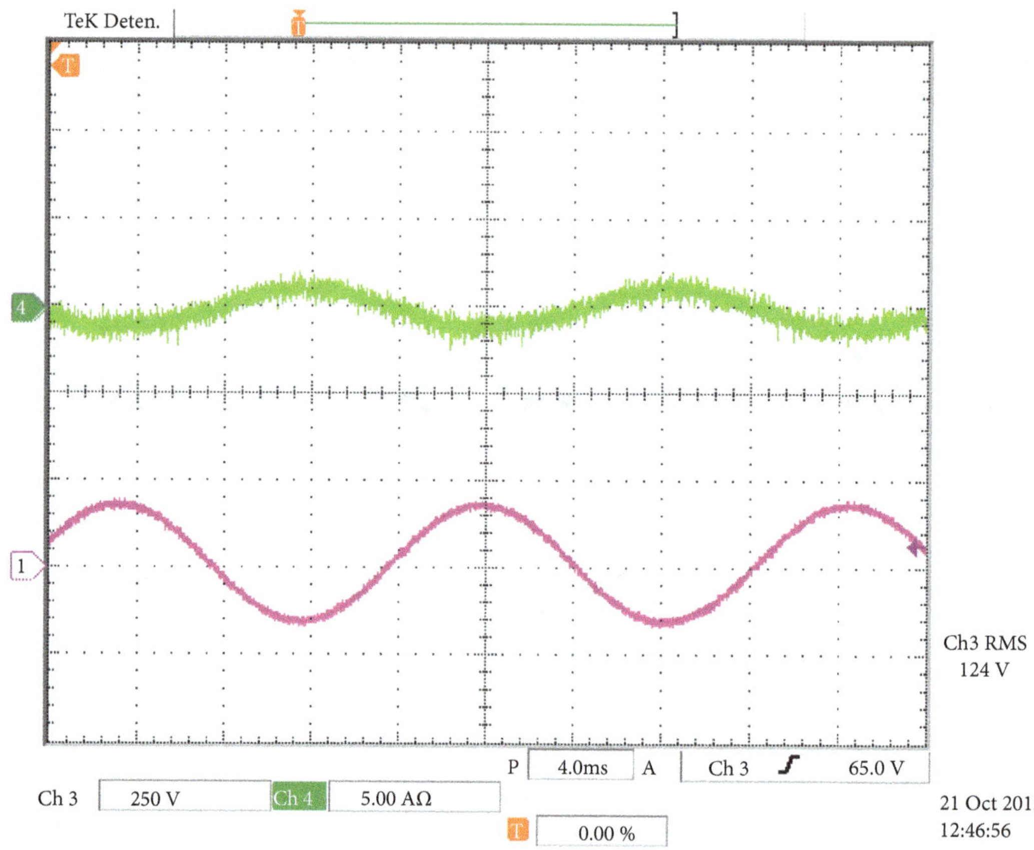

FIGURE 9: Experimental results when the topology only injects the active power. From top to bottom: grid current (5 A/div) and grid voltage (250 V/div).

the sliding surface, and then, the resultant differential equation system is evaluated for stability. This can be done by means of Lyapunov function [18] or by linearization.

2.4. Other Controller Blocks. For the proper operation of the system, other important control blocks must be taken into account, that is, the MPPT algorithm, the grid synchronization algorithm, and the voltage reference generator to control the capacitor voltage.

The constant k for the control of the capacitor voltage is determined by a PI controller, as in [19, 20]. To ensure a good steady-state operation of the system and thus a high power factor, this controller is tuned for a slow response.

The SOGI-FLL [21] is considered for the synchronization with the grid. This algorithm is capable of generating a good reference under distorted grid conditions.

Finally, other blocks, like blanking time and protections, are considered in the final implementation.

2.4.1. Digital Platform. As mentioned before, the platform considered for the control implementation is the CompactRIO from NI, together with the LabVIEW visual programming software. There have been programmed system protections and controllers for the entire system.

Figure 3 shows the control panel. The user can select several waveforms to be graphed in real time. Also, the FFT can be obtained, by selecting the appropriate tab on the control panel.

In Figure 4, some of the controllers implemented are shown. Specifically, the MPPT algorithm based on the P&O method is shown in Figure 4(a), and the SOGI-FLL controller employed for the synchronization with the grid is shown in Figure 4(b).

As part of the protections, overvoltage and overcurrent detection is included, but the blanking time of the inverter branch is also included in the program.

3. Simulation and Experimental Results

The system shown in Figure 1 was numerically simulated and physically implemented. The PSIM software was selected for simulations because of its ease of use and versatility. It also includes models for the PV panels. For the implementation of the switches, SiC devices were used. Table 1 shows the parameters considered for the system.

The nonlinear load considered is a traditional single-phase full-bridge diode rectifier plus a capacitive filter; the total harmonic distortion of this current is above 100%.

3.1. Simulation Results. The steady-state operation of the proposed system is illustrated in Figure 5. In this test, the load demands a high harmonic current content and the PV panels deliver more power than that absorbed by the load, so energy is injected to the grid. In Figure 5(a), the current demanded by the nonlinear load is shown. In Figure 5(b), the low distorted grid current can be observed; in this case, it should

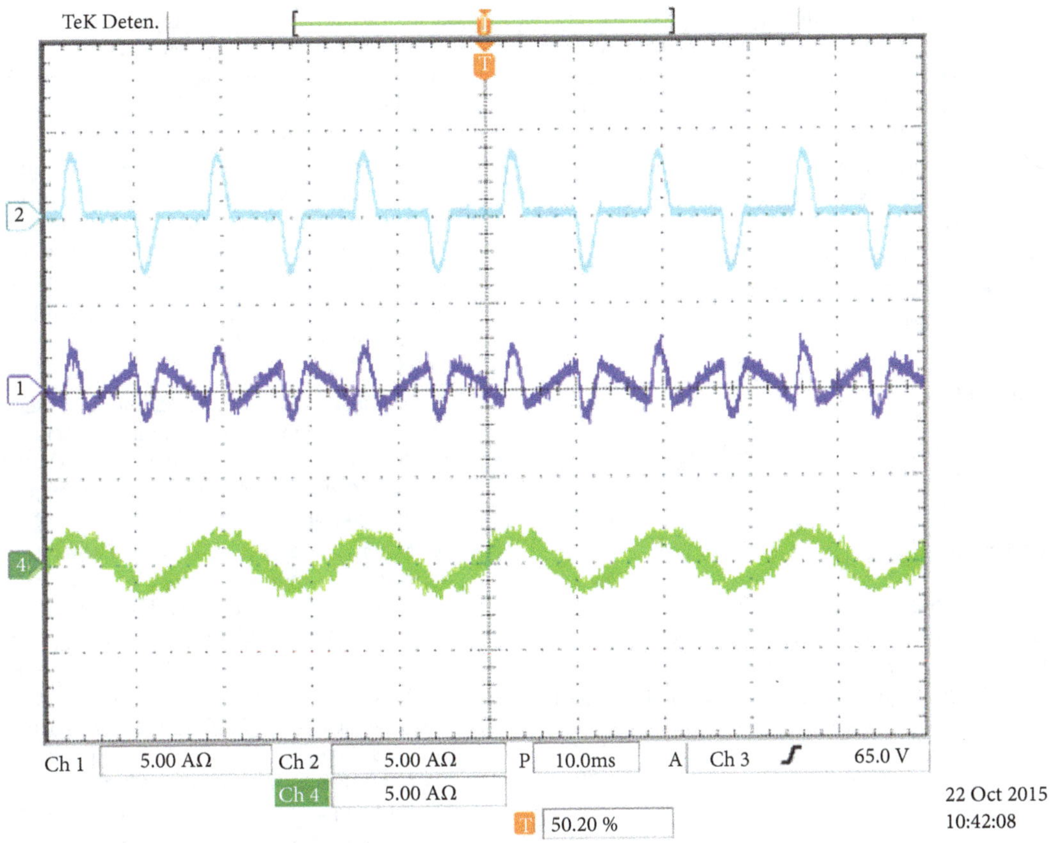

FIGURE 10: Experimental results when the converter operates only as an active filter. From top to bottom: load current (5 A/div), inverter current (5 A/div), and grid current (5 A/div).

be noted that this current is 180 degrees out of phase with the load current. In Figure 5(c), the inverter current is shown, where it can be observed that it is injecting both active power and harmonic currents. The power factor at the PCC is 0.99.

The operation of the system without nonlinear load is shown in Figure 6. In this case, the converter operates as a traditional grid-connected PV system. Only active power is injected into the grid. The current is 180 degrees out of phase with the grid voltage.

Additionally, the proposed system has the advantage of operating as a conventional active filter, as shown in Figure 7. In this case, the power from the PV panels is null. In Figure 7(a), the distorted load current can be observed. In Figure 7(b), the low distorted grid current is shown, which is in phase with the grid voltage. In Figure 7(c), the compensating current from the inverter can be observed.

3.2. Experimental Results. The steady-state operation of the proposed system is shown in Figure 8. In this test, the load demands a high harmonic current content and the PV panels provide more power than that absorbed by the nonlinear load, so energy is injected to the grid. At the top of the figure, the high distorted current demanded by the nonlinear load is shown (THD = 164%). At the bottom of the figure, the low distorted grid current can be observed (THD = 4.34%). Note that this current is 180 degrees out of phase with the nonlinear load current, as expected. At the center of the figure, the inverter current is shown, where it can be observed that it is injecting both active power and harmonic current components. The power factor at the PCC is 0.99.

The operation of the system without load is shown in Figure 9. In this case, the converter operates as a traditional grid-connected PV system. Only active power is injected into the grid. The current is 180 degrees out of phase with the grid voltage.

Additionally, the proposed system has the advantage of operating as a conventional active filter, as shown in Figure 10. In this case, the power from the PV panels is null. At the top of the figure, the distorted load current can be observed. At the bottom of the figure, the grid current is shown, which is in phase with the grid voltage. At the center of the figure, the compensating current of the inverter can be observed.

4. Conclusion

Grid-connected PV systems usually inject active power energy to the grid. They are mainly used to avoid the dependence of fossil fuels. Since the power stage used in grid-connected applications is an inverter, the functionality of this stage may be increased, acting as an active power filter and a power factor corrector.

In this paper, a multifunctional grid-connected PV system is analyzed and tested. The power stage consists of a

DC/DC boost converter and a single-phase inverter. The converter's protections and controllers are implemented in the CompactRIO digital platform from NI, which allows user friendly and easy tuning of the system. A sliding mode controller is employed that permits good operation in different modes: when the system delivers energy and compensates harmonics, when it only injects energy, or when it only acts as an active power filter.

The operation, design, and implementation of the system are presented. The simulation and experimental results confirm the feasibility of the proposal.

Conflicts of Interest

The authors declare that there is no conflict of interest regarding the publication of this paper.

Acknowledgments

This work has been sponsored by National Instruments Corporation through the 2016 NI Academic Research grant program and also was supported in part by the Comunidad de Madrid government (SINFOTON-CM) (S2013/MIT-2790).

References

[1] Renewable Energy Policy Network for the 21st Century, *Global Status Report 2014*, 2014, http://www.ren21.net/ren21activities/globalstatusreport.aspx.

[2] V. E. Wagner, J. C. Balda, D. C. Griffith et al., "Effects of harmonics on equipment," *IEEE Transactions on Power Delivery*, vol. 8, no. 2, pp. 672–680, 1993.

[3] N. R. Raju, S. S. Venkata, R. A. Kagalwala, and V. V. Sastry, "An active power quality conditioner for reactive power and harmonics compensation," in *1995 PESC '95 Record, 26th Annual IEEE Power Electronics Specialists Conference*, vol. 1, pp. 209–214, Atlanta, GA, USA, June 1995.

[4] A. M. Massoud, S. J. Finney, and B. W. Williams, "Seven-level shunt active power filter," in *2004 11th International Conference on Harmonics and Quality of Power*, pp. 136-141, Lake Placid, NY, USA, September 2004.

[5] S. Inoue, T. Shimizu, and K. Wada, "Control methods and compensation characteristics of a series active filter for a neutral conductor," *IEEE Transactions on Industrial Electronics*, vol. 54, no. 1, pp. 433–440, 2007.

[6] J. He, Y. Wei Li, F. Blaabjerg, and X. Wang, "Active harmonic filtering using current-controlled, grid-connected dg units with closed-loop power control," *IEEE Transactions on Power Electronics*, vol. 29, no. 2, pp. 642–653, 2014.

[7] P. C. Tan, Z. Salam, and A. Jusoh, "A single-phase hybrid active power filter using extension p-q theorem for photovoltaic application," in *2005 International Conference on Power Electronics and Drives Systems*, vol. 2, pp. 1250–1255, Kuala Lumpur, Malaysia, 2005.

[8] P. Neves, D. Goncalves, J. G. Pinto, R. Alves, and J. L. Afonso, "Single-phase shunt active filter interfacing renewable energy sources with the power grid," in *2009 35th Annual Conference of IEEE Industrial Electronics*, pp. 3264–3269, Porto, Portugal, November 2009.

[9] S. A. O. da Silva, L. P. Sampaio, and L. B. G. Campanhol, "Single-phase grid-tied photovoltaic system with boost converter and active filtering," in *2014 IEEE 23rd International Symposium on Industrial Electronics (ISIE)*, pp. 2502–2507, Istanbul, Turkey, June 2014.

[10] I. Méndez, N. Vázquez, J. Vaquero, J. Vázquez, C. Hernández, and H. López, "Multifunctional grid-connected photovoltaic-system controlled by sliding mode," in *IECON 2015 - 41st Annual Conference of the IEEE Industrial Electronics Society*, pp. 1339–1344, Yokohama, Japan, November 2015.

[11] M. C. Di Piazza and G. Vitale, "Static model," in *Photovoltaic Sources Modeling and Emulation*, pp. 56–67, Springer-Verlag, London, 2013.

[12] S. K. Dash, D. Verma, S. Nema, and R. K. Nema, "Comparative analysis of maximum power point (MPP) tracking techniques for solar PV application using MATLAB simulink," in *2014 Recent Advances and Innovations in Engineering (ICRAIE)*, pp. 1–7, Jaipur, India, May 2014.

[13] A. Morales-Acevedo, J. L. Diaz-Bernabe, and R. Garrido-Moctezuma, "Improved MPPT adaptive incremental conductance algorithm," in *IECON 2014 - 40th Annual Conference of the IEEE Industrial Electronics Society*, pp. 5540–5545, Dallas, TX, USA, October–November 2014.

[14] R. B. Roy, E. Basher, R. Yasmin, and M. Rokonuzzaman, "Fuzzy logic based MPPT approach in a grid connected photovoltaic system," in *2014 8th International Conference on Software, Knowledge, Information Management and Applications (SKIMA)*, pp. 1–6, Dhaka, Bangladesh, December 2014.

[15] N. Vázquez, Y. Azaf, I. Cervantes, E. Vázquez, and C. Hernández, "Maximum power point tracking based on sliding mode control," *International Journal of Photoenergy*, vol. 2015, Article ID 380684, 8 pages, 2015.

[16] A. A. Agrachev, A. S. Morse, E. D. Sontag et al., *Nonlinear and Optimal Control Theory*, Springer, Italy, 2004.

[17] J. Y. Hung, W. Gao, and J. C. Hung, "Variable structure control: a survey," *IEEE Transactions on Industrial Electronics*, vol. 40, no. 1, pp. 2–22, 1993.

[18] R. Sepulchre, M. Jankovic, and P. V. Kokotovic, *Constructive Nonlinear Control*, Springer Verlag, London, 1997.

[19] D. A. Torrey and A. M. A. M. Al-Zamel, "Single-phase active power filters for multiple nonlinear loads," *IEEE Transactions on Power Electronics*, vol. 10, no. 3, pp. 263–272, 1995.

[20] J. Matas, L. G. de Vicuna, J. Miret, J. M. Guerrero, and M. Castilla, "Feedback linearization of a single-phase active power filter via sliding mode control," *IEEE Transactions on Power Electronics*, vol. 23, no. 1, pp. 116–125, 2008.

[21] R. Villalobos Mendoza, *Design and Implementation of a Digital Circuit for Phase Synchronization of Applications with Interconnection with the Grid, (in Spanish), [M.S. thesis]*, Instituto Tecnológico de Celaya, Gto, Mexico, 2013.

12

Shockley's Equation Fit Analyses for Solar Cell Parameters from I-V Curves

Markus Diantoro,[1,2] **Thathit Suprayogi,**[1] **Arif Hidayat,**[1,2] **Ahmad Taufiq,**[1,2] **Abdulloh Fuad,**[1,2] **and Risa Suryana**[3]

[1]*Department of Physics, Faculty of Mathematics and Natural Sciences, Universitas Negeri Malang, Jl. Semarang 5, Malang 65146, Indonesia*
[2]*Central Laboratory for Minerals and Advanced Materials, Faculty of Mathematics and Natural Sciences, Universitas Negeri Malang, Jl. Semarang 5, Malang 65145, Indonesia*
[3]*Department of Physics, Faculty of Mathematics and Natural Sciences, Universitas Sebelas Maret, Solo, Jl. Ir. Sutami No. 36 A, Surakarta 57126, Indonesia*

Correspondence should be addressed to Markus Diantoro; mdiantoro@gmail.com

Academic Editor: Leonardo Palmisano

Some of the technical problems that appear are obtaining solar cell parameters from I-V curve measurement data. One simple method is using linear graphical fit at zero current or voltage conditions. Although the accuracy of the obtained values is acceptable, other problems may arise regarding the number of parameters which could be obtained. We report a comparison between manual or graphical fit and fit using Shockley's equation. The single I-V curve under the lighting was inferred to obtain the intrinsic parameters of the solar cells' performance. The fittings were performed using the nonlinear equation of Shockley by determining some initial values of fittings such as R_s, R_{sh}, n, I_0, I_{ph}, and T. In the case of the Shockley equation fit, the iteration was performed several times to obtain the least possible inferred parameters. We have successfully obtained a better result of nonlinear Shockley fitting compared to the manual linear fit.

1. Introduction

Rapid growth in science, technology, and industry has increased electric energy needs. About 87% of the world's energy consumption comes from fossils (oil, natural gas, and coal) which are a natural resource that cannot be renewed and lead to negative impact on the environment [1]. Therefore, developing electric energy from other natural resources which is eco-friendly is urgently required, such as the development of electric energy from solar energy [2]. Technically, solar energy can be converted into electric energy with a solar cell [3]. The solar cell is a device for green and clean electric power generation from solar energy. Solar cells have continued to evolve from single crystal silicon, thin film, organic solar cell, dye-sensitized solar cell (DSSC), and perovskite solar cell (PSC). Solar cell performance is continuously studied to raise conversion efficiency and capacity.

So far, the developing solar cell efficiency has been performed by improving the materials [4], construction [5, 6], contact systems [7], and characterizations [8, 9]. The solar cell optimization could also be optimized for analysis and modeling. One model for solar cell analysis is proposed based on the Shockley diode model. The analysis model of the solar cell from I-V characterization is with or without illumination.

One model for analyzing solar cell work is the single-diode model shown in Figure 1. At a given illumination, the

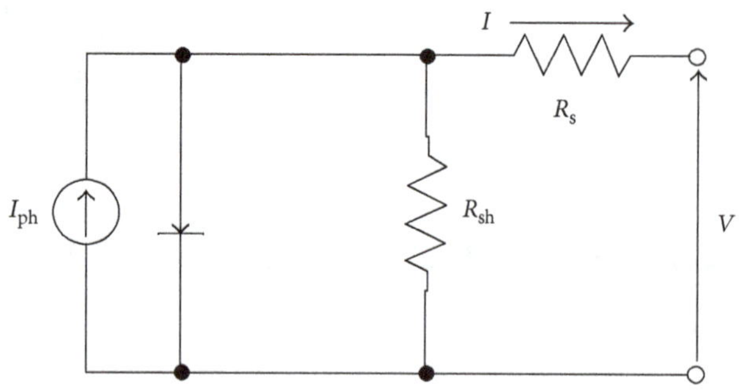

FIGURE 1: Single-diode model for a solar cell.

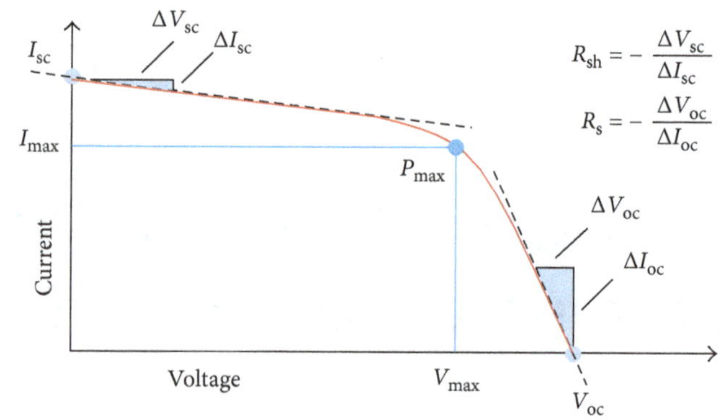

FIGURE 2: Series and shunt resistance extraction from a single I-V curve under illumination.

current/voltage relationship is given by (1) from the Shockley equation.

$$I = I_{ph} - I_0 \left[\exp\left\{ \frac{q(V + R_s I)}{nk_B T} \right\} - 1 \right] - \frac{V + R_s I}{R_{sh}}, \quad (1)$$

where I_{ph}, I_0, R_s, R_{sh}, q, n, k_B, and T are the photocurrent, the saturation current of the diode, the series resistance, the shunt resistance, the electron charge, the ideality factor, the Boltzmann constant, and the temperature, respectively.

The parameters in (1) correspond to the intrinsic characteristic of solar cell construction. The series resistance R_s represents the total resistance of the cell and is a composite of (1) the interfacial and active layer resistances, (2) the electrode resistances, and (3) the various contact and interconnect resistances [10]. The shunt resistance R_{sh} represents the p-n junction solar cell [11]. Parameters R_s and R_{sh} correspond to the fill factor (FF) of energy conversion in the solar cell.

$$\text{FF} = \frac{I_{max} V_{max}}{I_{sc} V_{oc}} \quad (2)$$

Decreasing R_s makes I_{sc} and V_{max} increase, and increasing R_{sh} makes V_{oc} and I_{max} increase [11, 12]. Many researchers develop solar cell trough variation R_s [10] and R_{sh} [13, 14]. The ideality factor n is related to the diode and its recombination [15]. In the case of $n = 1$, it represents a perfect crystalline structure with no structural defects [16]. The value of $n > 2$ is associated with the tunneling effect or the reduction of mobility due to the defects or less crystalline material [15].

The intrinsic solar cell parameters have been extracted from any method, such as a manual method with ideality factor approximation [17]. So, R_s and R_{sh} values correspond to the slope near V_{oc} and I_{sc}, respectively. By using this technique, many other parameters have not been found, such as I_{ph}, I_0, and n. Many researchers have done extraction of the solar cell parameter from the I-V curve. The parameters from (1) were extracted using the nonlinear fitting technique on a single I-V curve [18, 19]. However, the stability of the extraction results is difficult to achieve. Therefore, to extract the solar cell parameters like I_{ph}, I_0, R_s, R_{sh}, and n, the orthogonal distance regression with implicit equation model of the Shockley equation using origin lab was implemented.

2. Materials and Method

Data analysis technique on the I-V curve solar cell was needed to prove solar cell performance. Analysis of the I-V characteristic curve was done using the single-diode model equivalent circuit as shown in Figure 1 and (1). The intrinsic parameters of the solar cell were extracted using manual linear and nonlinear fitting methods using orthogonal distance regression.

The essential parameters of the solar cell based on the single-diode model was analyzed using the manual method with ideality factor value approximation [17]. Figure 2 shows how to determine the series resistance (R_s) and shunt resistance (R_{sh}) from an inverse slope near the V_{oc} and I_{sc}, respectively [17, 20].

The nonlinear fit was done using orthogonal distance regression using implicit model function (1) with origin lab. The initial value of the parameters was determined from the manual method for R_s and R_{sh}. The regression was done using about 500 iterations on the I-V curve until the best fit between the model and experimental data is obtained. By using this procedure, we found the intrinsic solar cell parameters of I_{ph}, I_0, R_s, R_{sh}, and n.

3. Results and Discussion

The comparison manual and nonlinear fittings of the I-V curve solar cell were applied to any solar cell generation. The comparison of two different equations was performed to know the effectivity for obtaining solar cell parameters based on the single-diode model. The intrinsic solar cell parameters were used to optimize the solar cell conversion efficiency.

3.1. Application to Silicon Solar Cell. The first result of the solar cell parameters of a silicon solar cell (R.T.C France) [21] using two different equation is presented. Figure 3 shows a comparison between the manual and nonlinear methods. It is observed that the linear method only fits in a particular area. Both the linear (solid blue line) and the nonlinear (solid red line) agree with experimental data

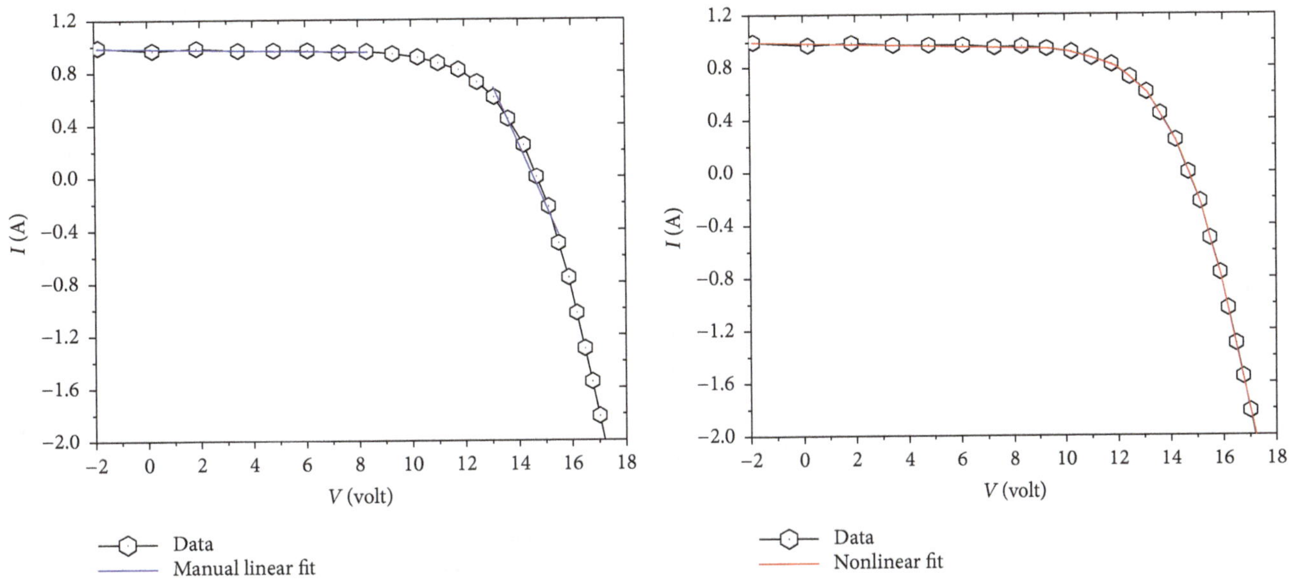

FIGURE 3: Curve fit on a silicon solar cell (R.T.C France) using the manual method (a) and the nonlinear curve fit using the Shockley equation (b).

TABLE 1: Comparison of manual and Shockley nonlinear fit on any solar cell. The Δ is from reference while χ^2 is from this work.

Number	Solar cell	V_{oc} (V)	I_{sc} (A)	I_{ph} (A)	I_0 (μA)	R_s (Ω)	R_{sh} (kΩ)	n	T (°C)	Δ/χ^2
1	Cell Si (R.T.C France) [21]	0.5728	0.7608	0.7608	0.3223	0.0364	0.0538	1.484	33	8.9×10^{-7}
	Manual	0.567	0.758	—	—	0.11	0.0324	—		—
	Previous work [18]	0.57	0.76	0.77	0.2	0.037	0.032	1.4		5.6×10^{-7}
	Nonlinear fit	0.568	0.758	0.7595	0.186	0.047	0.0355	1.418		2.9×10^{-6}
2	Module Si (Photowatt-PWP-201) [21]	16.778	1.03	1.0318	3.2876	1.2057	0.549	48.45	45	8.3×10^{-6}
	Manual	14.57	0.979	—	—	2.196	0.312	—		—
	Previous work [18]	17	1	1	2.3	1.3	0.83	47		6.3×10^{-6}
	Nonlinear fit	14.69	0.979	0.98	10	0.518	0.404	46.83		7×10^{-5}
3	Organic solar cell: ITO/CuPc/PTCBI/Ag [22]	—	—	—	—	—	—	—	20	—
	Manual	1.203	0.00449	—	—	78.59	1.222	—		—
	Previous work [18]	1.2	0.0046	0.0047	0.92	48	1.4	5.8		1.1×10^{-3}
	Nonlinear fit	1.204	0.00449	0.00468	0.918	48.34	1.232	5.74		3.13×10^{-10}
4	Organic solar cell-based p-type NiO: glass/ITO/NiO/P3HT:PCBM/LiF/Al [23]	—	—	—	—	—	—	—	27	
	Manual	0.638	0.0113	—	—	6.468	1.757	—		—
	Previous work [10]			0.0117	0.365	1.4	3.96	2.38		—
	Nonlinear fit	0.6406	0.0113	0.0113	0.0747	2.196	2.438	2.08		3.2×10^{-9}
5	Perovskite solar cell: mp-TiO_2/CH_3NH_3PbI [9]	—	—	—	—	—	—	—	27	—
	Manual	0.944	0.018	—	—	8.687	0.503	—		—
	Nonlinear fit	0.938	0.018	0.0182	0.0558	1.115	0.513	2.87		6.36×10^{-8}
6	DSSC [8]	0.704	0.00206	—	0.035	43.8	3.736	2.5	20	—
	Manual	0.708	0.002	—	—	93.27	3.532	—		—
	Nonlinear fit	0.702	0.00206	0.0021	0.0345	38.55	3.402	2.55		1.4×10^{-10}

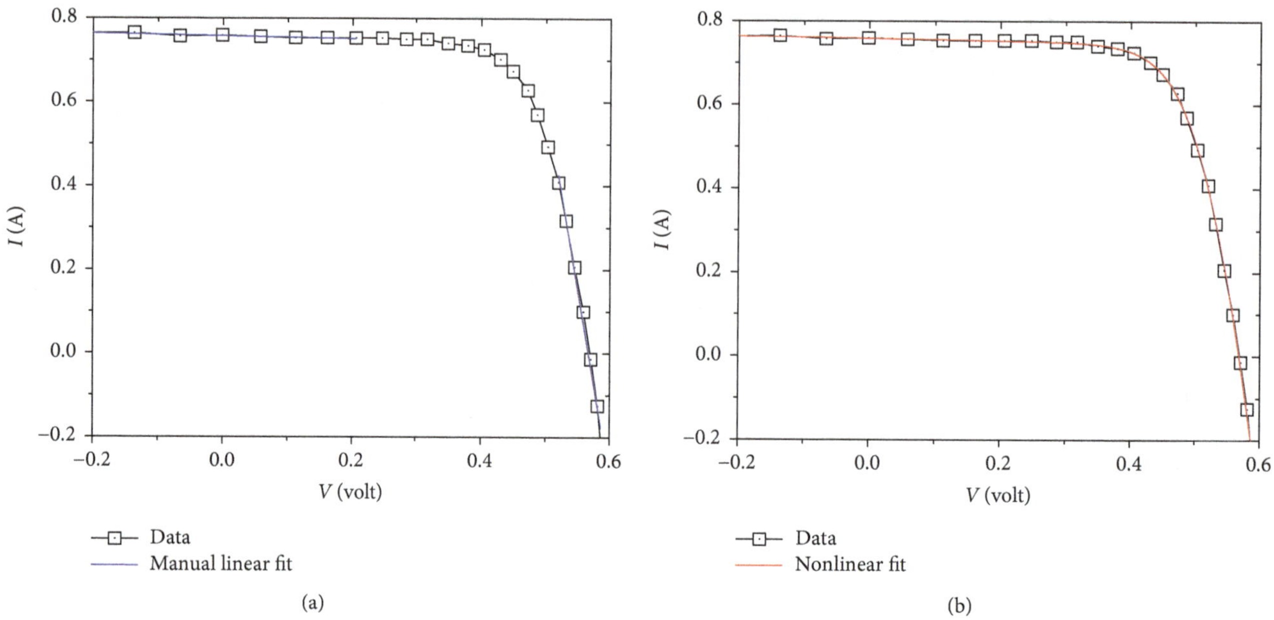

FIGURE 4: I-V curve fit on a silicon solar cell module (Photowatt-PWP-201) using the manual method (a) and nonlinear curve fit using the Shockley equation (b).

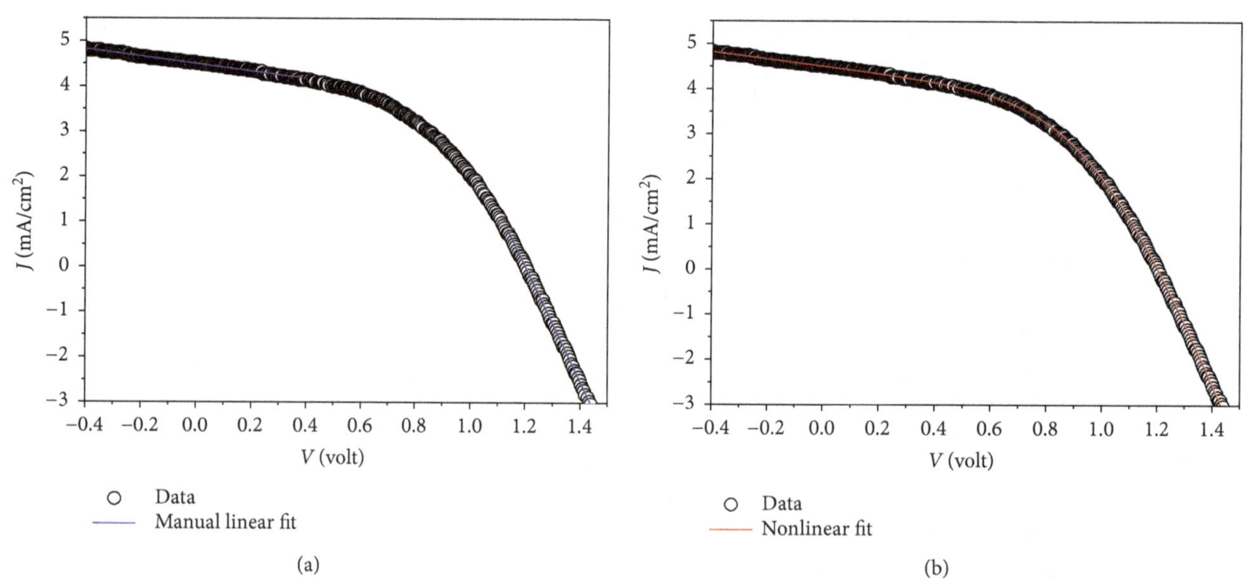

FIGURE 5: I-V curve fit on an organic solar cell based on ITO/CuPc/PTCBI/Ag: (a) manual method and (b) nonlinear curve fit.

of the I-V curve. Table 1 shows the result of the comparison of any extraction method.

Figure 4 shows the result of the I-V curve fit on a silicon solar cell module (Photowatt-PWP-201) [21]. It shows that the nonlinear fit method agrees with experimental data of the I-V curve and previous work.

3.2. Application to Organic Solar Cell. Comparison manual and nonlinear fit method on the I-V curve of an organic solar cell, ITO/CuPc/PTCBI/Ag [22], is shown in Figure 5. Not only silicon solar cell but also organic solar cell parameters can be extracted using nonlinear fit using orthogonal distance regression. Table 1 shows the result of comparison of any extracted method and the previous work.

The comparison manual and nonlinear curve fit applied to other organic solar cell-based p-type NiO, glass/ITO/NiO/P3HT: PCBM/LiF/Al [23], is shown in Figure 6. It shows suitability between the experimental data and nonlinear fit.

3.3. Application to Perovskite Solar Cell. Not only silicon and organic solar cells but also a perovskite solar cell I-V curve can be used to compare the manual and nonlinear curve fit based on the Shockley equation. Figure 7 shows the

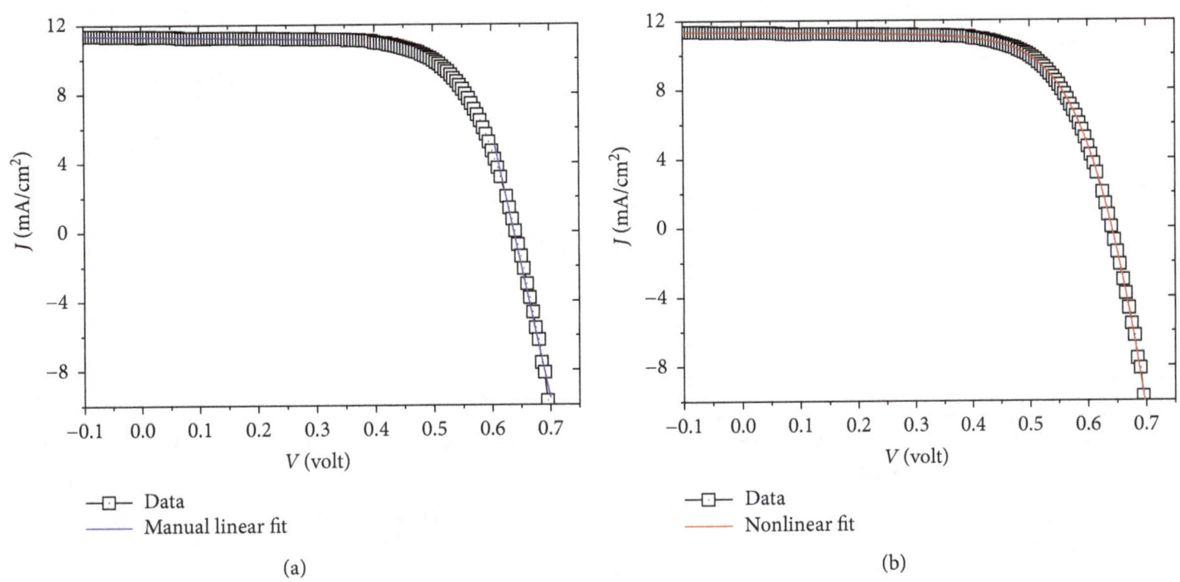

FIGURE 6: Organic solar cell based on p-type NiO: (a) manual method and (b) nonlinear curve fit.

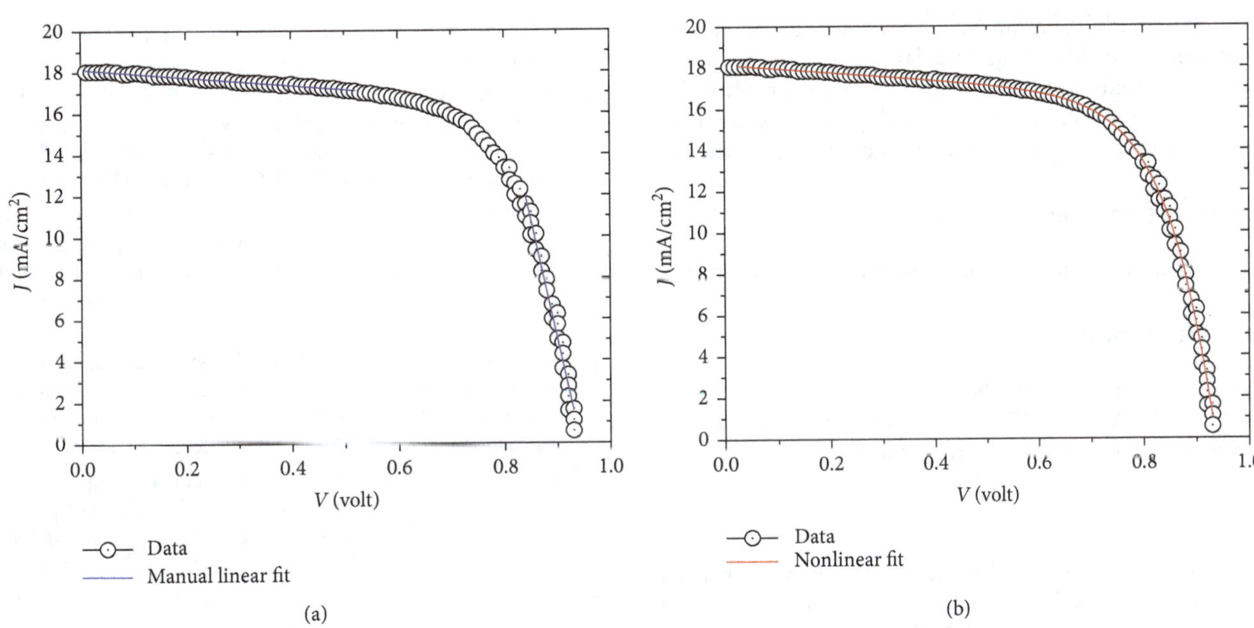

FIGURE 7: Fitting I-V curve perovskite solar cell based on mp-TiO_2 with CH_3NH_3PbI: (a) manual fit and (b) nonlinear curve fit.

comparison between experimental data and fitting method on a perovskite solar cell. Table 1 shows the result of the fitting method for perovskite solar cell parameters. The parameter of PSC can be used to upgrade solar cell conversion efficiency based on mesoporous TiO_2 and CH_3NH_3PbI [9].

3.4. Application to DSSC. The manual and nonlinear fit methods are applied to extract solar cell parameters on DSSC with TiO_2 as the photoanode and ruthenium as the sensitizer [8], and the results are shown in Figure 8. The I-V curve fit result of the DSSC parameter is shown in Table 1. The analyses of carboxyl effects on the TiO_2 photoanode has been studied using a single-diode model. The analyses were performed by determining DSSC intrinsic parameters in terms of carboxyl effects. It was found that the use of carboxyl acid gives rise to reducing the R_s value, which indicates the increase of electrical contact among TiO_2 particles [8].

4. Conclusion

The application of manual and nonlinear fit methods on any solar cell single I-V curve has been performed. The nonlinear

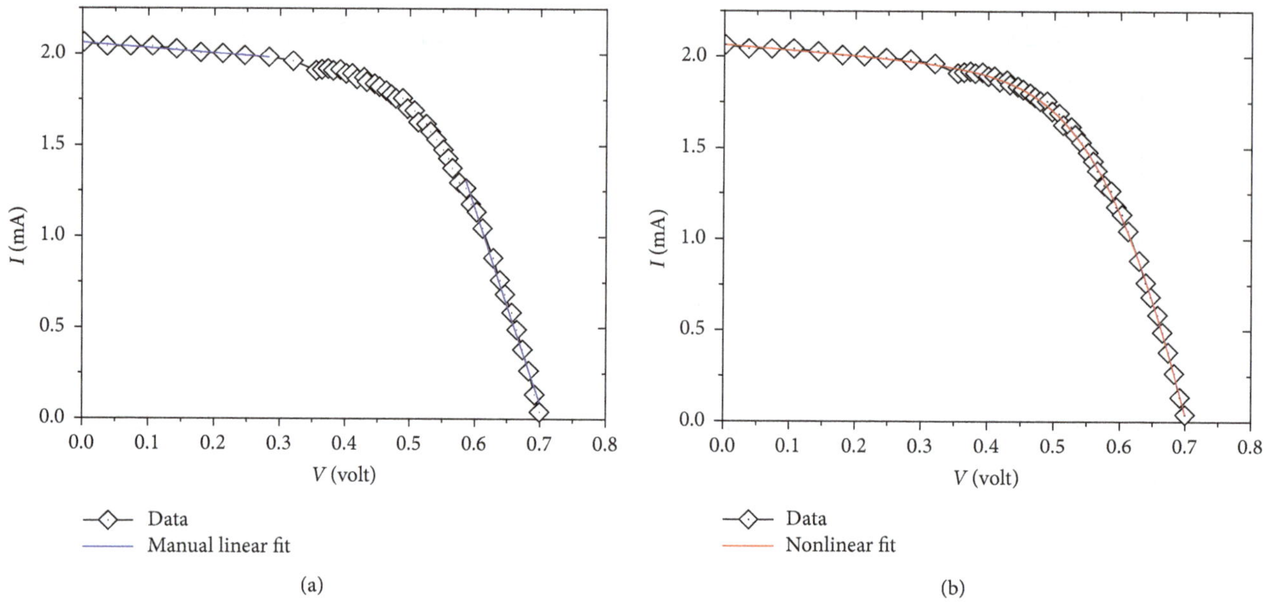

FIGURE 8: I-V curve fit of DSSC: (a) manual method and (b) nonlinear curve fit.

fit using Shockley's equation by employing orthogonal distance regression produces more solar cell parameters. It is also demonstrated that operating the nonlinear Shockley equation to several solar cell systems is highly agreeable with experimental data compared to the manual linear method.

Conflicts of Interest

The authors declare that they have no conflicts of interest.

Acknowledgments

This work was supported by the research grant of PUPT and INSINAS PRATAMA 2017 from Ministry of Research and Higher Education, Republic of Indonesia.

References

[1] E. Gençer and R. Agrawal, "A commentary on the US policies for efficient large scale renewable energy storage systems: focus on carbon storage cycles," *Energy Policy*, vol. 88, pp. 477–484, 2016.

[2] M. Arshad and S. Ahmed, "Cogeneration through bagasse: a renewable strategy to meet the future energy needs," *Renewable and Sustainable Energy Reviews*, vol. 54, pp. 732–737, 2016.

[3] M. A. Green, K. Emery, Y. Hishikawa, W. Warta, and E. D. Dunlop, "Solar cell efficiency tables (version 46)," *Progress in Photovoltaics: Research and Applications*, vol. 23, no. 7, pp. 805–812, 2015.

[4] S. Urnikaite, M. Daskeviciene, R. Send et al., "Organic dyes containing a hydrazone moiety as auxiliary donor for solid-state DSSC applications," *Dyes and Pigments*, vol. 114, pp. 175–183, 2015.

[5] H.-S. Ko, J.-W. Lee, and N.-G. Park, "15.76% efficiency perovskite solar cells prepared under high relative humidity: importance of PbI_2 morphology in two-step deposition of $CH_3NH_3PbI_3$," *Journal of Materials Chemistry A*, vol. 3, no. 16, pp. 8808–8815, 2015.

[6] M. Fathy, A. B. Kashyout, J. El Nady, S. Ebrahim, and M. B. Soliman, "Electrospun polymethylacrylate nanofibers membranes for quasi-solid-state dye sensitized solar cells," *Alexandria Engineering Journal*, vol. 55, no. 2, pp. 1737–1743, 2016.

[7] Y. Duan, Q. Tang, Y. Chen et al., "Solid-state dye-sensitized solar cells from poly (ethylene oxide)/polyaniline electrolytes with catalytic and hole-transporting characteristics," *Journal of Materials Chemistry A*, vol. 3, no. 10, pp. 5368–5374, 2015.

[8] M. Murayama and T. Mori, "Equivalent circuit analysis of dye-sensitized solar cell by using one-diode model: effect of carboxylic acid treatment of TiO_2 electrode," *Japanese Journal of Applied Physics*, vol. 45, no. 1B, pp. 542–545, 2006.

[9] H.-S. Kim and N.-G. Park, "Parameters affecting *I–V* hysteresis of $CH_3NH_3PbI_3$ perovskite solar cells: effects of perovskite crystal size and mesoporous TiO_2 layer," *Journal of Physical Chemistry Letters*, vol. 5, no. 17, pp. 2927–2934, 2014.

[10] J. D. Servaites, S. Yeganeh, T. J. Marks, and M. A. Ratner, "Efficiency enhancement in organic photovoltaic cells: consequences of optimizing series resistance," *Advanced Functional Materials*, vol. 20, no. 1, pp. 97–104, 2010.

[11] N. Koide, A. Islam, Y. Chiba, and L. Han, "Improvement of efficiency of dye-sensitized solar cells based on analysis of equivalent circuit," *Journal of Photochemistry and Photobiology A: Chemistry*, vol. 182, no. 3, pp. 296–305, 2006.

[12] H. J. Snaith, L. Schmidt-Mende, M. Grätzel, and M. Chiesa, "Light intensity, temperature, and thickness dependence of the open-circuit voltage in solid-state dye-sensitized solar cells," *Physical Review B*, vol. 74, no. 4, article 45306, 2006.

[13] J.-J. Huang, S.-P. Chiu, M.-J. Wu, and C.-F. Hsu, "Effect of titanium oxide compact layer in dye-sensitized solar cell prepared by liquid-phase deposition," *Applied Physics A*, vol. 122, no. 11, p. 971, 2016.

[14] A. Scuto, S. Lombardo, G. Di Marco et al., "Improvement of DSSC performance by voltage stress application," in *2016 IEEE*

International Reliability Physics Symposium (IRPS), Pasadena, CA, USA, 2016.

[15] A. J. Trindade and L. Pereira, "Bulk heterojunction organic solar cell area-dependent parameter fluctuation," *International Journal of Photoenergy*, vol. 2017, Article ID 1364152, 10 pages, 2017.

[16] Y. Roichman and N. Tessler, "Generalized Einstein relation for disordered semiconductors—implications for device performance," *Applied Physics Letters*, vol. 80, no. 11, pp. 1948–1950, 2002.

[17] D. S. H. Chan and J. C. H. Phang, "Analytical methods for the extraction of solar-cell single- and double-diode model parameters from I-V characteristics," *IEEE Transactions on Electron Devices*, vol. 34, no. 2, pp. 286–293, 1987.

[18] K. I. Ishibashi, Y. Kimura, and M. Niwano, "An extensively valid and stable method for derivation of all parameters of a solar cell from a single current-voltage characteristic," *Journal of Applied Physics*, vol. 103, no. 9, article 094507, 2008.

[19] C. Zhang, J. Zhang, Y. Hao, Z. Lin, and C. Zhu, "A simple and efficient solar cell parameter extraction method from a single current-voltage curve," *Journal of Applied Physics*, vol. 110, no. 6, article 64504, 2011.

[20] J. C. H. Phang, D. S. H. Chan, and J. R. Phillips, "Accurate analytical method for the extraction of solar cell model parameters," *Electronics Letters*, vol. 20, no. 10, pp. 406–408, 1984.

[21] T. Easwarakhathan, J. Bottin, I. Bouhouch, and C. Boutrit, "Nonlinear minimization algorithm for determining the solar cell parameters with microcomputers," *International Journal of Solar Energy*, vol. 4, no. 1, pp. 1–12, 1986.

[22] A. Yakimov and S. R. Forrest, "High photovoltage multiple-heterojunction organic solar cells incorporating interfacial metallic nanoclusters," *Applied Physics Letters*, vol. 80, no. 9, pp. 1667–1669, 2002.

[23] M. D. Irwin, D. B. Buchholz, A. W. Hains, R. P. H. Chang, and T. J. Marks, "p-Type semiconducting nickel oxide as an efficiency-enhancing anode interfacial layer in polymer bulk-heterojunction solar cells," *Proceedings of the National Academy of Sciences of the United States of America*, vol. 105, no. 8, pp. 2783–2787, 2008.

13

Grid-Connected Semitransparent Building-Integrated Photovoltaic System: The Comprehensive Case Study of the 120kWp Plant in Kunming, China

Yunfeng Wang,[1] **Ming Li**,[1] **Reda Hassanien Emam Hassanien**,[1,2] **Xun Ma**,[1] **and Guoliang Li**[1]

[1]*Solar Energy Research Institute, Yunnan Normal University, Kunming, China*
[2]*Agricultural Engineering Department, Cairo University, Giza 12613, Egypt*

Correspondence should be addressed to Xun Ma; maxun80313@126.com

Academic Editor: Gabriele Battista

A 120 kWp building-integrated photovoltaic (BIPV) system was installed on the south facade of the Solar Energy Research Institute building in Yunnan Normal University. The area of the curtain wall was 1560 m^2 (26 m × 60 m), which consisted of 720 semitransparent monocrystalline silicon double-glazing PV panels. This paper studied the yearly and monthly variations of power generation in terms of solar data and meteorological parameters. The total amount of power generation of the BIPV system measured from October 2014 to September 2015 was 64.607 MWh, and the simulation results with TRNSYS (Transient Systems Simulation Program) provided the 75.515 MWh predicted value of annual electricity production with the meteorological database of Meteonorm, while, based on the average value of the performance ratio (PR) of 60% and the life cycle assessment (LCA) of the system, the energy payback time (EPBT) of 9.38 years and the potential for pollutant emission reductions have been evaluated and the environmental cost is RMB ¥0.01053 per kWh. Finally, an economic analysis was carried out; the net present value (NPV) and the economic payback time of the BIPV system were estimated to be RMB ¥359,347 and 15 years, respectively.

1. Introduction

In the last 20 years, the world's energy consumption has sharply increased (40%), and it is expected to continue increasing by one-third in the period 2035 [1]. 20% to 40% of this global energy is consumed in the built environment [2]. In China, building energy consumption has been continuously increasing, which has been doubled during the last decades in China [3]. According to the growing trend of developed countries, the building energy consumption will account for at least 40% of all social total energy consumption in China [4]. In this situation, (nearly) zero-energy buildings (NZEB) have been considered the priority solutions for the architecture trend in the future. NZEB implies that all building-related operating energy is generated on the building site itself by renewable sources [5]. One of the most promising approaches to realize NZEB is by adopting solar energy-generating devices in the form of photovoltaic (PV) modules for electricity. As a kind of new energy, solar energy is the most abundant and inexhaustible, so several policies have been proposed by the Chinese government, such as the plan of the solar roof of buildings [6]. However, in modern urban cities, most buildings are high-rise and the roof area would be very limited for stand-alone PV system installation [7]. So the integration of PV systems into buildings becomes an imperative [8]. Building-integrated photovoltaics (BIPV) replacing part of the external walls with PV panels would be an appropriate alternative form of the PV system [9]. Integration of PV systems with distribution networks (grid-connected) could reduce the maximum demand charge and energy losses [10]. Moreover, BIPV can serve as a shading device for a window, a semitransparent glass facade, a building exterior cladding panel, a skylight, and a parapet unit or roofing system [11].

Li et al. [12] investigated the energy performance of a semitransparent a-Si PV facade. The simulation results revealed that the semitransparent PV facade was able to reduce the annual building energy use and peak cooling load by 1203 MWh and 450 kW, respectively. And in 2013, they conducted a study of the operational performance of a grid-connected BIPV system in Hong Kong. Annually, the electricity generation of 42,450 kWh was recorded and emissions of CO_2, SO_2, NO_x, and particulates could be reduced, respectively, by 32 tons, 43.3 kg, 41.8 kg, and 2.07 kg [13]. Furthermore, Hong et al. conducted a sensitivity analysis on how the impact factors of the rooftop PV system affect its electricity generation and developed a GIS-based optimization model for estimating the electricity generation of the rooftop PV system [14]. On a fixed PV array performance, its best electrical configuration had been researched by analyzing annual energy production with the surrounding obstacles causing partial shading conditions [15]. And the multiple calculations had been provided to determine whether solar energy has a viable strategic role in the global energy market [16]. The embodied energy payback period for a photovoltaic system integrated with the building was reported within the range of 12–13 years for the southern and western facades in the United Arab Emirates [17]. Besides, a lot of the previous studies researched the energy harvested and evaluated the performance of BIPV systems [18–21].

However, there is still a lack of research on the performance of semitransparent BIPV facades compared with conventional glazing in low-latitude and high-altitude cities, such as the city of Kunming. This experiment was the first and largest BIPV application installed on the building facade of the Solar Energy Research Institute of Yunnan Normal University. Thus, the aim of this paper is to analyze the PV power generation performance and evaluate the economic benefits of a grid-connected semitransparent BIPV system. The energy output and meteorological data were recorded and analyzed, and a TRNSYS (Transient Systems Simulation Program) model was built; the net energy benefit, carbon emission reduction, and cost savings as well as energy, carbon emissions, and cost payback were evaluated by using life cycle assessment (LCA). The results can be used to evaluate the performance and provide the reference data for more applications of semitransparent BIPV under the actual operating environments.

2. BIPV System Description

The BIPV system was built in Kunming City in June 2014; then, it was put into operation in October 2014. Kunming City at the southwest of China is a special plateau city, which receives an annual total solar radiation at the horizontal plane of ~1530.24 kWh/m^2. It also has the mildest climates characterized by short, cool dry winters with mild days and crisp night, and long, warm, and humid summers. Therefore, the climate and location of Kunming are suitable for the development of solar energy, and a lot of PV systems have been built and running in Kunming. The first and largest grid-connected semitransparent BIPV system of Kunming in this study (Figure 1) was installed outside of

FIGURE 1: Yunnan Normal University BIPV system.

the south facade of the building of the Solar Energy Research Institute in Yunnan Normal University. The BIPV system was 60 m in length and 26 m in height at the south face of the 5-storey building with a total area of 1560 m^2, including 720 monocrystalline silicon double-glazing PV modules. The modules were installed with 85° tilt angles and were 6 m apart from the building for the ventilation. The PV arrays were comprised of 30 parallel strings where each string has 24 modules, and the peak power and voltage of each string were 4080 Wp and 796 V, respectively. The specifications of the semitransparent PV modules are given in Table 1. Each PV module with dimensions of 1985 × 1038 mm consisted of 64 pieces of the solar cell and a 6 mm transparent glass cover, a 6 mm glass backside, and a certain space between the PV cells to transmit light in order to ensure a 47% light transmittance as shown in Figure 2. Solar cells adopted in the PV module were monocrystalline silicon wafer cells with an 8.25% efficiency, and the dimensions of each cell were 125 × 125 mm. The reason why the efficiency of the PV module was lower than that of the conventional monocrystalline silicon module was that the areas of the 64 pieces of solar cells in a module were just 48.5% of the total area of a PV module. It was based on the cost of sacrificing the efficiency of the PV module to ensure partial light transmission. The laminated glass forms were installed on both the west and east directions and on the rooftops of the PV curtain wall (Figure 3). And some ventilation blinds were placed at the top and the bottom of the PV curtain wall for proper ventilation (Figure 4). The PV module temperature will be controlled or affected according to the adjustment of these ventilation blinds by opening or closing it in different seasons.

The 720 solar modules were configured to 25 (parallel) × 24 (series) and 5 (parallel) × 24 (series). The 600 modules were connected to the inverter (Growatt CP100 with a matching maximum DC voltage of 1000 V and a maximum input current of 250 A, maximum efficiency of 97.1%, and maximum DC input power of 115 kW), and the remaining 120 modules were connected to another inverter (Growatt 20000UE with a matching maximum DC voltage of 1000 V and a maximum input current of 20 A, maximum efficiency of 98%, and maximum DC input power of 20.8 kW). The 120 kWp power generated was synchronized and supplied

Table 1: The PV module specifications under standard test conditions (STCs).

Number	Item	Parameter
1	Dimensions of one module (mm)	1985 × 1038 × 13.52
2	Weight of one module (kg)	63
3	Light transmittance	47%
4	Nominal power of one module (Wp)	170
5	Short-circuit current (A)	5.65
6	Open-circuit voltage (V)	39.6
7	Efficiency	8.25%
8	Maximum power point current (A)	5.12
9	Maximum power point voltage (V)	33.2
10	Layer structure	Toughened glass (6 mm) + PVB (2.28 mm) + (125 × 125 mm) sc-Si cells + toughened glass (6 mm)

Figure 2: The semitransparent double-glazing PV module.

Figure 3: The inside view of the PV curtain wall.

to the 400 V campus grid through a grid export conditioner. The Maximum Power Point Tracker (MPPT) was used to adjust the voltage of the PV generator to operate at its maximum power point.

Figure 5 shows the schematic diagram of the electrical system and the monitoring system, and each data is collected by 2 photovoltaic inverters connected with a power-monitoring module bus to investigate the generating characteristics of the BIPV system. A pyrheliometer, thermometers, a watt-hour meter, and so forth are also installed to collect the data needed for a detailed analysis of the BIPV system.

3. BIPV System Performance Analysis

3.1. Experimental Analysis of Power-Generating Performance of BIPV. Figure 6 shows the amount of power generation in the building from October 2014 to September 2015 monthly. The monthly average amount of generation was 5383.9 kWh, while the cumulative amount for 12 months was 64,607.5 kWh. The month with the least generated amount (740.2 kWh) was September. This is because of the installation tilt and operating temperature. The BIPV system was installed with an 85° tilt angle facing towards the south, due to the solar zenith angle being high in summer, so that the solar irradiation received by the PV modules decreased significantly in September. Besides, the generated amount (1664.9 kWh) was also less in March. This is because February and March in Kunming's climate correspond to the coldest season. This was especially true, and it even snowed in 2015, so the solar radiation sharply lowered with the extended severe climate and affected the power generation; however, this kind of extreme weather was just for the year 2015. The month with the highest generated amount (11,025 kWh) was December, up to 14.9 times compared with that in September.

However, the operating temperature (or the panel temperature) of PV modules seems to be a little different between summer and winter for Kunming City, as shown in Figure 7(a), which shows the average operating temperature of PV modules in the sunny days of June 14 and December 15. It was found that the average operating temperature of summer and winter before 8:29 in the morning was identical, while in the noon, the temperature in winter was about 10°C higher than the operating temperature of summer. The operating temperatures of both days increased rapidly from 9:00, and about 3 hours later, the temperatures reached the maximum temperature and then gradually decreased and got close to each other. One of the reasons why the average operating temperature was lower in summer was that the ventilation blinds on the top and bottom of the PV curtain wall as shown in Figure 4 were opened in order to decrease the air temperatures and reduce the cooling load of the building. Although the operating temperature of the PV module in December was up to

(a) (b)

FIGURE 4: The ventilation blinds at the top and bottom of the PV curtain wall.

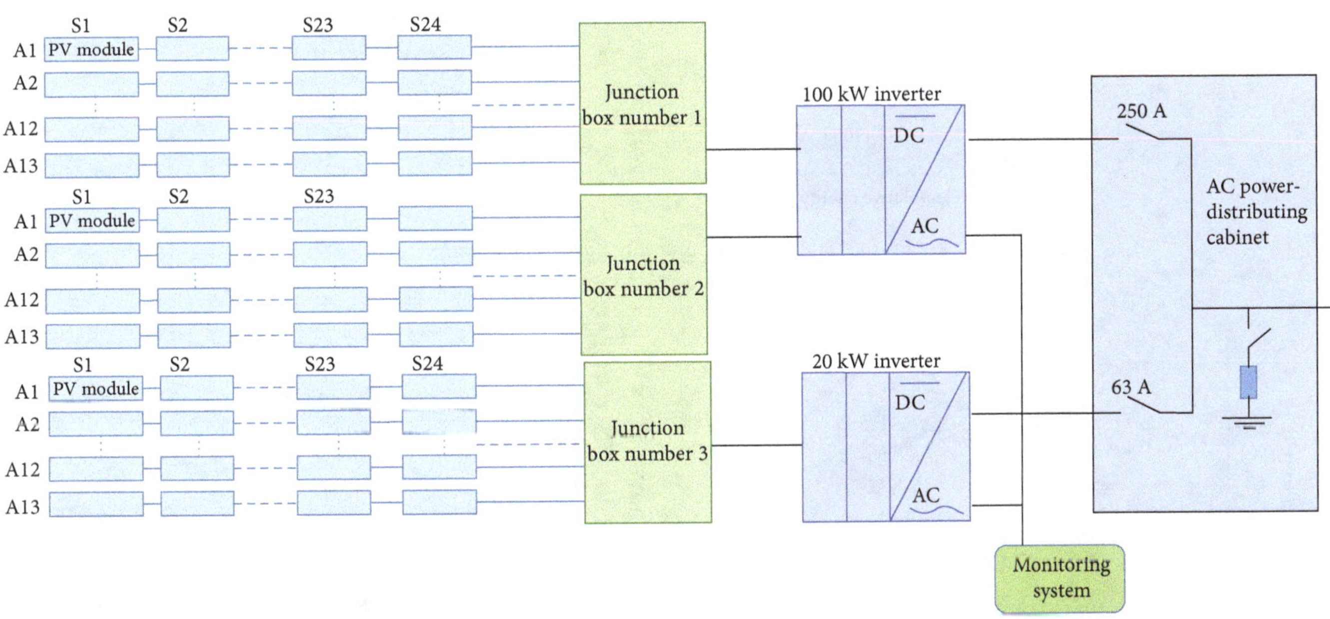

FIGURE 5: The connection schematic diagram of the system.

55.2°C, the air temperature in the gap between the PV curtain wall and the exterior wall of the building was just 32°C, and the average of the 1st–5th floor temperatures eventually reached around 22°C, which was 6°C higher than the ambient temperature as shown in Figure 7(b). Because the ventilation blinds on the top and bottom of the PV curtain wall were closed, the increased air temperature could be used to keep the rooms warm without any auxiliary heating in winter. Another reason was that the PV modules can receive more solar irradiation in winter because of the installed inclination angles of the PV curtain wall. The solar radiation intensity on the surface of the PV curtain facing south with an 85° tilt angle in the two typical sunny days of June 2015 and December 2014 is shown in Figure 8. The highest and average solar radiation intensities on the vertical surface of the PV modules (957 W/m^2 and 417 W/m^2) in a typical day of December were both up to 2 times higher than those in June (495 W/m^2 and 233 W/m^2), respectively. It seems that the solar radiation would have the largest effect on the power generation. Furthermore, the daily power generation in December of 2014 and in the two identical months of 2014 and 2015 is shown in Figures 9 and 10, respectively. It indicates that the daily power generation may fluctuate greatly even in the same month of different years, due to the solar radiation variation which was the typical climate in Kunming.

3.2. TRNSYS Simulation Results and Discussion. TRNSYS is a transient system energy modeling software developed to solve complex energy system problems. The software uses individual components referred to as types connected to each other with each representing one part of the overall system. Figure 11 shows the TRNSYS model used to simulate the power generation of the PV system, which has a weather database to predict the behavior of PV installations in many locations such as Kunming. In this work, a five-parameter

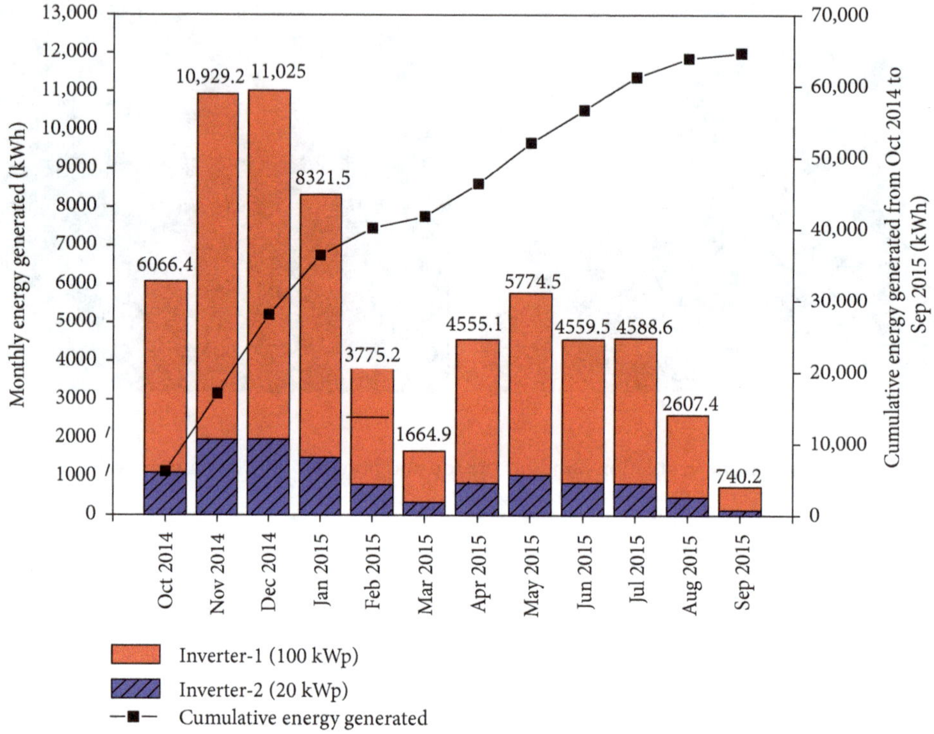

FIGURE 6: Amount of monthly power generation.

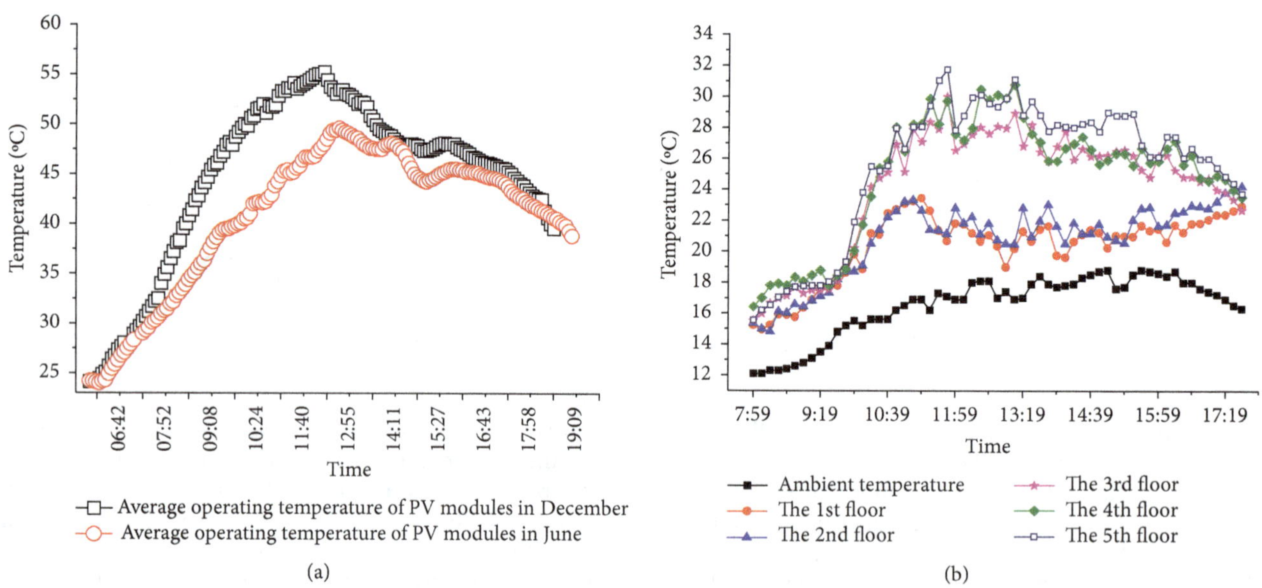

FIGURE 7: (a) The average operating temperature of PV modules and (b) the air temperature in the gap between the PV curtain wall and the exterior wall of the building.

model (Type 194 "photovoltaic array") has been used to determine the maximum power point from *I-V* curves in the operating conditions which are given by the meteorological inputs. The inverter information is given as a text file to the PV component in Type 194.

3.2.1. TRNSYS Model Validation. The PV system model was created using TRNSYS 16 with known photovoltaic cell envelope characteristics. The model was then validated with the data collected using pyrheliometers, thermometers, and so forth. The meteorological inputs used in the Type 109-User were the in-field monitoring data. From the results of the data collection, a curve was developed illustrating the power output of one day with respect to the solar radiation. This curve was used to validate the TRNSYS model by matching the TRNSYS power outputs at various solar

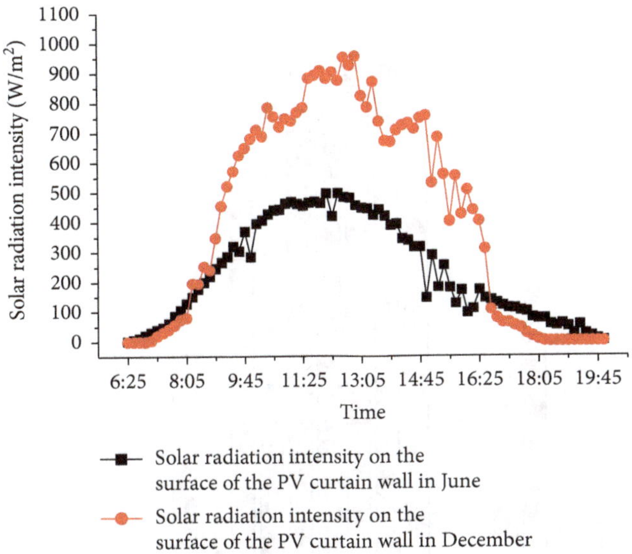

FIGURE 8: Solar radiation intensity on the vertical surface of the PV curtain wall.

radiations. Figure 12 demonstrates the power output of the 20 kW inverter array parts at a monitoring weather condition which was matched with the TRNSYS model power output. The weather condition and PV power output of the day of November 25, 2015, had been chosen to verify the accuracy of the TRNSYS model.

It could be found from Figure 12 that there were some differences between the hourly measured power output and the simulation results with the TRNSYS model on a sunny day. Maybe the reason was that the temperature provided to Type 194 was the ambient temperature during the test periods and not the real temperature of the PV curtain wall; in addition, the solar radiation intensity provided to Type 109 was the total radiation on the horizontal. Type 194 used temperature data from the standard NOCT (nominal operating cell temperature) measurements to compute the module temperature T_c at each time step. The NOCT temperature is the operating temperature of the module with a wind speed of 1 m/s, no electrical load, and a certainly specified insolation and ambient temperature. Besides, Type 194 uses the NOCT data to determine the ratio of the module transmittance-reflectance product to the module loss coefficient [22]. And indeed, the literature shows that the choice of the temperature model is relevant only on a daily basis but has a small effect on the annual predicted PV output [23]. Besides, the calculation of the radiation on a tilted surface derives from the radiation on the horizontal surface. This calculation involves the use of a global diffuse correlation and of a tilted surface radiation model to estimate the direct, diffuse, and albedo components of the radiation on the tilted surface. The literature has shown that the Perez model (model 4 in TRNSYS) can be a good choice to predict the energy output. However, the relative error of the hourly power generation was less than 17%, and the relative error of the average hourly power generation was just 6.25%. The calculated result agreed with the measured result on the amount of power generation of the day, 98.6 kWh and 104.4 kWh, respectively, and the relative error between the measured result and simulation result was 5.55%. In general terms, the five-parameter model (Type 194) provides a simple tool to predict the PV production accurately.

3.2.2. The BIPV Energy Output according to the TRNSYS Model. The yearly energy PV production has been calculated with TRNSYS, using the meteorological database of Meteonorm and Type 109 and assuming that the PV installation is connected to the grid with no failures. The annual PV power generation for the BIPV system is shown in Figure 13. The results reveal that the calculated annual electricity production of the BIPV is about 75.515 MWh. On the other hand, the monthly peak production was reached in November with 9.67 MWh and the lowest was found in June with 3.789 MWh.

4. Analysis and Discussion

4.1. Performance Assessment. In order to identify the performance of the BIPV system, the performance ratio (PR) was calculated. The PR, which is a crucial parameter for the PV performance evaluation, describes the relationship between the actual and theoretical energy outputs of the PV array. It represents the energy losses such as loss due to temperature, wiring loss, shadowing, soiling, inverter, mismatch loss, and loss across the bypass diodes and shows the incomplete utilization of incoming solar radiation and proportion of energy available at the grid after losses in grid-connected PV systems. It is the most widely used performance parameter for comparative analysis of different PV panel technologies and is also used for comparing grid-connected PV systems irrespective of their location, power capacity, and mounting structure [24]. The main equation was used to investigate the PR as follows [25]:

$$PR = \frac{E_{y\text{-}s}}{(H/1000) \times P_{n\text{-}s}}, \quad (1)$$

where $E_{y\text{-}s}$ is the energy generated by the PV system during the year [Wh], H is the annual solar irradiation on the PV modules [Wh/m^2·year], and $P_{n\text{-}s}$ is the nominal power of the system measured at STC [Wp].

According to the test data of the first year of installation (the year 2015), the PR of the BIPV system is calculated as being equal to 64.3%, which includes some extreme weather events like snows in this year. The BIPV system has just run for two years since it was completed; more precise data for calculating the performance ratio of the system have not been acquired during its working period. However, such PR value can be considered within the normal range; literature showed that PV arrays installed in Italy have a PR value varying across the range from 62% to 81% [26], though it has a different tilt angle and different system structure. Some research shows that the climatic conditions have a great impact on the PR [27]. For example, the year that had the greatest number of days with bad weather (rain, snow, or fog) had the lower PR; on the contrary, the highest PR

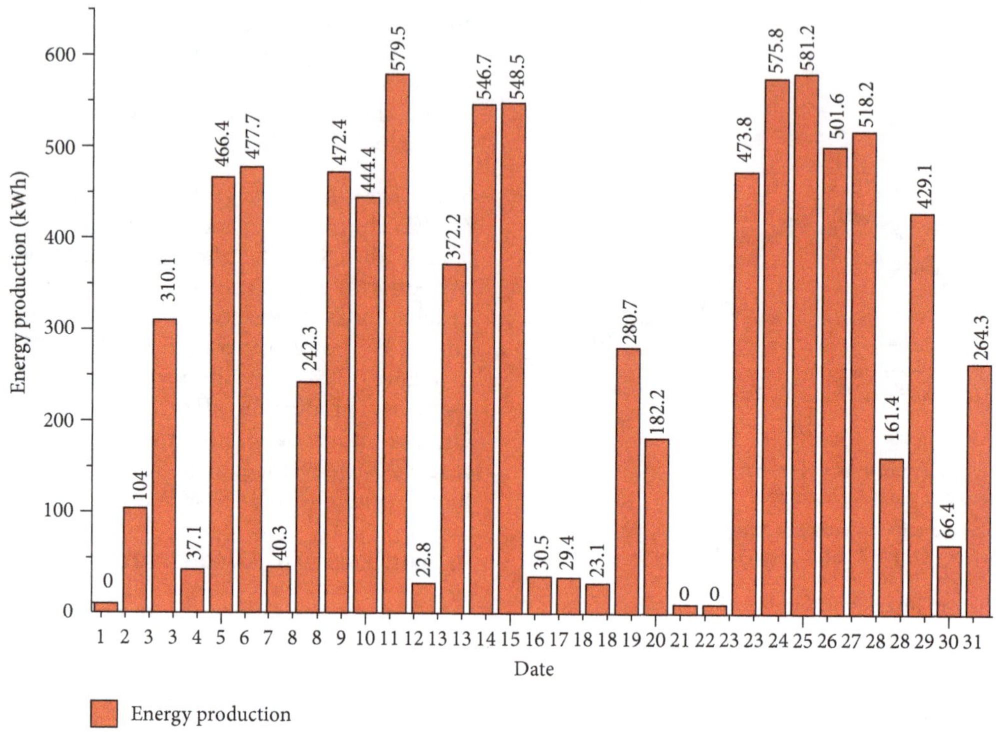

FIGURE 9: Daily December power generation.

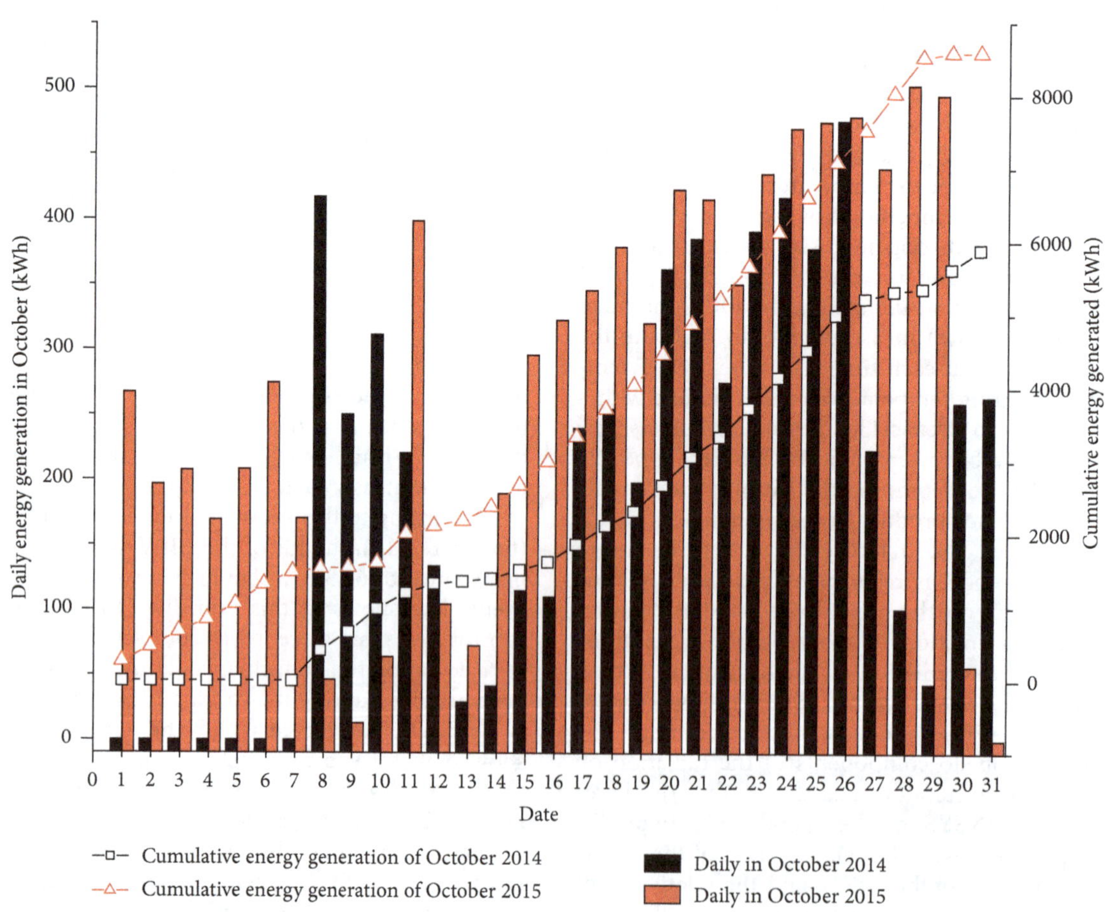

FIGURE 10: Power generation compared in the same month of different years.

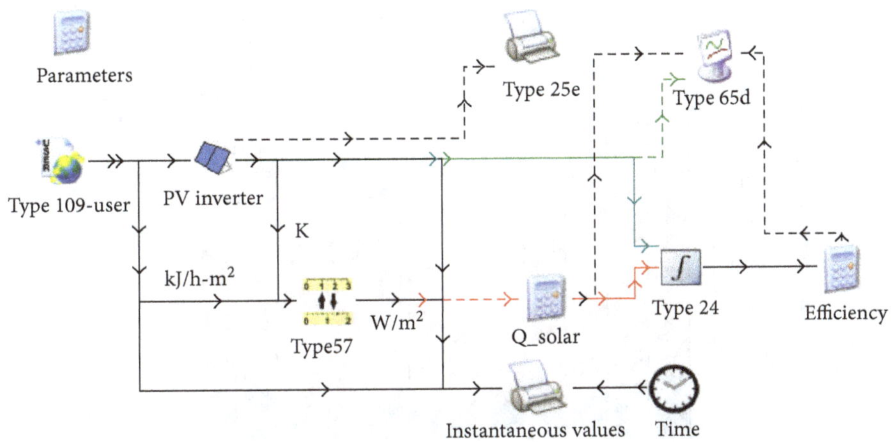

FIGURE 11: Scheme of the TRNSYS PV model.

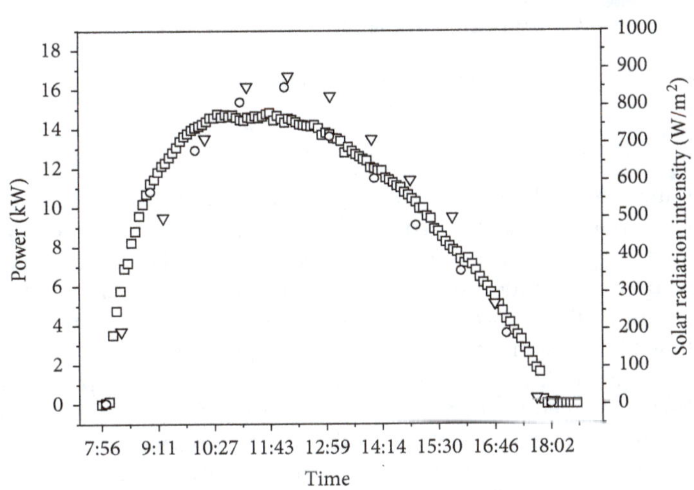

▽ Measured solar radiation intensity on the title surface of PV curtain wall
□ Measured power output of the 20 kW array
○ Power output with TRNSYS model of 20 kW inverter array

FIGURE 12: TRNSYS PV model validation for the power output of a day.

occurred in the year with the maximum recorded number of clear days [25]. So if there are no extreme weather events in a year, the PR of the BIPV system could be higher than 65%. While considering the typical degradation rate of PV modules (equal to 1%–2%/year) [24], the average value of PR could be 60% according to the simulated data of the TRNSYS model during the BIPV system with an expected lifetime of 25 years. The reason for the low PR of the BIPV system is that the area of solar cells in a module was just 48.5% of the total area of a PV module; however, the solar irradiation H in (1) includes all solar irradiation received by a module. In order to ensure partial light transmission and building daylighting, it is the only choice to sacrifice the efficiency of the PV module for the moment. From this view, it can be supposed that different types of solar cells and different designed schemes of the envelope curtain wall could improve the PR of a BIPV system; for example, a new kind of solar module which has a lower light transmittance, known as a higher efficiency, would be employed to replace the existing one.

4.2. Environmental Assessment. It is generally known that the PV system is widely recognized as one of the most clean technologies for power generation. However, it consumes energy during its life cycle, particularly in the manufacturing processes, which may be larger than its energy output during its whole life. In order to thoroughly investigate the life cycle environmental effects and energy payback performance of the BIPV system, life cycle assessment (LCA) is used to measure its sustainability. LCA is usually used as a technique to compare and analyze various aspects associated with the development of a product and its potential impact throughout the product's life [28]. The goal of an LCA is to quantify nonrenewable (fossil) primary energy use and global warming potential in electricity generation. According to LCA, the life cycle primary energy use is considered the sum of

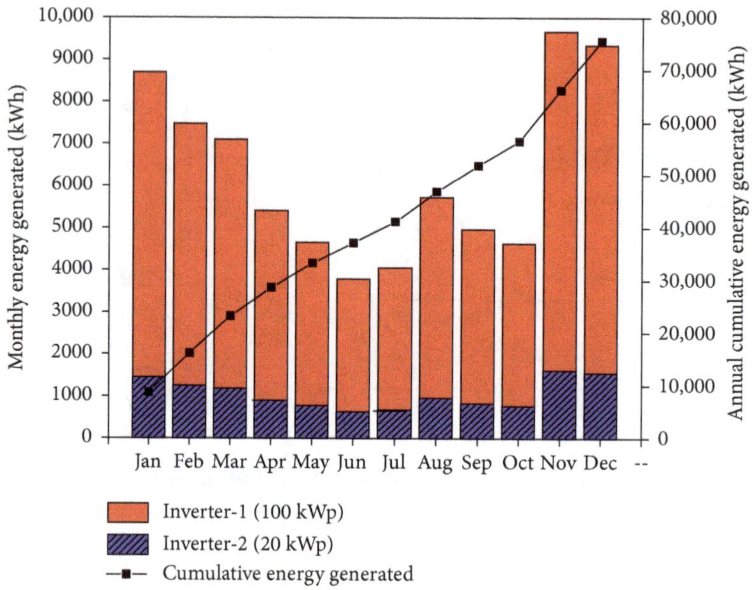

FIGURE 13: Predicted monthly BIPV power output using the TRNSYS model.

the energy consumed in three life cycle phases, which includes energy consumed in exploration, extraction, processing, manufacturing, decommissioning, and disposal of all the materials associated with the power generation system [29]. Besides, the energy payback time (EPBT) would be used, in order to find out whether the BIPV system can bring a net gain of energy for the user during its whole lifetime, and if so then to what extent. The EPBT is defined as the years required for a PV system to generate a certain amount of energy (converted into equivalent primary energy) for compensation of the energy consumption over its life cycle, including energy requirements in PV modules' manufacturing, assembly, transportation, system installation, operation and maintenance, and system decommissioning or recycling [30]. The equation of EPBT is presented as follows:

$$\text{EPBT} = \frac{E_{\text{input}} + E_{\text{BOS,E}}}{E_{\text{output}}}, \tag{2}$$

where E_{input} is the primary energy input of the PV module during the life cycle, which includes energy requirements in module manufacturing, transportation, installation, operation and maintenance, and module decommissioning or recycling [MWh]. $E_{\text{BOS,E}}$ is the energy requirement of the balance of system (BOS) components, which include support structures, cabling, electronic and electrical components, inverters, and batteries [MWh]. E_{output} is the annual primary energy savings due to electricity generation by the PV system [MWh].

There were very few papers using the actual energy yield to conduct an LCA study on BIPV systems in China, especially on a BIPV system of mono-Si modules, on account of many unavailable data of mono-Si production from the manufacturers, in particular, the energy consumption data of the Czochralski procedure. Besides, different manufacturers have different energy requirements for silicon purification and crystallization processes, which increases the difficulty of the research. Lastly, there is a rapid technological improvement in the PV industry and cell production in China, and it is difficult to accurately estimate the energy requirements of mono-Si module production by experience. So this paper conducts the life cycle assessment study for the mono-Si BIPV system and discusses the energy payback performance just according to the data from literature [31–36].

The equivalent primary energy consumption for manufacturing solar mono-Si double-glazing PV modules is estimated as 4.67 MWh/kWp. And the BOS component is estimated as 1.12 MWh/kWp in this system, because the supporting structure is steel which is built outside the building exterior wall, and the battery can be saved in consideration of the special structure of the double-glazing module and the grid-connected scheme. And the energy consumption of the other components, including transportation, installation, operation and maintenance, and module decommissioning or recycling, would be 2-3 orders of magnitude lower than the energy consumption of PV module manufacturing and the BOS component; thus, they could be negligible when the life cycle energy requirement is calculated in the equation of EPBT. The EPBT of the BIPV system is calculated to be 9.38 years, so during the PV module's lifetime (normally expected to be 25 years), the BIPV system could still generate a substantial amount of electricity and gain additional benefits.

According to relevant Chinese documents of the National Bureau of Statistics, from the year 2006, the electricity consumption converted into the standard coal coefficient should employ the equivalent coefficient of 0.1229 kg/kWh, which means that 0.1229 kg of standard coal generates 1 kWh of power. The environmental benefits

Table 2: Environmental benefit measurement and relevant parameters for the BIPV system.

The major emission pollutants	Emission rate of pollutants (kg/kg$_{ce}$)	Emission reduction density (kg/kWh)	Emission density (kg/kWh)	Environmental values of pollutants (RMB yuan/kg)	Environmental cost of generation (RMB yuan/kWh)	Environmental benefits of emission reduction (RMB yuan/kWh)
CO_2	1.731	2.127×10^{-1}	7.986×10^{-2}	2.3×10^{-2}	0.184×10^{-3}	4.893×10^{-3}
SO_2	0.018	2.212×10^{-3}	8.304×10^{-4}	6	4.983×10^{-3}	1.327×10^{-2}
NO_x	0.008	9.832×10^{-4}	3.691×10^{-4}	8	2.953×10^{-3}	7.866×10^{-3}
CO	0.00026	3.195×10^{-5}	1.199×10^{-5}	1	1.199×10^{-5}	3.195×10^{-5}
Particulate	0.11	1.351×10^{-2}	5.075×10^{-3}	1.2×10^{-1}	6.090×10^{-4}	1.622×10^{-3}
Slag	0.03	3.687×10^{-3}	1.384×10^{-3}	1.0×10^{-1}	1.384×10^{-4}	3.687×10^{-4}
Total	/	/	/	/	1.053×10^{-2}	2.805×10^{-2}

Table 3: The economic parameters [39].

Life span (years)	Annual output yield (MWh)	PV electricity tariff (RMB yuan/kWh)	Nominal interest rate	Inflation rate
25	75.515	0.6	4%	3%

of the BIPV system can be calculated as shown in Table 2, where the emission reduction density is calculated by the following formula [37]:

$$D_{er} = 0.1229 \times R_e, \tag{3}$$

where D_{er} is the emission reduction density while generating 1 kWh of power [kg/kWh] and R_e is the emission rate of pollutants per ton standard coal.

And the emission density is expressed using the following formula:

$$D_e = \frac{E_{c,l} \times D_{er}}{E_{g,l}}, \tag{4}$$

where D_e is the emission density [kg/kWh], $E_{c,l}$ is the total energy consumption of the BIPV system throughout the lifecycle [kWh], and $E_{g,l}$ is the total power generation of the entire life span [kWh]. The environmental cost of generation and the environmental benefits of emission reduction are calculated by the following, respectively:

$$\begin{aligned} C_g &= V_p \times D_e, \\ B_r &= R_e \times 0.1229 \times V_p, \end{aligned} \tag{5}$$

where C_g is the environmental cost of generating each kilowatt-hour of electricity [RMB/kWh], V_p is the environmental value of per ton of pollutants [RMB/kg], and B_r is the environmental benefit of emission reduction for each kilowatt-hour of electricity generated [RMB/kWh].

Because of the grid-connected BIPV system, the annual average emissions of CO_2, SO_2, NO_x, CO, particulates, and slag could be reduced by 1974 kg, 20.53 kg, 9.12 kg, 0.29 kg, 125.46 kg, and 34.22 kg, respectively. By calculation, as shown in Table 2, the environmental cost of the BIPV power generation system is RMB ¥0.01053 per kWh. Without the BIPV system, the environmental cost of power generation would be RMB ¥0.02805 per kWh through the thermal power, which means that the environmental cost could be reduced by RMB ¥0.01752 per kWh due to the benefit of the BIPV system. Given that most parts of China have been affected by fog and haze, the environmental benefits should be a major consideration once such BIPV systems are widely applied in China.

4.3. Economic Analysis. This section shows the procedure followed to calculate the payback time for the BIPV system. The net present value (NPV) is used and can be expressed as follows [6, 38]:

$$\text{NPV} = \sum_{t=0}^{p} (\text{CI} - \text{CO})(1 + i)^{-t}, \tag{6}$$

where CI is the cash inflows, CO is the cash outflows, i is the nominal interest rate, p is the life span of PV, in years, and t is the projected age limit. The higher the NPV, the better for the BIPV system, and when NPV is equal to 0, the payback period would be acquired. The initial cost of PV modules and the inverters in this system are RMB ¥20/Wp and RMB ¥1/Wp in the year of 2014, respectively. If including the supplying and fixing of steel framing, wires, and the data acquisition system, the total cost was equal to RMB ¥3,070,400. Fortunately, the public grant covered 80% of the total cost of the BIPV system, with an amount equal to RMB ¥2,500,000. The operation and maintenance (O&M) costs are approximately equal to 0.05%/year of the total cost. According to the real market prices and the expected electricity cost trend for the next years, the average commercial electric tariff for Kunming is expected to be RMB ¥0.6 (≈US $0.1) per kWh; the annual average saving in electricity charge is computed to be RMB ¥45,309 (≈US $7551.5). The other main economic parameters are shown in Table 3. On the basis of these considerations, the payback period is 15 years with government subsidy, and the NPV is RMB ¥359,347 at the end of a lifetime.

4.4. Energy Capacity Fraction of the Building. The building of the BIPV system in this study (Figure 1) is the Solar Energy

Table 4: The calculated results with different mounting tilt angles.

Tilt angles	90°	85°	80°	70°	60°
Annual power production (MWh)	68.685	75.515	82.359	95.617	107.311
EPBT (years)	10.32	9.38	8.60	7.41	6.60
Environmental cost of generation (RMB yuan/kWh)	1.158×10^{-2}	1.053×10^{-2}	0.965×10^{-2}	0.832×10^{-2}	0.741×10^{-2}
Payback period (years)	17	15	14	12	11
NPV (RMB yuan)	272,820	359,347	446,053	614,015	762,164

Research Institute in Yunnan Normal University; actually, much experimental equipment is placed in the building, including some high-power and high-energy-consuming facilities, for instance, the amorphous silicon thin-film solar cell production equipment, reheat furnaces, and wind tunnel testing facility. So it is meaningful to discuss the energy capacity fraction according to the Chinese national standard "Design standard for energy efficiency of public buildings" GB50189. Especially, because of the climate matter of Kunming, there is hardly any air-conditioning in the building. The lowest energy generation during the year is in June and July according to the results above; fortunately, this is the Chinese university summer vacation period, and so the energy consumption of the building is also the lowest. By the way, the previous research [40] found that due to the chimney effect of the PV curtain wall, the temperature of rooms covered by the PV curtain wall can keep warm without any heating facilities in winter; it can be regarded as another benefit of the BIPV to reduce the energy consumption of the building. Referring to the records of electricity consumption from 2014 up to now and comparing the annual energy consumption with annual cumulative power generation, it can be estimated that the average amount of energy consumed in the building is approximately 57,684 kWh each year in consideration of two long vacations and not any more new equipment, which is less than the amount produced from the photovoltaic system, which totaled 75,515 kWh. According to the definition of the capacity fraction (CF) in [41]

$$\mathrm{CF} = \frac{E_{\mathrm{PV}}}{E_{\mathrm{load}}}, \tag{7}$$

where CF means the energy capacity fraction of the building, E_{PV} is the annual electricity produced by the PV array, and E_{load} is the annual electricity consumed by the load. Therefore, the final energy capacity fraction is 130.91%, and the energy surplus is 17,831 kWh, which means that the goal of the zero-energy building is achieved. However, few facilities have not been put into full operation in the building, and obviously, the total energy consumption of the building could not be covered by the PV system at all in the future, while taking into account so many experimental facilities fully running. In order to improve the energy capacity fraction, it is possible to optimize the PV panel tilt angle or reduce the operating temperature in order to maximize the electricity production.

5. Discussion

As discussed above, it can be found that the energy payback time (EPBT), the environmental cost, the NPV, and so on, as well as all the performance parameters of the BIPV system are closely related to the amount of power generation. The higher the power generation, the more advantages the BIPV system has, and the main influencing factors to power generation are the solar radiation intensity and the efficiency of the PV modules. The efficiency of the semitransparent PV modules could not be as high as that of the traditional PV modules with opaque panels, since partial light transmission and building daylighting must be ensured, leading to certain areas in a module that have to be sacrificed. However, with the technologies developing, maybe the efficiency of the semitransparent PV modules used in a BIPV system would be improved significantly in the future. At present, the simple and efficient method to increase the power generation is to change the mounting tilt angles of the PV curtain wall. The yearly PV power generations have been calculated with different tilt angles by the TRNSYS models; the results are shown in Table 4.

With the mounting tilt angle decreasing from 90° to 60°, the annual power production increases from 68.685 MWh to 107.311 MWh gradually, because the PV modules could receive more solar radiation at the 60° tilt angle than at 90° for a full year. Though more power production can be achieved with a smaller tilt angle, it is impossible for a BIPV system, especially the PV curtain wall integrated with a building, to design a tilt angle smaller than 60 degrees, except in a rooftop PV system. The EPBT, environmental cost of generation, and payback period would reduce with the decrease in tilt angles, and it is beneficial for a designer to decide which angle would be chosen, according to different emphasis points. The payback period and the NPV also are affected by other important factors, such as the total cost of the BIPV system, the nominal rate of interest, and the rate of inflation. In this work, a lot of state subsidies are provided; thus, the payback period can be shortened and benefits can be gotten in its life span. If the nominal interest rate and the inflation rate change in the next 30 years and the government subsidy still remains to be RMB ¥2,500,000, the payback period would be extended. Moreover, the amount of government subsidies plays an essential part in the payback period and NPV of this kind of BIPV systems. The effect of different interest rates and the amount of government subsidy on the payback period and NPV is calculated in Table 5. If the government subsidy reduces to less than

TABLE 5: The variation of the payback period and NPV with the interest rate.

Government subsidy (RMB yuan)	2,500,000	2,500,000	2,500,000	2,200,000	2,400,000	2,600,000
Nominal interest rate	4.00%	4.50%	4.25%	4.00%	4.00%	4.00%
Inflation rate	3.00%	2.00%	1.50%	3.00%	3.00%	3.00%
Real rate of interest [39]	0.97%	2.45%	2.71%	0.97%	0.97%	0.97%
Payback period (years)	15	17	18	24	18	13
NPV (RMB yuan)	359,347	211,565	189,551	62,133	260,210	458,287

RMB ¥2,150,000, the cost of investment could not be recovered during the 25-year life span. Nevertheless, the environmental effects should be beyond the economic benefits; more and more, the similar BIPV system would be employed with the decrease in PV costs. And all of the results above show the comprehensive analysis procedure of a BIPV system, including the methods of evaluation and prediction, which can be referenced for system designing.

6. Conclusions

A study of the operational performance of a grid-connected building-integrated photovoltaic (BIPV) system, a 120 kWp monocrystalline, in Kunming, China, was conducted. The experimental results from October 2014 to September 2015 have been analyzed and have been compared with a TRNSYS model, which provides good agreement with the experimental results. Moreover, the TRNSYS model has been built to predict the amount of power generation of the BIPV system under the meteorological database of Meteonorm, and the annual electricity production of 75.515 MWh was estimated.

The PR of the BIPV system was calculated with the test data of the first year of installation (the year 2015), which was equal to 64.3%. However, according to the simulated data of the TRNSYS model during the BIPV system expecting a lifetime of 25 years, the average value of PR could be 60%. Based on the PR and LCA of the system, the EPBT is calculated to be 9.38 years, which shows that the BIPV system could still generate a substantial amount of electricity in its expected lifetime. Besides, the annual average emissions of CO_2, SO_2, NO_x, CO, particulates, and slag could be reduced by 1974 kg, 20.53 kg, 9.12 kg, 0.29 kg, 125.46 kg, and 34.22 kg, respectively, after the adoption of the BIPV power generation system, the environmental cost of which is RMB ¥0.01053 per kWh and with the environmental economic benefit of RMB ¥0.01752 per kWh. Consequently, the net present value (NPV) and the economic payback time of the BIPV system were estimated. The NPV is equal to approximately RMB ¥359,347 at the end of the life span, and the payback is 15 years. These results mean that the system has both good environmental and economic benefits. Not only that, the BIPV system is also effective in helping the building to achieve the zone energy consumption goal; in other words, nearly all of the existing building's energy consumption could be covered by the power generation of the installed BIPV system so far, even a surplus. The results can be used as the reference for future studies related to improving the performance of the BIPV system or be effectively extended to the investigation of other BIPV systems.

Conflicts of Interest

The authors declare that they have no conflicts of interest.

Acknowledgments

This work is supported by a grant from the Yunnan Province Science Foundation for Youths (2014FD015).

References

[1] *IEA World Energy Outlook 2013*, International Energy Agency, 2013.

[2] L. Pérez-Lombard, J. Ortiz, and C. Pout, "A review on buildings energy consumption information," *Energy and Buildings*, vol. 40, no. 3, pp. 394–398, 2008.

[3] X. Kong, S. Lu, P. Gao, N. Zhu, W. Wu, and X. Cao, "Research on the energy performance and indoor environment quality of typical public buildings in the tropical areas of China," *Energy and Buildings*, vol. 48, pp. 155–167, 2012.

[4] S. Kadoshin, T. Nishiyama, and T. Ito, "The trend in current and near future energy consumption from a statistical perspective," *Applied Energy*, vol. 67, no. 4, pp. 407–417, 2000.

[5] M. J. Ritzen, Z. A. E. P. Vroon, R. Rovers, and C. P. W. Geurts, "Comparative performance assessment of a non-ventilated and ventilated BIPV rooftop configurations in the Netherlands," *Solar Energy*, vol. 146, pp. 389–400, 2017.

[6] W. Wang, Y. Liu, X. Wu et al., "Environmental assessments and economic performance of BAPV and BIPV systems in Shanghai," *Energy and Buildings*, vol. 130, pp. 98–106, 2016.

[7] L. Y. Seng, G. Lalchand, and G. M. Sow Lin, "Economical, environmental and technical analysis of building integrated photovoltaic systems in Malaysia," *Energy Policy*, vol. 36, no. 6, pp. 2130–2142, 2008.

[8] L. Aelenei, R. Pereira, A. Ferreira, H. Gonçalves, and A. Joyce, "Building integrated photovoltaic system with integral thermal storage: a case study," *Energy Procedia*, vol. 58, pp. 172–178, 2014.

[9] A. G. Hestnes, "Building integration of solar energy systems," *Solar Energy*, vol. 67, no. 4-6, pp. 181–187, 1999.

[10] G. C. Bakos and M. Soursos, "Technical feasibility and economic viability of a grid-connected PV installation for low cost electricity production," *Energy and Buildings*, vol. 34, no. 7, pp. 753–758, 2002.

[11] B. Norton, P. C. Eames, T. K. Mallick et al., "Enhancing the performance of building integrated photovoltaics," *Solar Energy*, vol. 85, no. 8, pp. 1629–1664, 2011.

[12] D. H. W. Li, T. N. T. Lam, W. W. H. Chan, and A. H. L. Mak, "Energy and cost analysis of semi-transparent photovoltaic in office buildings," *Applied Energy*, vol. 86, no. 5, pp. 722–729, 2009.

[13] D. H. W. Li, S. K. H. Chow, and E. W. M. Lee, "An analysis of a medium size grid-connected building integrated photovoltaic (BIPV) system using measured data," *Energy and Buildings*, vol. 60, pp. 383–387, 2013.

[14] T. Hong, C. Koo, J. Park, and H. S. Park, "A GIS (geographic information system)-based optimization model for estimating the electricity generation of the rooftop PV (photovoltaic) system," *Energy*, vol. 65, pp. 190–199, 2014.

[15] B. Celik, E. Karatepe, S. Silvestre, N. Gokmen, and A. Chouder, "Analysis of spatial fixed PV arrays configurations to maximize energy harvesting in BIPV applications," *Renewable Energy*, vol. 75, pp. 534–540, 2015.

[16] F. Cucchiella and I. D'Adamo, "Estimation of the energetic and environmental impacts of a roof-mounted building-integrated photovoltaic systems," *Renewable and Sustainable Energy Reviews*, vol. 16, no. 7, pp. 5245–5259, 2012.

[17] H. Radhi, "Energy analysis of façade-integrated photovoltaic systems applied to UAE commercial buildings," *Solar Energy*, vol. 84, no. 12, pp. 2009–2021, 2010.

[18] R. Eke and A. Senturk, "Monitoring the performance of single and triple junction amorphous silicon modules in two building integrated photovoltaic (BIPV) installations," *Applied Energy*, vol. 109, pp. 154–162, 2013.

[19] S. Wittkopf, S. Valliappan, L. Liu, K. S. Ang, and S. C. J. Cheng, "Analytical performance monitoring of a 142.5kW$_p$ grid-connected rooftop BIPV system in Singapore," *Renewable Energy*, vol. 47, pp. 9–20, 2012.

[20] S. Kang, T. Hwang, and J. T. Kim, "Theoretical analysis of the blinds integrated photovoltaic modules," *Energy and Buildings*, vol. 46, pp. 86–91, 2012.

[21] S. Li, J. Joe, J. Hu, and P. Karava, "System identification and model-predictive control of office buildings with integrated photovoltaic-thermal collectors, radiant floor heating and active thermal storage," *Solar Energy*, vol. 113, pp. 139–157, 2015.

[22] *Solar Energy Laboratory User Manual: TRNSYS 17 a TRaNsient SYstem Simulation Program - Volume 4 - Mathematical Reference*, vol. 4, 2009.

[23] B. Quesada, C. Sánchez, J. Cañada, R. Royo, and J. Payá, "Experimental results and simulation with TRNSYS of a 7.2 kWp grid-connected photovoltaic system," *Applied Energy*, vol. 88, no. 5, pp. 1772–1783, 2011.

[24] M. Kumar and A. Kumar, "Performance assessment and degradation analysis of solar photovoltaic technologies: a review," *Renewable and Sustainable Energy Reviews*, vol. 78, pp. 554–587, 2017.

[25] N. Aste, C. Del Pero, and F. Leonforte, "The first Italian BIPV project: case study and long-term performance analysis," *Solar Energy*, vol. 134, pp. 340–352, 2016.

[26] S. Li Causi, S. Castello, and F. De Lia, "ENEA role in the Italian roof-top programme," in *Proceedings of 3rd World Conference on Photovoltaic Energy Conversion, 2003*, vol. 3, pp. 2644–2647, Osaka, Japan, May 2003.

[27] N. Aste, C. Del Pero, and F. Leonforte, "PV technologies performance comparison in temperate climates," *Solar Energy*, vol. 109, pp. 1–10, 2014.

[28] A. F. Sherwani and J. A. Usmani, "Life cycle assessment of solar PV based electricity generation systems: a review," *Renewable and Sustainable Energy Reviews*, vol. 14, no. 1, pp. 540–544, 2010.

[29] R. Kannan, K. C. Leong, R. Osman, and H. K. Ho, "Life cycle energy, emissions and cost inventory of power generation technologies in Singapore," *Renewable and Sustainable Energy Reviews*, vol. 11, no. 4, pp. 702–715, 2007.

[30] J. Peng, L. Lu, and H. Yang, "Review on life cycle assessment of energy payback and greenhouse gas emission of solar photovoltaic systems," *Renewable and Sustainable Energy Reviews*, vol. 19, pp. 255–274, 2013.

[31] M. J. (Mariska) de Wild-Scholten, "Energy payback time and carbon footprint of commercial photovoltaic systems," *Solar Energy Materials & Solar Cells*, vol. 119, pp. 296–305, 2013.

[32] Y. Jia, J. Wang, Z. Han, Y. Pang, and P. An, "Analysis on environmental load of wind, PV and coal-fired power generation based on life cycle assessment," *Journal of Chinese Society of Power Engineering*, vol. 36, pp. 1000–1009, 2016.

[33] L. Pei, *The study on eco-design of high-rise residential buildings in Wuhan based on energy simulation and life cycle assessment*, Huazhong University of Science and Technology, Wuhan, China, 2015.

[34] Y. Li, *Life cycle assessment of crystalline silicon modules in China*, Shanghai Jiao Tong University, Shanghai, China, 2015.

[35] J. X. Zhang and G. F. Zhu, "Comparative studies of photolvoltaic power generation and coal-fired power generation base on life cycle assessment," *Environmental Science and Management*, vol. 39, pp. 86–90, 2014.

[36] Y. He, J. Zhang, Q. Wang, and J. Xie, "Environmental impact assessment on multicrystalline wafer production based on life cycle assessment method," *Sichuan Environment*, vol. 32, pp. 83–90, 2013.

[37] Y. Sun, R. Wang, L. Xiao, J. Liu, Y. Yu, and X. Zhuang, "Economical and environmental analysis of grid-connected photovoltaic system in China," *China Population, Resources and Environment*, vol. 21, pp. 88–94, 2011.

[38] Q. Zhu, L. Si, and T. Jiang, "Economical and environmental analysis of building photovoltaic systems with different installation styles," *Acta Energiae Solaris Sinica*, vol. 33, pp. 24–29, 2012.

[39] G. Evola and G. Margani, "Renovation of apartment blocks with BIPV: energy and economic evaluation in temperate climate," *Energy and Buildings*, vol. 130, pp. 794–810, 2016.

[40] Y. F. Wang, R. H. E. Hassanien, M. Li, G. X. Xu, and X. Ji, "An experimental study of building thermal environment in building integrated photovoltaic (BIPV) installation," *Bulgarian Chemical Communications*, vol. 48, pp. 158–164, 2016.

[41] A. Gallo, B. T. Molina, M. Prodanovic, J. G. Aguilar, and M. Romero, "Analysis of net zero-energy building in Spain. Integration of PV, solar domestic hot water and air-conditioning systems," *Energy Procedia*, vol. 48, pp. 828–836, 2014.

14

Performance Optimization of Unglazed Nanofluid Photovoltaic/Thermal System: Energy and Exergy Analyses

M. Imtiaz Hussain[1] **and Jun-Tae Kim**[2]

[1]*Green Energy Technology Research Center, Kongju National University, Cheonan, Republic of Korea*
[2]*Department of Architectural Engineering, Kongju National University, Cheonan, Republic of Korea*

Correspondence should be addressed to Jun-Tae Kim; jtkim@kongju.ac.kr

Academic Editor: Kei Ohkubo

The focus of this paper is to predict the transient response of a nanoengineered photovoltaic thermal (PV/T) system in view of energy and exergy analyses. Instead of a circular-shaped receiver, a trapezoidal-shaped receiver is employed to increase heat transfer surface area with photovoltaic (PV) cells for improvement of heat extraction and thus achievement of a higher PV/T system efficiency. The dynamic mathematical model is developed using MATLAB® software by considering real-time heat transfer coefficients. The proposed model is validated with experimental data from a previous study. Negligible discrepancies were found between measured and predicted data. The validated model was further investigated in detail using different nanofluids by dispersing copper oxide (CuO) and aluminum oxide (Al_2O_3) in pure water. The overall performance of the nanoengineered PV/T system was compared to that of a PV/T system using water only, and optimal operating conditions were determined for maximum useful energy and exergy rates. The results indicated that the CuO/water nanofluid has a notable impact on the energy and exergy efficiencies of the PV/T system compared to that of Al_2O_3/water nanofluid and water only cases.

1. Introduction

One of the major challenges facing the world today is the increase in concentrations of greenhouse gases in the environment, which is generally considered as being caused by the substantial use of fossil fuels. Due to the limited availability of fossil fuels and rising air pollution, the replacement of fossil resources by renewable energy has become inevitable. Among various types of renewable energy sources, sun-based energy is the most promising in terms of accessibility, affordability, and cleanliness. The direct conversion of sunlight into electrical energy can be carried out with photovoltaic (PV) cells [1]. Continuous exposure of PV surfaces to the solar radiation in high ambient temperature results in a decrease in device electrical efficiency. Therefore, accumulated solar heat across the PV module is removed through circulation of heat transfer fluids [2, 3]. This integration of a PV module and a thermal collector is called a photovoltaic/thermal (PV/T) system; such a system can viably generate heat and electricity simultaneously.

Over the years, a substantial number of studies have been carried out to find a suitable heat transfer fluid that can effectively extract heat from the PV module. Various heat transfer fluids such as water, glycol, ethylene, and acetone have been tried with the purpose of acquiring high collector efficiency [4, 5], but the low thermal conductivity values of these fluids always remain main hurdles to achieving this goal [6]. To cope with this problem, nanofluid as a heat transfer fluid has been introduced. A colloidal solution (nanofluid) can be obtained by dispersion of nanoparticles in base fluid. Generally, the thermal conductivity and heat transfer coefficient of a nanofluid are higher than those of other available fluids [7]. Therefore, colloidal solutions are expected to improve the overall performance of the PV/T system compared to the case of using water only. Tiwari et al. [8] discussed the effects of volume concentration of Al_2O_3 in the base fluid and variations in the mass flow rate of nanofluid on the performance of a flat plate solar collector. Rejeb et al. [9] presented a combined experimental and numerical study to evaluate the influence of alumina and copper nanoparticles

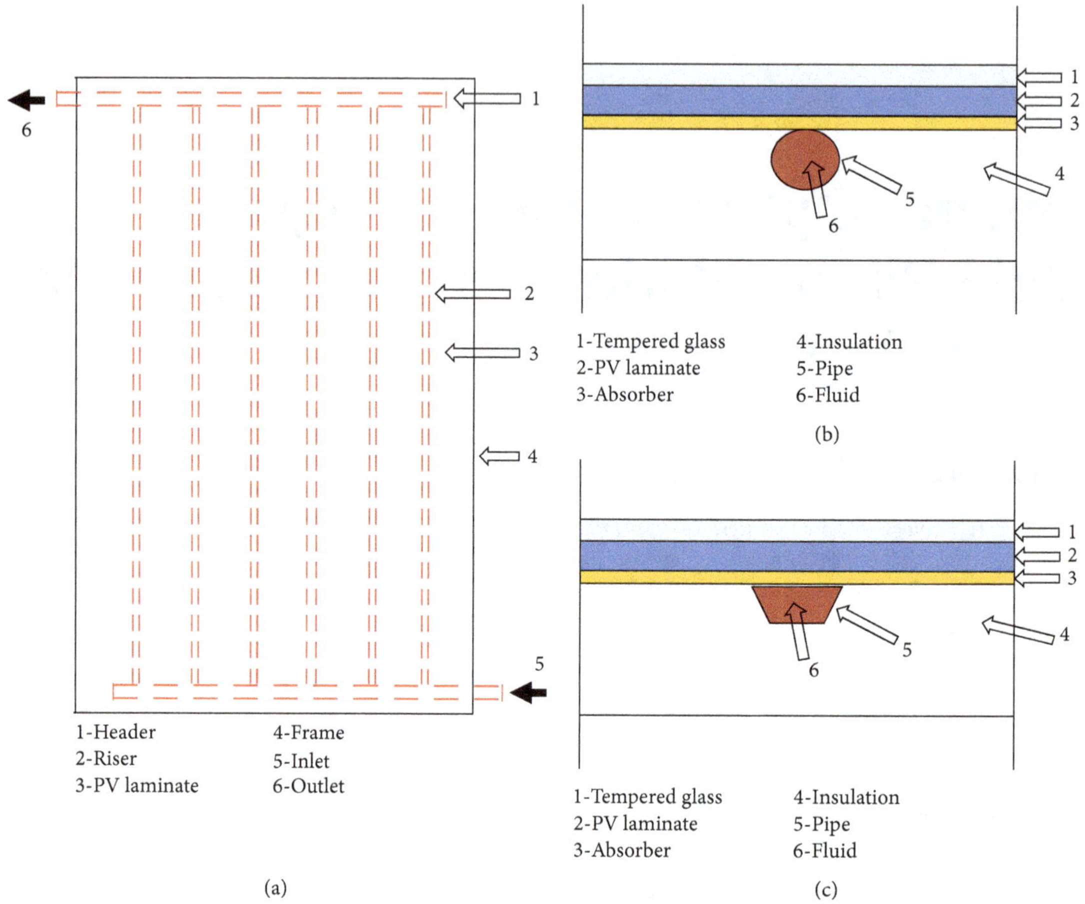

FIGURE 1: PV/T collector: (a) schematic details, (b) cross section with a circular receiver, and (c) cross section with a trapezoidal receiver.

in different base fluids (water and ethylene glycol) on overall PV/T system performance. It was observed that water as a base fluid gives a higher performance than does ethylene glycol. Hasan et al. [10] proposed a jet impingement cooling system for the PV/T system, in which nozzles directly inject nanofluid (coolant) to the back surface of the PV module. Using jet array nanofluid impingement in the PV/T system, the maximum power of the proposed collector increased by 62.5% in comparison with the conventional PV module. Simultaneous application of nanofluid as a coolant and an organic paraffin wax as a phase change material (PCM) on the exergy and energy efficiencies of the PV/T system is investigated experimentally by Hosseinzadeh et al. [11]. According to the results, the PCM in the nanofluid-based PV/T system immensely enhances the output thermal exergy of the collector. Shahsavar et al. [12] analyzed the exergy performance of a naturally ventilated PV/T air collector in which a thin metal sheet is introduced in the middle of an air passage to improve the heat exchanger performance and thus achieve lower PV temperature.

A considerable number of studies available in the literature indicate that the exergy and energy analyses of the PV/T system have been performed considering steady-state heat transfer conditions. To the best of the authors' knowledge, a few attempts have been made to calculate the comprehensive transient behavior of various PV/T components for the exergy analysis. As pointed out by Jones and Underwood [13], a steady-state model cannot accurately predict the PV temperature responses under rapidly fluctuating solar radiation. To address the aforementioned existing research gaps, a dynamic mathematical model of a PV/T system using nanofluid as a coolant is developed and validated using experimental data from the literature. Interdependence transient processes of the PV/T system were predicted using MATLAB software considering real-time calculation of heat transfer coefficients. To maximize the energy and exergy efficiencies, an optimal mode of operation is determined using different heat transfer fluids and geometrical conditions.

2. Collector Design and Simulation Details

In this study, a sheet and tube PV/T is considered that comprises an off-the-shelf PV module, a thermal absorber, a fluid tube, and thermal insulation. A schematic diagram and a cross-section view of the PV/T system are shown in Figure 1. A monocrystalline silicon cell with ideal conversion efficiency of 15–20% is employed in the PV module. The PV module and the absorber plate are fixed together by means of an adhesive. The fluid carriers (copper tubes) are bonded to the absorber plate at equal spacing in a parallel-flow arrangement throughout the plate width. The absorber plate serves as heat exchanger fins for the fluid tubes. The top and bottom

TABLE 1: Thermophysical properties of nanoparticles.

Nanoparticles	Properties Specific heat (J/kg·K)	Thermal conductivity (W/m·K)	Density (kg/m^3)
Copper oxide	551	32.9	6310
Aluminum oxide	773	30	3890
Pure water	4179	0.613	997.1

ends of the tubes are connected with common headers to maintain balanced fluid flow and distance between adjacent tubes. The sides and the rear surface of the panel are covered with the thermal insulation to reduce losses due to absorbed solar energy.

Nanoparticles of aluminum oxide (Al_2O_3) and copper oxide (CuO) with a fixed concentration of 0.7 wt% in the base fluid (pure water) were considered in this study. Temperature-dependent thermophysical properties for all utilized heat transfer fluids were determined so that, compared to water as a thermal fluid, the improvements in the heat transfer rate with nanofluid could be evaluated. The properties of the employed heat transfer fluids are shown in Table 1.

Explicit energy flow analysis was performed by modifying the dynamic energy balance equation for the PV/T system given by Chow [14]. The MATLAB ODE solver was used to derive the transient temperatures from the instantaneous energy balance equations. Using the finite difference method [15], a control volume (node) was created, within which the energy and mass balance equations were applied. The parameters considered for the simulation are given in Table 2. Each component of the PV/T system is denoted by node. Therefore, the given collector model consists of a total of five nodes. Figures 1(b) and 1(c) show the temperature nodes for the different PV/T collector components.

3. Mathematical Model

A mathematical model was developed to describe the interdependence of the temperature responses across the different components of the PV/T system. The heat balance across different components of the PV/T system was derived by considering the vicinity of a single copper tube, as all the tubes are arranged in parallel fashion with fixed spacing between them. As indicated by Chow [14], in the parallel-tube arrangement, the temperature and fluid flow rate conditions in all tubes can be assumed to be the same. For the model development, the following assumptions are considered:

(i) The material properties of all components are considered to be constant.

(ii) Considering the same fluid flow rate obtains through all tubes.

(iii) The edge losses are assumed to be negligible.

(iv) The heat transfer coefficients are calculated in real time.

TABLE 2: Data for simulation.

PV module	Length and width	1.62 m and 0.98 m
	Absorptivity (α_p)	0.9
	Emissivity (ε_p)	0.88
	Specific heat (C_p)	900 J/kg·K
	Packing factor	0.8
	Temperature coefficient (β_r)	0.0045/°C
	Reference PV efficiency (%)	17.3
Copper tube	Inner diameter (D_i)	0.008 m
	Thickness (δ_t)	0.0012 m
	Specific heat (C_t)	903 J/kg·K
	Density (ρ_t)	2702 kg/m^3
	Number of tubes	9
	Tube spacing	0.11 m
Absorber plate	Density (ρ_b)	2702 kg/m^3
	Emissivity (ε_b)	0.87
	Thermal conductivity (K_b)	310 W/m·K
	Thickness (δ_b)	0–1 mm
Insulation	Density (ρ_i)	20 kg/m^3
	Specific heat (C_i)	670 J/kg·K
	Thermal conductivity (K_i)	0.034 W/m·K

3.1. Energy Balance for PV Cells

$$M_p C_p \left(\frac{dT_p}{dt}\right) = G\alpha_p - E - h_{wind} A_{pa} (T_p - T_a) - h_{ps} A_{ps} (T_p - T_s) - h_{pb} A_{pb} (T_p - T_b) - h_{pt} A_{pt} (T_p - T_t), \quad (1)$$

$$E = GP\eta_{el}, \quad (2)$$

$$\eta_{el} = \eta_r [1 - \beta_r (T_p - T_r)], \quad (3)$$

where E is used for electrical power. P is the packing factor and η_e is the electrical efficiency. T_r, η_r, and β_r are the reference PV cell temperature, efficiency, and temperature coefficient, respectively. h_{wind} is the convective heat transfer coefficient due to wind [16]. G and α_p are the solar radiation and absorptivity of the PV cells, respectively; M_p, C_p, and T_p are the mass, specific heat, and temperature of the PV cells, respectively; T_b, T_t, and T_a are the absorber plate, tube, and ambient air temperature, respectively. h_{pa} is the radiative heat transfer coefficient between PV and the sky (T_s). h_{pb} and h_{pt} are the conductive heat transfer coefficients between PV and the absorber and PV and the tube, respectively.

3.2. Energy Balance for Absorber Plate

$$M_b C_b \left(\frac{dT_b}{dt}\right) = h_{pb} A_{pb} (T_p - T_b) - h_{bt} A_{bt} (T_b - T_t) - h_{bi} A_{bi} (T_p - T_i), \quad (4)$$

where T_i is the insulation temperature. M_b and C_b are the mass and specific heat of the absorber plate, respectively. h_{bt} and h_{bi} are the conductive heat transfer coefficients between the absorber and the tube and the absorber and the insulation, respectively.

3.3. Energy Balance for Tube

$$M_t C_t \left(\frac{dT_t}{dt}\right) = h_{pt} A_{pt} (T_p - T_t) + h_{bt} A_{bt} (T_b - T_t) - A_{t,fl} h_{t,fl} (T_t - T_{fl}) - A_{ti} h_{ti} (T_t - T_i), \tag{5}$$

where T_{fl} is the circulating fluid temperature. M_t and C_t are the mass and specific heat of the tube, respectively. h_{ti} is the conductive heat transfer coefficient between the tube and the insulation.

3.4. Energy Balance for Insulation

$$M_i C_i \left(\frac{dT_i}{dt}\right) = h_{bi} A_{bi} (T_p - T_i) - A_{ti} h_{ti} (T_t - T_i) - h_{ia} A_{ia} (T_i - T_a), \tag{6}$$

where M_i and C_i are the mass and specific heat of the insulation material, respectively. h_{ia} is the convective heat transfer coefficient between the insulation and the ambient air.

3.5. Energy Balance for Fluid

$$M_{fl} C_{fl} \left(\frac{dT_{fl}}{dt}\right) = \dot{m}_{fl} C_{fl} (T_o - T_{in}) + A_{t,fl} h_{t,fl} (T_t - T_{fl}), \tag{7}$$

where $\dot{m}_{fl}$ is the fluid mass flow rate. T_{in} and T_o are the inlet and outlet temperatures of the fluid, respectively. $h_{t,fl}$ is the convective heat transfer coefficient between the tube and the fluid.

3.6. Heat Transfer Coefficients. The convection heat transfer coefficient between PV laminate and ambient air due to wind can be written as

$$h_{wind} = 3u_a + 2.8, \tag{8}$$

where h_{wind} is the convection heat transfer coefficient due to wind and u_a is the wind velocity. The radiation heat transfer coefficient between the PV laminate and the sky is given as follows [17]:

$$h_{ps} = \varepsilon_p \sigma (T_p + T_s) (T_p^2 + T_s^2), \tag{9}$$

where ε_p is the emissivity of the PV plate and σ is the Stefan-Boltzman constant. The conduction heat transfer coefficient between the PV laminate and the absorber plate can be calculated using the following equation:

$$h_{pb} = \frac{k_{ad}}{\delta_{ad}}, \tag{10}$$

where k_{ad} and δ_{ad} are the thermal conductivity and thickness, respectively, of the adhesive layer. The conduction heat transfer coefficient between the absorber plate and insulation can be determined as follows [9]:

$$h_{bi} = \frac{2k_i}{\delta_i}, \tag{11}$$

where k_{ad} and δ_{ad} are the thermal conductivity and thickness, respectively, of the insulation. The conductive heat transfer coefficient between the PV laminate and the tube can be expressed using the following correlation [18]:

$$h_{pt} A_{pt} = \frac{\delta_i L}{(x_p/2k_p) + (\delta_{ad}\delta_{ad}/k_{ad} D_o)}, \tag{12}$$

$$x_p = \frac{W}{4}, \tag{13}$$

where k_p is the thermal conductivity of the PV plate and W is the tube spacing. D_o and L are the outer diameter and length, respectively, of the tube. The conductive heat transfer coefficient between the absorber plate and the tube can be expressed by the following correlation [14]:

$$h_{bt} = \frac{2k_b}{x_b}, \tag{14}$$

$$x_b = \frac{(W - D_o)}{4}, \tag{15}$$

where k_b is the thermal conductivity of the absorber plate.

The maximum output relative to the ambient temperature can be achieved using exergy analysis. The thermal, electrical, and overall exergy efficiencies given by Yazdanifard et al. [19] for a solar hybrid collector were modified and used for the simulation.

$$\eta_{ex,th} = \frac{\dot{E}_{th}}{\dot{E}_{in}} = \frac{\dot{m}_{fl} C_{fl} (T_o - T_{in})(1 - 293/(293 + (T_o - T_a)))}{\dot{E}_{in}}, \tag{16}$$

$$\eta_{ex,el} = \frac{\dot{E}_{el}}{\dot{E}_{in}} = \frac{GP\eta_r [1 - \beta_r (T_p - T_r)]}{\dot{E}_{in}}, \tag{17}$$

$$\eta_{ex,PVT} = \frac{\dot{E}_{th} + \dot{E}_{el}}{\dot{E}_{in}} = \eta_{ex,th} + \eta_{ex,el}, \tag{18}$$

where η_r, T_r, and β_r are the reference efficiency, temperature, and coefficient, respectively; $\eta_{ex,th}$, $\eta_{ex,el}$, and $\eta_{ex,PVT}$ are the thermal, electrical, and overall exergy efficiencies, repectively; $\dot{E}_{th}$ and $\dot{E}_{el}$ are the thermal and electrical exergy rates, respectively. Exergy of the absorbed solar radiation can be calculated using the following expression [19, 20]:

$$\dot{E}_{in} = \left[1 + \frac{1}{3}\left(\frac{T_a}{T_{sun}}\right)^4 - \frac{4T_a}{3T_{sun}}\right] GA_c, \tag{19}$$

where T_{sun} and A_c are the sun temperature and collector area, respectively. The equivalent thermal efficiency can be calculated as

$$\eta_{PVT} = \eta_{th} + \frac{\eta_{el}}{\eta_{pp}}, \quad (20)$$

where η_{PVT}, η_{el}, η_{th}, and η_{pp} are the equivalent thermal efficiency, the electrical and thermal efficiency, and the electrical power generation efficiency of the power plant, respectively. Model calibration for short-term performance evaluation is performed using root mean square error (RMSE) analysis and other statistical indicators, as follows [21, 22]:

$$\text{RMSE}_1 = \left(\frac{1}{n}\sum_{i=1}^{n} d_i^2\right)^{1/2}, \quad (21)$$

$$\text{RMSE}_2 = \left(\frac{1}{n}\sum_{i=1}^{n}\left(\frac{d_i}{Y_i}\right)^2\right)^{1/2}, \quad (22)$$

$$r^2 = \left[\frac{n\left(\sum_{i=1}^{n} X_i Y_i\right) - \left(\sum_{i=1}^{n} X_i\right)\left(\sum_{i=1}^{n} Y_i\right)}{\sqrt{\left[n\sum_{i=1}^{n} X_i^2 - \left(\sum_{i=1}^{n} X_i\right)^2\right]\left[n\sum_{i=1}^{n} Y_i^2 - \left(\sum_{i=1}^{n} Y_i\right)^2\right]}}\right]^2, \quad (23)$$

where d_i is the difference between the ith predicted and measured values, and n is the number of data points. Y_i and X_i are the ith measured and predicted values.

3.7. Model Validation. Authentication of the proposed dynamic model is carried out by comparing predicted results with the published test data of the uncovered PV/T system given by Rejeb et al. [9]. The predicted and measured values of the mean nanofluid temperature were found to be 42.18°C and 41.73°C, respectively. To further determine the reliability, the developed model is validated through comparing the predicted PV cell temperature and the experimental data. The daily PV cell temperature measured by Rommel et al. [23] and predicted results from the mathematical model are presented in Figure 2. For comparison purposes, similar design and operational parameters were considered for each case, as suggested by [9, 23]. It can be seen that the predicted results matched well with the experimental data. There was no apparent deviation or bias between the predicted and the measured results. For model calibration, different statistical indicators were employed [18]. Using RMSE1, RMSE2, and R-squared indicators, the obtained values were found to be 0.129, 0.014, and 0.303, respectively. A conclusion can be drawn from the above comparison that the suggested model is reliable at predicting the performance of an unglazed PV/T system even under extreme weather fluctuations.

4. Results and Discussions

It is important to study the influence of nanoparticle concentration on the thermophysical properties, when attempting to evaluate the performance of a nanofluid-based PV/T system. Figure 3 shows the thermal conductivity ratio (k_{nf}/k_{bf}) and

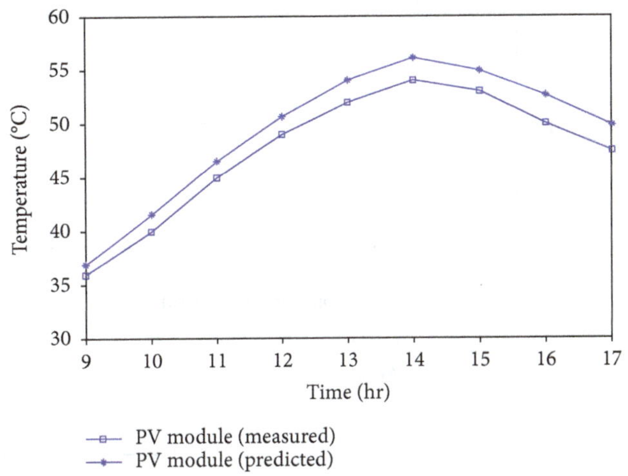

FIGURE 2: Comparison of measured and predicted PV temperature.

density ratio (ρ_{nf}/ρ_{bf}) enhancement of CuO nanofluid with increasing particle weight fraction. nf and bf stand for nanofluid and base fluid, respectively. The results for Al_2O_3 nanofluid with increasing weight fraction of Al_2O_3 also exhibit a similar trend. However, from the comparison of the thermophysical properties of CuO and Al_2O_3, it is anticipated that a slightly better performance can be obtained when CuO/water is used instead of Al_2O_3/water. On the contrary, an increase in volume fraction of CuO and Al_2O_3 results in a decrease in the marginal specific heat ratio (C_{nf}/C_{bf}) of both nanofluids. The typical value of thermal conductivity for CuO, Al_2O_3, and water are 32.9 W/m·K, 30 W/m·K, and 0.613 W/m·K, respectively. Therefore, colloidal solutions can be anticipated to exhibit notably improved thermophysical properties relative to pure water [24].

The pumping power was calculated from (24) for the trapezoidal and circular receivers and plotted against different concentrations of nanofluids. Figure 4 shows that pressure drop and thus pumping power increase with increasing concentration of the nanofluids. It is observed from the results that pumping powers for both nanofluids are nearly the same. However, CuO nanofluid with a trapezoidal receiver shows the maximum pumping power required than does Al_2O_3 nanofluid with trapezoidal and circular receivers. Compared to other cases considered in this study, the marginally higher pumping power for CuO nanofluid was probably due to the high viscosity [25] and noncircular shaped of the receive. The pumping power can be expressed as follows:

$$\text{Pumping power} = \left(\frac{\dot{m}_{fl}}{\rho_{fl}}\right) * \Delta p, \quad (24)$$

where ρ_f is the density of circulating fluid, and nanofluid density can be defined as [26]

$$\rho_{nf} = \rho_{np}(\phi) + \rho_{bf}(1-\phi), \quad (25)$$

where ϕ is the concentration of nanoparticles into the base

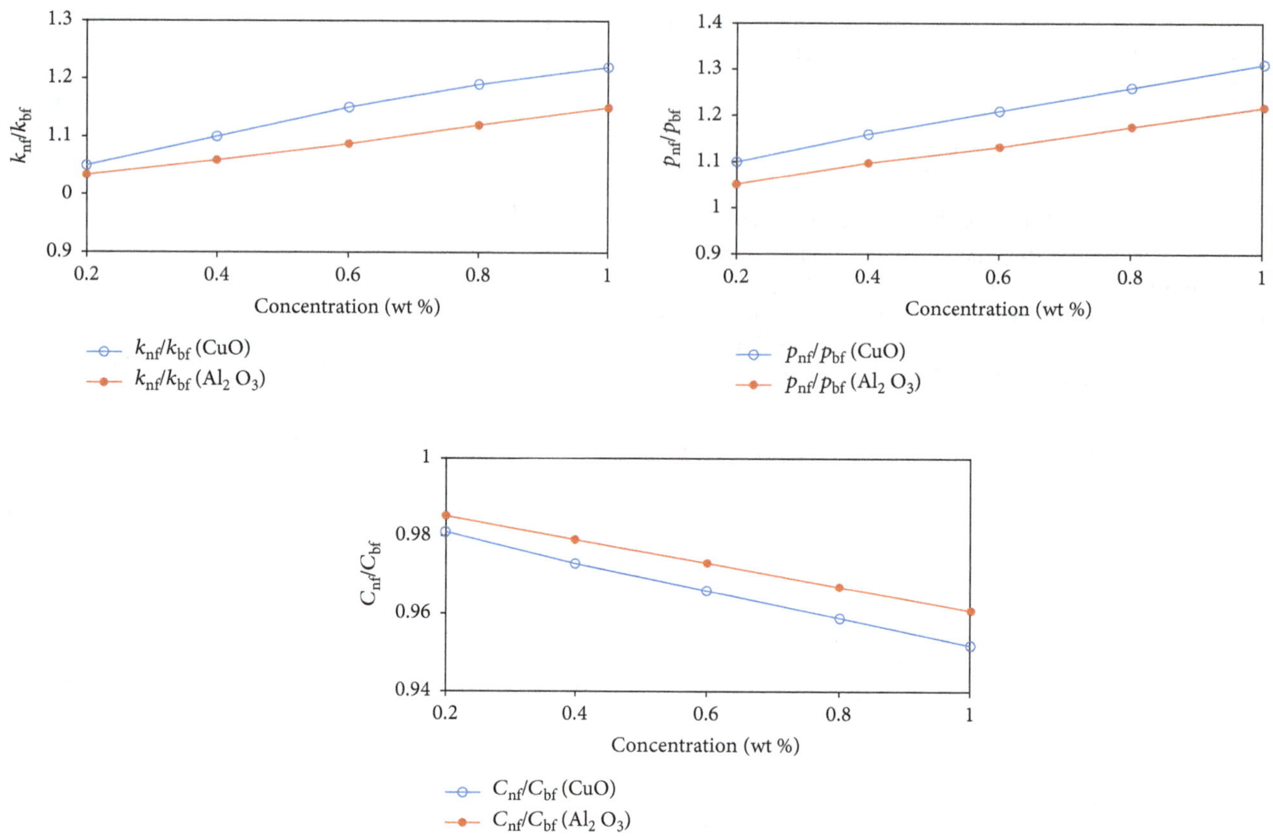

FIGURE 3: Variations of thermal physical properties of Al_2O_3/water and CuO/water with different concentrations.

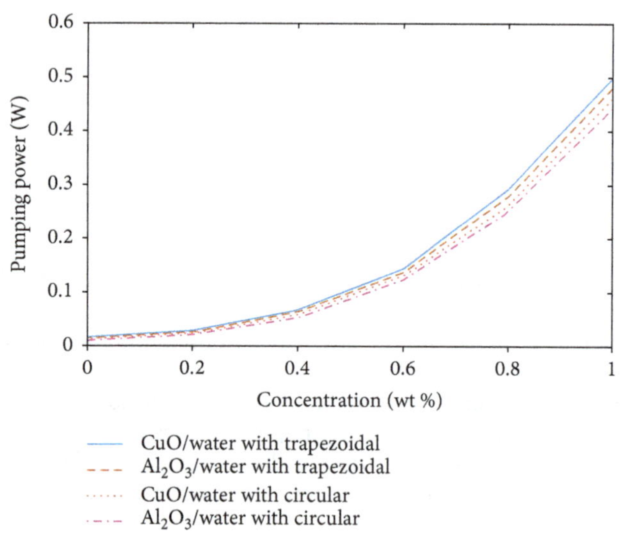

FIGURE 4: Variation of pumping power with the influence of nanoparticle concentration.

fluid. Pressure drop (Δp) can be calculated using the following correlation [27]:

$$\Delta p = f \frac{\rho_{fl} V^2}{2} \frac{L}{D_i} + K \frac{\rho_{fl} V^2}{2}, \tag{26}$$

where K is the loss coefficient taken according to [27] and V is the fluid velocity in m/s. The friction factor (f) for the laminar and turbulent flow can be expressed as

(1) For laminar flow

$$f = \frac{64}{R_e}. \tag{27}$$

(2) For turbulent flow

$$f = \frac{0.079}{R_e^{0.25}}. \tag{28}$$

R_e is the Reynolds number.

In a tube-and-fin type PV/T collector, the role of the absorber plate, which is sandwiched between the PV module and the fluid pipe, cannot be neglected in the heat transfer process. In fact, variations in the thickness of the absorber plate directly influence the overall performance of the PV/T system. Figure 5 depicts the variations of the thermal and electrical efficiencies of the PV/T system for different thicknesses of the absorber plate. Decreasing the thickness of the absorber plate leads to a decrease in the thermal resistance between the PV module and the circulating fluid. For this reason, the thermal and electrical efficiencies increased. Heat transfer occurs at a faster rate across the thinner absorber plate than it does across the thicker plate. It is worth noting

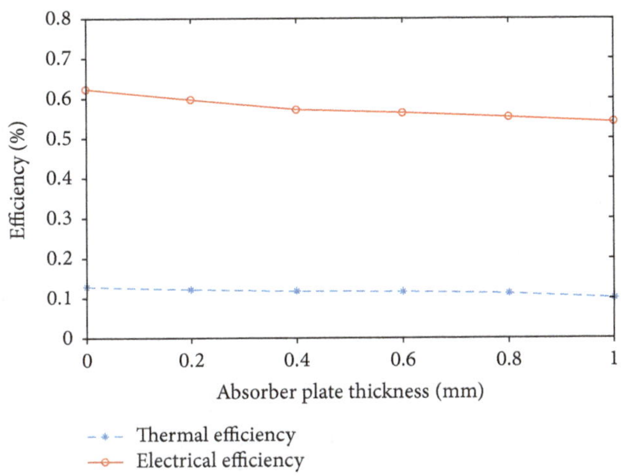

FIGURE 5: Variations of thermal and electrical efficiency with absorber thickness.

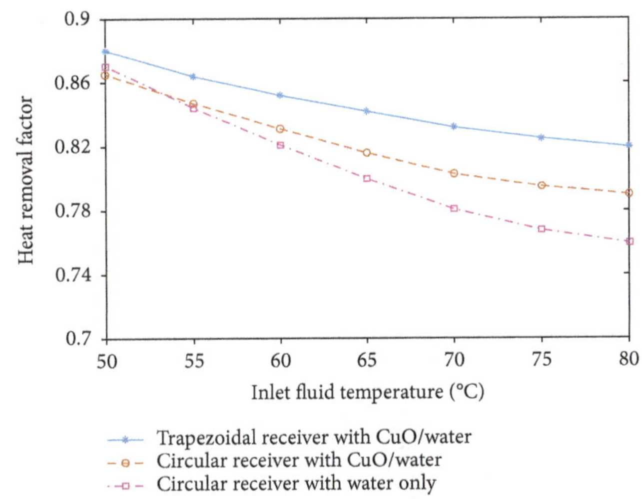

FIGURE 6: Variations of heat removal factor with inlet fluid temperature using different receiver shapes and fluids.

that the PV/T system without absorber plate (which means zero thickness) shows better overall performance than the PV/T with the absorber plate. Increases of the thermal and electrical efficiency are due to direct contact of the fluid-carrying tubes with the PV rear surface.

Figure 6 shows the relationship between the heat removal factor and the fluid inlet temperature, obtained using different heat transfer fluids and shapes of receivers. For purposes of comparison, the similar surface areas of both the circular- and trapezoidal-shaped receivers were considered. As can be seen, when the inlet fluid temperature increased, the heat removal factors decreased for all cases, independent of the receiver shapes and fluid types. However, the decrease in heat removal factor with the trapezoidal-shaped receiver is much smaller than the decrease when using the circular-shaped receiver. In addition, the trapezoidal receiver with CuO/water as a heat exchanger has a higher heat removal factor than the circular receiver with CuO/water and water. When the inlet fluid temperature increased from 50°C to 80°C, heat removal factors decreased from 0.88 to 0.82, 0.86 to 0.79, and 0.87 to 0.76 for the trapezoidal receiver with nanofluid, circular receiver with nanofluid, and water, respectively. Though the overall heat loss coefficient increased considerably with the fluid temperature, the heat transfer efficiency of the nanofluid did not decrease much because of this material's superior thermophysical properties compared to those of water. This may be attributed to the high thermal conductivity of the nanofluid, even at higher operating temperature. It is also observed that initially at a lower operating temperature the water depicts a marginally higher heat removal factor than nanofluids this was due to low density and the high specific heat of water. As the fluid temperature increases, the density of the nanofluids decreases and contact of nanoparticles with the surface of absorber increases [28], which leads to improvement in heat exchanger performance. Therefore, the heat removal factor for the nanofluid-based PV/T system decreases at a slower rate with increase in inlet fluid temperature.

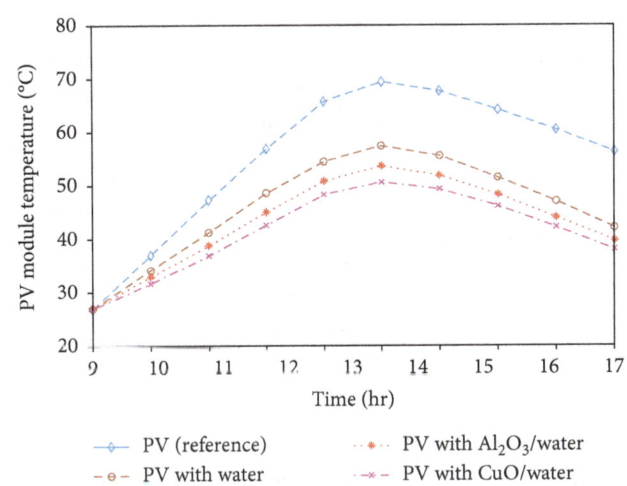

FIGURE 7: Daily variations of PV temperature with CuO/water, Al_2O_3/water, and water.

The effects of nanofluid as a coolant on the temperature evolution of PV cells are investigated in comparison with the case of pure water. The daily PV cell temperature is evaluated using pure water, Al_2O_3/water, and CuO/water, with results shown in Figure 7. For the purpose of comparative analysis, the PV cell temperature is evaluated with and without coolant fluid. A consensus can be drawn from these results that adding nanoparticles to the water notably reduces the PV cell temperature compared to the sole use of water as the working fluid. Furthermore, the addition of nanoparticles to pure water results in an increase of the overall thermal conductivity of the working fluid and, hence, is the apparent cause of the high outlet fluid temperature. It is noted that the daily maximum reference PV cell temperature obtained without cooling fluid is 69.4°C, while, using water only, Al_2O_3/water, and CuO/water, the values were 57.4°C, 53.7°C, and 50.6°C, respectively. These results show the clear dominance of CuO/water as a heat transfer fluid over Al_2O_3/water in the context of heat removal performance.

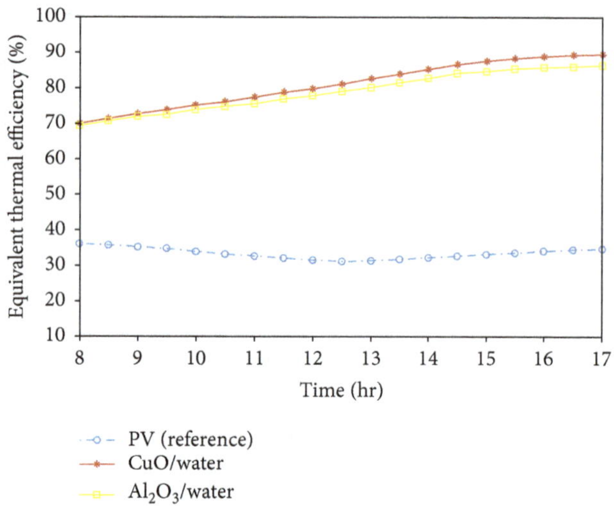

FIGURE 8: Daily variations of equivalent thermal efficiency with reference PV, CuO/water, and Al_2O_3/water.

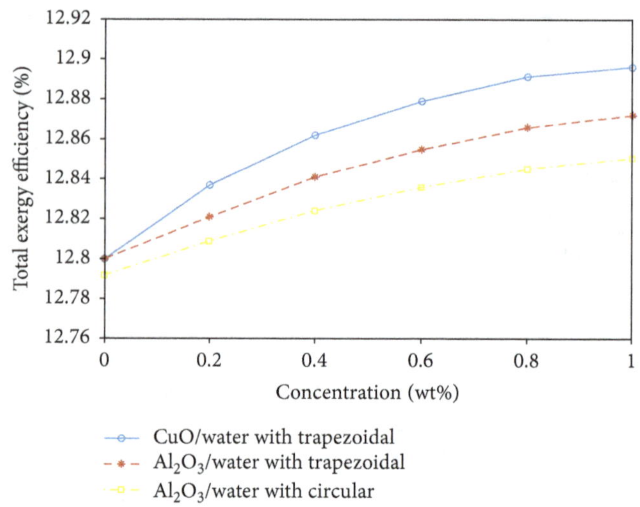

FIGURE 9: Variations of total exergy efficiency with concentrations (CuO/water and Al_2O_3/water).

Generally, primary energy saving or equivalent thermal efficiency is used as a performance metric for solar hybrid systems. Daily variation of primary energy savings of the PV and the PV/T systems can be obtained by converting electrical energy to equivalent thermal efficiency. Figure 8 shows the daily variation of the equivalent thermal efficiency of the PV and the PV/T systems. The average equivalent thermal efficiency for the PV (reference) is 35.53%; values for the PV/T system with CuO/water and Al_2O_3/water are found to be 80.94 and 78.83%, respectively. In the case of the reference PV cells, the decrease in equivalent thermal efficiency is due to the higher operating temperature that arises from the accumulation of extra solar radiation at the PV surface, in particular during solar noon hours. Primarily, the power output and electrical efficiency of the PV cells are influenced by the operating temperature. It should be noticed that the PV/T system with CuO/water in particular has shown better results than the PV system without coolant and the PV/T system with Al_2O_3/water.

The variation of the exergy efficiency as a function of different concentrations of CuO and Al_2O_3 nanoparticles in the base fluid is presented in Figure 9. Total exergy efficiency values (thermal and electrical) of CuO/water and of Al_2O_3/water were estimated using (18). According to Bejan's narrative [29], the maximum power output is inversely associated with the entropy generation of the solar system. This means that maximization of the power output is equivalent to minimization of the entropy generation rate. It is observed that, independent of nanofluid type, the exergy efficiency of a solar collector increases with increasing nanoparticle concentration in the base fluid. Generally, thermal energy efficiency varies as a function of the outlet temperature and the heat capacity of the thermal fluid. On the other hand, the exergy efficiency is more related to the fluid outlet temperature than to the heat capacity [19]. Therefore, the collector with CuO/water with a trapezoidal receiver showed higher exergy efficiency than did Al_2O_3/water with both trapezoidal- and circular-shaped receivers. This can be explained by the fact that CuO/water as a working fluid offers a higher heat transfer coefficient, which corresponds to a lower PV cell temperature and a higher outlet nanofluid temperature. However, Al_2O_3/water with a trapezoidal receiver performed marginally better than in the case of using Al_2O_3/water with a circular receiver. This is because the direct attachment of a flat top of a trapezoidal receiver onto the PV rear surface increases the total heat transfer surface area and hence improves the heat exchange performance between PV module and the circulating fluid.

Figure 10 shows the total exergy efficiency of the collector as a function of the daily sunlight hours when using Al_2O_3/water, CuO/water, and water as the working fluid. Based on the available solar radiation, the exergetic efficiencies for different working fluids were estimated. It is important to show the relationship between solar radiation and exergy efficiency because irreversibility varies directly with incident solar radiation. It can be clearly seen that the exergy efficiencies when using different fluids follow a pattern similar to the one for daily solar radiation. It is also observed that during the peak sun intensity hours (noon), when the sunlight is strongest, the irreversibility is maximized. Higher solar radiation intensity results in higher temperature difference between the collector surface and the ambient air. A larger difference reflects or may cause extra heat loss from the collector surface and, hence, lowers the exergy efficiency. Therefore, the temperature difference is a key point to consider when attempting to improve exergy efficiency. It is concluded that CuO/water as a heating medium has a profound impact on the collector performance, better than that of Al_2O_3/water and water only.

5. Conclusion

In this study, a transient mathematical model of the nanoengineered PV/T system is proposed based on real-time calculation of all heat transfer coefficients. CuO/water and Al_2O_3/water nanofluids were considered as the heat

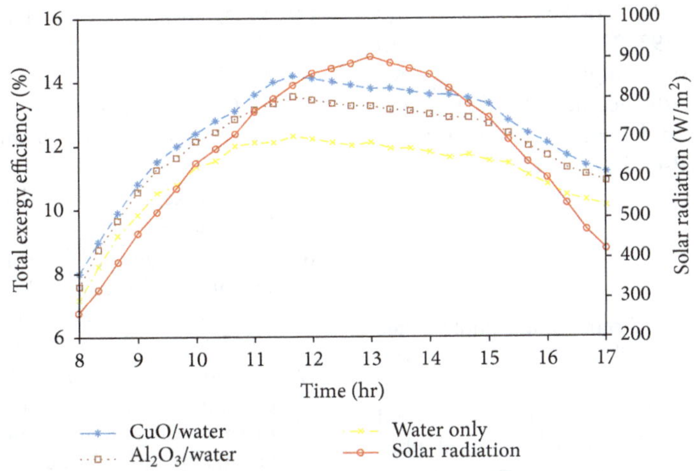

FIGURE 10: Daily variations of total exergy efficiency with solar radiation (CuO/water and Al_2O_3/water).

transfer fluids. Instead of a circular pipe, a trapezoidal-shaped absorber pipe was used with the intention of improving the surface area and, subsequently, the heat extraction from the PV cells. Experimental validation of the proposed model was performed using outdoor test results from the literature. The optimal energy and exergy efficiencies were located considering different operating conditions. Applying a nanofluid improves the energy and exergy efficiencies of the PV/T system; it was found that the CuO/water nanofluid exhibited better performance than that of the Al_2O_3/water or water-only cases. The highest predicted PV cell temperature without cooling fluid was 69.4°C, whereas, with water, Al_2O_3/water, and CuO/water this value dropped to 57.4°C, 53.7°C, and 50.6°C, respectively. It was observed that the irreversibility of the system increases with increasing solar radiation (at solar noon in particular), which leads to acceleration of heat losses to ambient air and thus results in a considerable exergy loss. The PV/T system with CuO/water nanofluid showed higher exergy efficiency than did the system with Al_2O_3/water nanofluid. The higher exergy efficiency is caused more by the outlet temperature of the CuO/water nanofluid than by the heat capacity.

Nomenclature

M: Mass (kg)
C: Specific heat (J/kg·°C)
T: Temperature (°C)
A: Surface area (m^2)
h: Heat transfer coefficient (W/m^2·°C)
u_a: Wind velocity (m/s)
E: Electrical power (W)
P: Packing factor
G: Solar radiation (W/m^2)
k: Thermal conductivity (W/m·°C)
W: Tube spacing (m)
L: Length (m)
D_i and D_o: Inner and outer tube diameters (m)
$\dot{m}_{fl}$: Mass flow rate (kg/s)
$\dot{E}$: Exergy rate.

Greek

α: Absorptivity
η: Efficiency
η_{PVT}: Primary energy saving efficiency
β_r: Solar cell temperature coefficient (1/K)
ε: Emissivity
σ: Stefan-Boltzmann constant ($W \cdot m^{-2} \cdot K^{-4}$)
δ: Thickness (m)
ρ: Density (kg/m^3)
ϕ: Nanoparticle concentration.

Subscripts

p: PV cells
b: Absorber plate
t: Tube
i: Insulation
fl: Circulating fluid
nf: Nanofluid
np: Nanoparticle
bf: Base fluid
a: Ambient air
pp: Power plant
s: Sky
el: Electrical
ex: Exergy
r: Reference
o and in: Outlet and inlet
th: Thermal.

Conflicts of Interest

The authors declare that there is no conflict of interest regarding the publication of this paper.

Acknowledgments

This work was supported by the Korea Research Fellowship Program through the National Research Foundation of

Korea (NRF) funded by the Ministry of Science, ICT, and Future Planning (2016H1D3A1938222); Research Grant from Kongju National University through the Korea Agency for Infrastructure Technology Advancement funded by the Ministry of Land, Infrastructure and Transport of the Korean government (Project no. 17TBIP-C124966-01).

References

[1] J. M. Gordon, "Optimal sizing of stand-alone photovoltaic solar power systems," *Solar Cells*, vol. 20, no. 4, pp. 295–313, 1987.

[2] A. S. Joshi, A. Tiwari, G. N. Tiwari, I. Dincer, and B. V. Reddy, "Performance evaluation of a hybrid photovoltaic thermal (PV/T)(glass-to-glass) system," *International Journal of Thermal Sciences*, vol. 48, no. 1, pp. 154–164, 2009.

[3] M. Imtiaz Hussain, A. Ali, and G. H. Lee, "Multi-module concentrated photovoltaic thermal system feasibility for greenhouse heating: model validation and techno-economic analysis," *Solar Energy*, vol. 135, pp. 719–730, 2016.

[4] S. Chol, *Enhancing Thermal Conductivity of Fluids with Nanoparticles*, vol. 231, ASME-Publications-Fed, 1995.

[5] R. Prasher, P. Bhattacharya, and P. E. Phelan, "Thermal conductivity of nanoscale colloidal solutions (nanofluids)," *Physical Review Letters*, vol. 94, no. 2, 2005.

[6] Z. Said, R. Saidur, M. A. Sabiha, A. Hepbasli, and N. A. Rahim, "Energy and exergy efficiency of a flat plate solar collector using pH treated Al 2 O 3 nanofluid," *Journal of Cleaner Production*, vol. 112, pp. 3915–3926, 2016.

[7] C. H. Li and G. P. Peterson, "Experimental investigation of temperature and volume fraction variations on the effective thermal conductivity of nanoparticle suspensions (nanofluids)," *Journal of Applied Physics*, vol. 99, no. 8, article 084314, 2006.

[8] A. K. Tiwari, P. Ghosh, and J. Sarkar, "Solar water heating using nanofluids–a comprehensive overview and environmental impact analysis," *International Journal of Emerging Technology and Advanced Engineering*, vol. 3, no. 3, pp. 221–224, 2013.

[9] O. Rejeb, M. Sardarabadi, C. Ménézo, M. Passandideh-Fard, M. H. Dhaou, and A. Jemni, "Numerical and model validation of uncovered nanofluid sheet and tube type photovoltaic thermal solar system," *Energy Conversion and Management*, vol. 110, pp. 367–377, 2016.

[10] H. A. Hasan, K. Sopian, A. H. Jaaz, and A. N. al-Shamani, "Experimental investigation of jet array nanofluids impingement in photovoltaic/thermal collector," *Solar Energy*, vol. 144, pp. 321–334, 2017.

[11] M. Hosseinzadeh, M. Sardarabadi, and M. Passandideh-Fard, "Energy and exergy analysis of nanofluid based photovoltaic thermal system integrated with phase change material," *Energy*, vol. 147, pp. 636–647, 2018.

[12] A. Shahsavar, M. Ameri, and M. Gholampour, "Energy and exergy analysis of a photovoltaic-thermal collector with natural air flow," *Journal of Solar Energy Engineering*, vol. 134, no. 1, article 011014, 2012.

[13] A. D. Jones and C. P. Underwood, "A thermal model for photovoltaic systems," *Solar Energy*, vol. 70, no. 4, pp. 349–359, 2001.

[14] T. T. Chow, "Performance analysis of photovoltaic-thermal collector by explicit dynamic model," *Solar Energy*, vol. 75, no. 2, pp. 143–152, 2003.

[15] J. A. Clarke, *Energy Simulation in Building Design*, Routledge, London, 2007.

[16] M. Imtiaz Hussain, G. H. Lee, and J.-T. Kim, "Experimental validation of mathematical models of identical aluminum and stainless steel engineered conical solar collectors," *Renewable Energy*, vol. 112, pp. 44–52, 2017.

[17] M. Imtiaz Hussain and G. H. Lee, "Thermal performance evaluation of a conical solar water heater integrated with a thermal storage system," *Energy Conversion and Management*, vol. 87, pp. 267–273, 2014.

[18] S. Bhattarai, J. H. Oh, S. H. Euh, G. Krishna Kafle, and D. Hyun Kim, "Simulation and model validation of sheet and tube type photovoltaic thermal solar system and conventional solar collecting system in transient states," *Solar Energy Materials and Solar Cells*, vol. 103, pp. 184–193, 2012.

[19] F. Yazdanifard, E. Ebrahimnia-Bajestan, and M. Ameri, "Performance of a parabolic trough concentrating photovoltaic/thermal system: effects of flow regime, design parameters, and using nanofluids," *Energy Conversion and Management*, vol. 148, pp. 1265–1277, 2017.

[20] R. Petela, "Exergy of heat radiation," *Journal of Heat Transfer*, vol. 86, no. 2, p. 187, 1964.

[21] R. J. Stone, "Improved statistical procedure for the evaluation of solar radiation estimation models," *Solar Energy*, vol. 51, no. 4, pp. 289–291, 1993.

[22] D. H. Kim, B. M. Jenkins, T. R. Rumsey, M. W. Yore, and N. J. Kim, "Simulation and model validation of a horizontal shallow basin solar concentrator," *Solar Energy*, vol. 81, no. 4, pp. 463–475, 2007.

[23] M. Rommel, D. Zenhäusern, A. Baggenstos, O. Türk, and S. Brunold, "Development of glazed and unglazed PVT collectors and first results of their application in different projects," *Energy Procedia*, vol. 70, pp. 318–323, 2015.

[24] M. S. Liu, M. C. C. Lin, I. T. Huang, and C. C. Wang, "Enhancement of thermal conductivity with CuO for nanofluids," *Chemical Engineering & Technology*, vol. 29, no. 1, pp. 72–77, 2006.

[25] R. S. Vajjha and D. K. Das, "A review and analysis on influence of temperature and concentration of nanofluids on thermophysical properties, heat transfer and pumping power," *International Journal of Heat and Mass Transfer*, vol. 55, no. 15-16, pp. 4063–4078, 2012.

[26] S. R. Shamshirgaran, M. K. Assadi, H. H. al-Kayiem, and K. V. Sharma, "Investigation of thermal behaviour, pressure drop, and pumping power in a Cu nanofluid-filled solar flat-plate collector," *MATEC Web of Conferences*, vol. 131, article 01003, 2017.

[27] Z. Said, R. Saidur, N. A. Rahim, and M. A. Alim, "Analyses of exergy efficiency and pumping power for a conventional flat plate solar collector using SWCNTs based nanofluid," *Energy and Buildings*, vol. 78, pp. 1–9, 2014.

[28] V. Salamon, D. Senthil Kumar, and S. Thirumalini, "Experimental investigation of heat transfer characteristics of automobile radiator using tio2-nanofluid coolant," *IOP Conference Series: Materials Science and Engineering*, vol. 225, article 012101, 2017.

[29] H. F. Oztop and K. Al-Salem, "A review on entropy generation in natural and mixed convection heat transfer for energy systems," *Renewable and Sustainable Energy Reviews*, vol. 16, no. 1, pp. 911–920, 2012.

15

Physics-Based Modeling and Experimental Study of Si-Doped InAs/GaAs Quantum Dot Solar Cells

A. P. Cédola,[1] D. Kim,[2] A. Tibaldi,[3] M. Tang,[2] A. Khalili,[1] J. Wu,[2] H. Liu,[2] and F. Cappelluti[1]

[1]*Department of Electronics and Telecommunications, Corso Duca degli Abruzzi 24, 10129 Torino, Italy*
[2]*Department of Electronic and Electrical Engineering, University College London, Torrington Place, London WC1E 7JE, UK*
[3]*Istituto di Elettronica e di Ingegneria dell'Informazione e delle Telecomunicazioni (IEIIT), Consiglio Nazionale delle Ricerche (CNR), Corso Duca degli Abruzzi 24, 10129 Torino, Italy*

Correspondence should be addressed to F. Cappelluti; federica.cappelluti@polito.it

Academic Editor: Urs Aeberhard

This paper presents an experimental and theoretical study on the impact of doping and recombination mechanisms on quantum dot solar cells based on the InAs/GaAs system. Numerical simulations are built on a hybrid approach that includes the quantum features of the charge transfer processes between the nanostructured material and the bulk host material in a classical transport model of the macroscopic continuum. This allows gaining a detailed understanding of the several physical mechanisms affecting the photovoltaic conversion efficiency and provides a quantitatively accurate picture of real devices at a reasonable computational cost. Experimental results demonstrate that QD doping provides a remarkable increase of the solar cell open-circuit voltage, which is explained by the numerical simulations as the result of reduced recombination loss through quantum dots and defects.

1. Introduction

Quantum dots (QDs) are being widely investigated since several years as a promising material for advanced concepts such as multiple junction [1], intermediate band [2, 3], and hot carrier [4] solar cells. The basic building block of such devices is typically made by an almost standard diode structure embedding a region with QDs. Electronic effects and charge transfer mechanisms in the QDs and between the QD localized states and the continuum states of the host semiconductor present quantum mechanical features, while a large part of the device is made by bulk regions working in the semiclassical transport regime. Thus, device level simulation models of QD solar cells inherently require multiscale approaches which shall be able to combine the micro- and macroscale description at an affordable computational cost [5].

In the hierarchy of multiscale models for the simulation of QD solar cells (QDSCs), one of the simplest approaches is provided by spatially resolved rate equation-based models, which couple microscopic calculated quantities such as the QD electronic structure, optical properties, and scattering rates to macroscopic equations for transport in the extended states. Several of the QD model parameters can also be extracted from routinary experimental data, making such approach suitable for both design and interpretation purposes. Modelling approaches able to reproduce the behavior of realistic devices and analyze the interplay of the involved physical mechanisms allow gaining insight into experimental results and provide useful feedback to the technology development. In this respect, while detailed balance theory of QDSCs has received a lot of attention and is useful to indicate the long-term target efficiency [6–8], less effort has been devoted to models based on semiconductor transport equations [9–13] and to the implications of QD carrier dynamics on the photovoltaic performances [14]. On the other hand, previous work by some of the present authors has shown that introducing a proper description of interband and intersubband QD dynamics is a crucial asset to address some of the most critical issues encountered in QDSCs, such as poor carrier collection efficiency and degradation of the open-circuit voltage [15, 16].

In this work, we apply the QD-*aware* physics-based model firstly proposed in [15] to the study of InAs/GaAs QDSC exploiting selective Si doping. QD doping is extensively investigated as an attractive means to control photocarrier dynamics and improve QDSC performance (see, e.g., the recent review in [17] and the references therein). The present work extends the analysis already proposed in [18] on the interplay between doping and recombination processes in QDSCs and provides an experimental-based validation to the conclusions in [18]. To ensure an accurate description of the cell behavior, the model in [15, 18] has been complemented by an accurate electromagnetic model for carrier photogeneration. In the following, we present the basic ideas of the model, highlighting its thermodynamic consistency and discussing the identification of microscopic parameters from quantum models and experimental characterization, and we show with a real case study the valuable insight into device behavior that can be gained with this simulation approach.

2. Model

QD-based solar cells usually exploit a $p-n$ or $p-i-n$ structure with a region including a periodic stack of QD layers. Figure 1 schematically shows the energy band diagram of an In(Ga)As QD surrounded by GaAs barriers and the carrier transfer processes that are included in the model, as discussed in the following. From the electronic standpoint, the 3D confinement gives rise to a deeply confined energy level, the ground state (GS), one or more excited states (ES), both of them with 0D density of states, and a quasi-two-dimensional (2D) energy state associated with the wetting layer (WL) [19]. At device level, the WL subband is described as a 0D level with high degeneracy factor, accounting for the 2D density of states of the WL and the weakly confined excited states.

We assume negligible coupling between the QD layers; thus, the exchange of carriers between localized states in different layers is always mediated by the barrier extended states. Under such hypothesis, the QDSC operation is as follows: electrons and holes are photogenerated in the barrier and QD states by above- and below-gap photons, respectively. Charge transfer between continuum and bound states is characterized by capture and escape processes through the WL level. In the barrier, carriers move by drift-diffusion, and when they arrive within the interaction range of the QD region (a few nanometers), they can be either emitted through the QD layer or captured in the QD states. Dually, confined carriers may escape from the QD states through thermal emission to the WL and from the WL to the barrier. Only thermal escape is considered in this work, because in the QDs under study, at room temperature, photon-assisted escape is negligible [20]; moreover, due to the shallow confinement, thermal emission is so efficient that the inclusion of an electric field-assisted mechanism does not provide any significant change to the overall picture. In [15], a good correlation between simulated and measured open-circuit voltage of cells with QD ground state emission between 1000 nm and 1300 nm has been demonstrated. On the other hand, deeper QDs may be significantly affected by field-assisted tunneling. Thus, the model formulation described in the following shall be applied only to the study of relatively shallow QDs at room temperature, hereinafter referred as thermally limited operation.

The electrical problem is formulated in terms of electrostatics and free carrier continuity equations, accounting for charge localization due to the QDs and charge transfer mechanisms between barrier and bound states. At the QD layer, Poisson's equation and electron continuity equation (for the sake of brevity, we omit hole equations) read as

$$\frac{\partial}{\partial x}\left(\epsilon\frac{\partial\phi}{\partial x}\right) = -q\left[p - n + \sum_{\gamma}\left(p_{\gamma} - n_{\gamma}\right) + N_D^+ - N_A^-\right],$$
$$\frac{\partial n}{\partial t} = \frac{1}{q}\frac{\partial J_n}{\partial x} - U_{\mathrm{B}} + G_{\mathrm{B}} - U_{n,\mathrm{CAP}}, \tag{1}$$

where ϕ is the electrostatic potential, n and p are the electron and hole densities, respectively, in the barrier, n_γ and p_γ are the electron and hole densities, respectively, in the QD state γ, and N_D^+ and N_A^- are the ionized donor and acceptor density, respectively. In the continuity equation, J_n is the electron drift-diffusion current density, U_{B} the net recombination rate, G_{B} the band-to-band photogeneration rate, and $U_{n,\mathrm{CAP}}$ the WL-mediated net capture rate from the barrier into the QD states, that is, the difference between the capture rate from barrier to WL and the escape rate from WL to barrier. U_{B} includes both radiative and nonradiative processes. Radiative recombination is modeled as

$$U_{\mathrm{B,r}} = B_{\mathrm{op}}\left(pn - n_i^2\right), \tag{2}$$

B_{op} being the GaAs spontaneous emission coefficient and n_i the thermal equilibrium carrier density. Nonradiative recombination follows the Shockley-Read-Hall theory

$$U_{\mathrm{B,nr}} = \frac{pn - n_i^2}{\tau_p\left(n + n_1\right) + \tau_n\left(p + p_1\right)}, \tag{3}$$

where $\tau_{n(p)}$ is the SRH recombination lifetime and n_1 and p_1 are the electron and hole densities, respectively, when Fermi level corresponds to the trap energy level, here assumed at midgap.

Capture and cascaded relaxation processes are described by a set of rate equations establishing a detailed balance, for each QD level, among all the interband and intersubband charge transfer mechanisms. At each QD layer, the rate equation describing the electron population in the subband γ reads as

$$\frac{\partial n_\gamma}{\partial t} = U_{n,\mathrm{CAP}}^{\gamma+1\to\gamma} - U_{n,\mathrm{CAP}}^{\gamma\to\gamma-1} - U_\gamma + G_\gamma, \tag{4}$$

where $U_\gamma \approx f_{n,\gamma} f_{p,\gamma}/\tau_{r,\gamma}$ and G_γ are the net interband radiative recombination and photogeneration rates of the γ subband

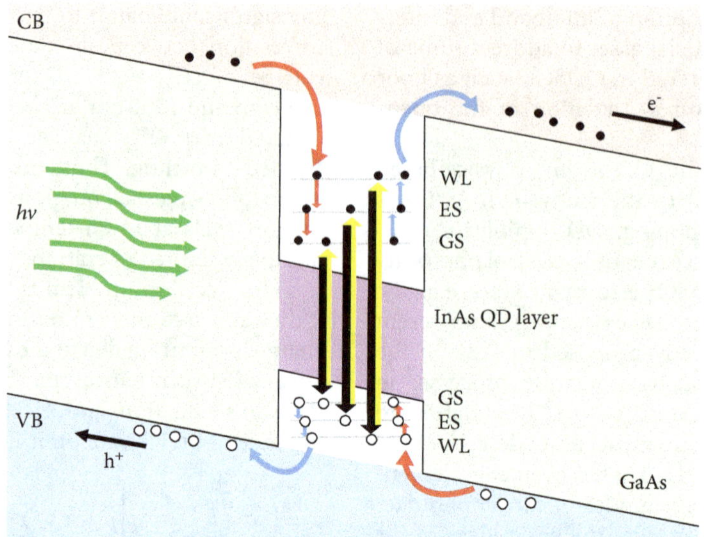

FIGURE 1: Energy band diagram of one InAs QD layer embedded in GaAs including all the carrier processes considered in the model: capture/relaxation (red), escape (blue), photogeneration (yellow), and recombination (black). WL is treated as a 0D state.

state (γ = WL, ES, GS), respectively. For the sake of simplicity, we consider only the radiative transitions associated to electron and hole states in the same subband. A radiative lifetime of 1 ns is assumed for all the QD states. The term $U_{n,\text{CAP}}^{\gamma\to l}$ is the net capture rate from the intersubband state k to the l one. Note that capture and relaxation processes are allowed to exist only between adjacent energy levels; thus, for γ = WL, $\gamma+1$ state indicates the barrier, whereas for γ = GS the terms involving $\gamma-1$ must be ignored. To make the formulation treatable, we assume in the following Maxwell-Boltzmann statistics for the continuum states and Fermi-Dirac statistics for the QD states. Moreover, we neglect QD inhomogeneous broadening; thus, the QD density of states is a Dirac delta function.

The capture or relaxation process of an electron from state γ to state $\gamma-1$ can be formulated as

$$R_{\text{CAP}} = c' N_\gamma f_\gamma N_{\gamma-1}\left(1 - f_{\gamma-1}\right), \tag{5}$$

where c' is the scattering rate ($\text{cm}^3\cdot\text{s}^{-1}$) of the transition and N_k is the effective density of states in the k band, with distribution function f_k. The reverse escape process reads as

$$\begin{aligned} R_{\text{ESC}} &= e' N_{\gamma-1} f_{\gamma-1} N_\gamma\left(1 - f_\gamma\right) \\ &= e' N_{\gamma-1} N_\gamma f_\gamma\left(1 - f_{\gamma-1}\right) e^{\left(E_{Fn}^{\gamma-1} - E_{Fn}^{\gamma}\right)/k_B T} e^{\left(E_\gamma - E_{\gamma-1}\right)/k_B T}, \end{aligned} \tag{6}$$

where the last term is derived taking advantage of the following identity, which holds regardless of the specific statistics, Boltzmann or Fermi [21],

$$f_\gamma\left(1 - f_{\gamma-1}\right) e^{\left(E_{Fn}^{\gamma-1} - E_{Fn}^{\gamma}\right)/k_B T} = f_{\gamma-1}\left(1 - f_\gamma\right) e^{\left(E_{\gamma-1} - E_\gamma\right)/k_B T}. \tag{7}$$

In (6), E_k and E_{Fn}^k are the energy of the k state and the quasi-Fermi level of the corresponding electron population, respectively. For barrier electrons, obviously, E_k coincides with the minimum of the conduction band. The detailed balance at thermal equilibrium imposes

$$c' = e' e^{\left(E_\gamma - E_{\gamma-1}\right)/k_B T}, \tag{8}$$

yielding the net capture rate $U_{n,\text{CAP}} = R_{\text{CAP}} - R_{\text{ESC}}$

$$U_{n,\text{CAP}}^{\gamma\to\gamma-1} = \frac{1}{\tau_{\text{cap}}^{\gamma-1}} n_\gamma \left(1 - \frac{n_{\gamma-1}}{N_{\gamma-1}}\right)\left(1 - e^{\left(E_{Fn}^{\gamma-1} - E_{Fn}^{\gamma}\right)/k_B T}\right), \tag{9}$$

$n_k = N_k f_k$ being the free electron density in the k state and $\tau_{\text{cap}}^{\gamma} = 1/c' N_{\gamma-1}$. Equation 9 highlights the correlation between the onsets of capture ($U_{n,\text{CAP}} > 0$) or escape ($U_{n,\text{CAP}} < 0$) and the nonequilibrium condition between electron populations in the states E_γ and $E_{\gamma-1}$.

The above formulation can be generalized so as to include the detailed energy dependence of the density of states of the different subbands (e.g., to include a more physical description of the QD states and WL states), Fermi-Dirac statistics in the continuum bands, as well as the energy dependence of the microscopic scattering matrix element $M^{\gamma,\gamma-1}$ describing the interaction between the two states [21]:

$$\begin{aligned} R_{\text{CAP}} &= \int_{E_\gamma} dE'_\gamma \int_{E_{\gamma-1}} g_\gamma\left(E'_\gamma\right) f_\gamma\left(E'_\gamma\right) M^{\gamma,\gamma-1} g_{\gamma-1}\left(E'_{\gamma-1}\right) \\ &\quad \cdot \left(1 - f_{\gamma-1}\left(E'_{\gamma-1}\right)\right) dE'_{\gamma-1}, \\ R_{\text{ESC}} &= \int_{E_\gamma} dE'_\gamma \int_{E_{\gamma-1}} g_\gamma\left(E'_\gamma\right)\left(1 - f_\gamma\left(E'_\gamma\right)\right) M^{\gamma-1,\gamma} g_{\gamma-1} \\ &\quad \cdot \left(E'_{\gamma-1}\right) f_{\gamma-1}\left(E'_{\gamma-1}\right) dE'_{\gamma-1}. \end{aligned} \tag{10}$$

Introducing the energy averaged scattering element $\langle M^{\gamma,\gamma-1}\rangle$, averaged over the entire ensemble of possible initial and final states, and after some analytical manipulation, one derives an expression of $U_{n,\mathrm{CAP}}$ completely analogous to 8, with $\tau_{\mathrm{cap}}^{\gamma-1} = 1/\langle M^{\gamma,\gamma-1}\rangle N_{\gamma-1}$, $n_\gamma = \int_E g_\gamma f_\gamma dE$, $N_{\gamma-1} = \int_E g_{\gamma\text{-}1} dE$.

Capture and relaxation in self-assembled In(Ga)As/GaAs QDs happen through complicated dynamics involving emission of longitudinal optical (LO) phonon emission, polaron (electron-phonon coupling) decay, defect-mediated relaxation, carrier-carrier scattering, and so forth (see, e.g., [22–24] and the references therein). The scattering time τ_{cap} derived from quantum models is typically a function of the carrier density in the WL and QD states, which in turn depends on injection/photogeneration of carriers according to the cell operating conditions and on design parameters such as doping. At low carrier density, electron-LO-phonon interaction and Auger electron-hole scattering act as very efficient relaxation channels, providing relaxation times on the scale of tenths of ps to tens of ps [22, 23, 25]. At high carrier density, carrier-carrier scattering may become significant, yielding again relaxation times on the ps scale [24]. In particular, for n-doped (p-doped) QD hole (electron), the population is extremely low; thus, electron-hole scattering is negligible and relaxation occurs through a combination of electron-electron (hole-hole) scattering and carrier-LO-phonon relaxation. These predictions are confirmed by several experimental data on undoped and doped samples: in [26], relaxation times from WL to QD ground state were found to be 2, 3, and 6 ps for p-doped, n-doped, and undoped samples, respectively. Further experimental data of similar sign are summarized in [23]. Overall, theoretical and experimental results on In(Ga)As/GaAs QDs show that net capture/relaxation rates—on a ps scale—are extremely fast with respect to the characteristic QD band-to-band and intraband radiative lifetimes (about 1 ns and 100 ns, resp. [25]), making the QD carrier lifetime largely dominated by the QD radiative lifetime. Moreover, the open-circuit voltage (V_{oc}) penalty in thermally limited QDSCs is dominated by the ratio between carrier lifetime in the barrier and carrier lifetime in QDs [27], the last one being the net result of the competing processes of capture/relaxation/recombination through the QDs—from the one hand—and escape from the QD bound states and electric-field-driven sweep out through the extended states—from the other hand. Thus, at least in In(Ga)As/GaAs QDs and under nonconcentration operation, capture/relaxation times can be reasonably treated as constant parameters, neglecting their carrier density dependence and the possible electric-field dependence due to tunneling mechanisms. In this respect, it may be remarked that the electrical field across the interdot layers plays however an important role in the resulting net capture rate in QDs: at high field, free-carrier transit time is high and the probability of capture from the barrier into the QDs remains low (such situation is representative of the short-circuit condition, where QD recombination turns to be negligible). In [28], carrier capture from the GaAs barrier was found to be fully quenched in fields of only 15 kV/cm. When the electrical field is significantly screened (e.g., as forward bias approaches the maximum power point and open-circuit voltage), the transit time decreases and the net capture rate in the QDs increases; that is, QDs turn into radiative trap centers (see [15, 16] for a detailed discussion).

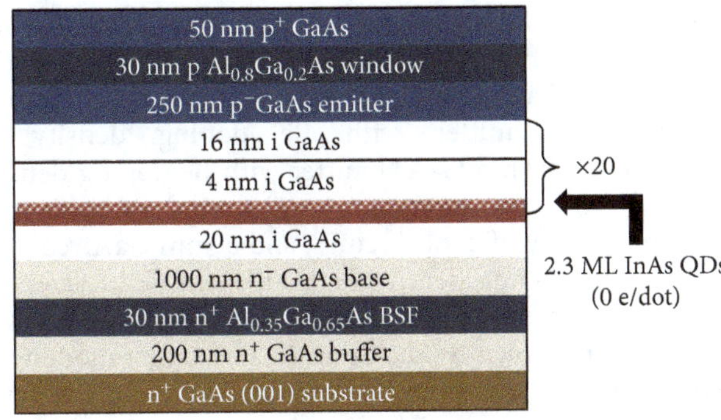

Figure 2: Sketch of the epilayer structure of the 20 × QD solar cell (R3 and R4 samples).

In the following simulations, we use empirical data extracted in [15] from the interband-pump-intraband-probe spectroscopy experiments reported in [29]. To this aim, we assume very fast capture into the WL, which occurs through one-LO-phonon interaction on a 0.1 ps time scale, followed by cascaded relaxation of carriers into the ES and GS states. Fitting the time-resolved IR absorption of QDs similar to those in the present work, we estimated electron time constants of about 1 ps [15] for WL-ES and ES-GS relaxation. Due to their large effective mass, holes form closely spaced energy levels, characterized by fast phonon-mediated capture and relaxation mechanisms, whose time constants are set to 100 fs.

Concerning the photogeneration rates, the simulation calculates the (optical) electric field profile across the multilayer structure, with each material characterized by the complex refractive index $\tilde{n} = n + i\kappa = n + i\alpha\lambda/(4\pi)$ (α being the absorption coefficient), exploiting a scattering matrix formalism for coherent multilayers [30]. From the optical electric field profile $E_{\mathrm{op}}(x)$, the absorbed photon density (at each wavelength) and the photogeneration rate (assuming unitary quantum yield) are then calculated through the divergence of the Poynting vector, yielding

$$G = \frac{\sigma}{2h\nu} \left| E_{\mathrm{op}} \right|^2, \tag{11}$$

where $h\nu$ is the photon energy and $\sigma = n\alpha/\mu_0 c$ the electrical conductivity (μ_0 and c being the vacuum magnetic permeability and light velocity, resp.).

The model numerical implementation is based on the discretization of the electrical equations through the Scharfetter-Gummel scheme and the self-consistent solution of the resulting system of nonlinear equations through Newton-based methods [31].

3. Experimental

All the epitaxial structures were grown using a solid-source molecular beam epitaxy (MBE) on n^+ GaAs (100) substrates. InAs/GaAs QDSC structures based on those previously demonstrated by Kim et al. [32] were used, as shown in Figure 2. The structures consist of a 200 nm GaAs buffer layer with Si doping density of 1×10^{18} cm^{-3}, 30 nm $Al_{0.35}Ga_{0.65}As$ back surface field (BSF) with Si doping density of 1×10^{18} cm^{-3}, 1000 nm GaAs base with Si doping density of 1×10^{17} cm^{-3}, 250 nm GaAs emitter with Be doping density of 2×10^{18} cm^{-3}, 100 nm GaAs emitter with Be doping density of 5×10^{18} cm^{-3}, 30 nm $Al_{0.75}Ga_{0.25}As$ window layer with Be doping density of 2×10^{18} cm^{-3}, and 50 nm GaAs contact layer with Be doping density of 1×10^{19} cm^{-3}. 20 layers of InAs QDs were grown in the intrinsic region of the SCs. The QDs were grown by the Stranski-Krastanov mode with the InAs coverage of 2.1 monolayers (ML) at a substrate temperature of ~500°C. A high growth temperature was used during the growth of the GaAs spacer layers to suppress the formation of dislocations [33–35]. No cap layer was applied between the InAs QDs and the GaAs space layers. Direct Si doping with doping densities of 14 e/dot was applied to the QDs in the Si-doped QDSCs [32]. Whereas the InAs QD layers in all other samples were separated by 20 nm GaAs spacer, the spacer layers in the QDSC with additional spacer layers (R5) were 25 nm each, which gave additional 100 nm in total. For the post-growth sample cleaning, the SCs were ultrasonicated in acetone and isopropanol for 10 min each at room temperature. To remove the oxide on the surface of the samples, the SCs were immersed in diluted ammonia solution (1 : 19) for 30 s. A Au-Zn alloy (95% Au, 5% Zn) was deposited in grid patterns to form 200 nm thick *p*-type electrodes using a metal shadow mask and a thermal evaporator.

The morphology of the QD layers was studied using a Veeco Nanoscope V atomic force microscope (AFM). 532 nm excitation from a diode-pumped solid-state laser was used for photoluminescence (PL) measurements. The sample temperature during the PL measurements was controlled using a He-cooled cryostat. Current density versus voltage ($J-V$) measurements was performed using an LOT-calibrated solar simulator with a xenon lamp under one-sun air mass (AM) 1.5 G illumination at 25°C. The devices were connected to a Keithly 2400 sourcemeter via a 4-point probe station, and ReRa Tracer 3 software was used to collect the data. Photocurrent measurements were obtained using a Halogen lamp chopped to 188 Hz through a Newport monochromator. A 4-point probe connected with a lock-in amplifier was used to collect data. The monochromatic beam was then calibrated using a Si photodiode, and the data was analysed with Photor QE 3.1 software to produce the external quantum efficiency (EQE).

4. Results and Discussion

An overview of the photovoltaic characteristics of the undoped and Si-doped QD solar cells and of an undoped reference cell (same geometry and doping but without the inclusion of QDs) is presented in Figure 3 and Table 1. The target performances for the REF cell were $V_{oc} = 1.04$ V, short circuit current density (J_{sc}) of 14.2 mA/cm^2, and fill factor (FF) of 86.5%. All the cells, included in the REF one, show quite low FF and unusual rounded shape of the $J-V$ characteristic, whose origin is attributed to issues in the realization of the metal grid. The observed behavior cannot be fitted by a

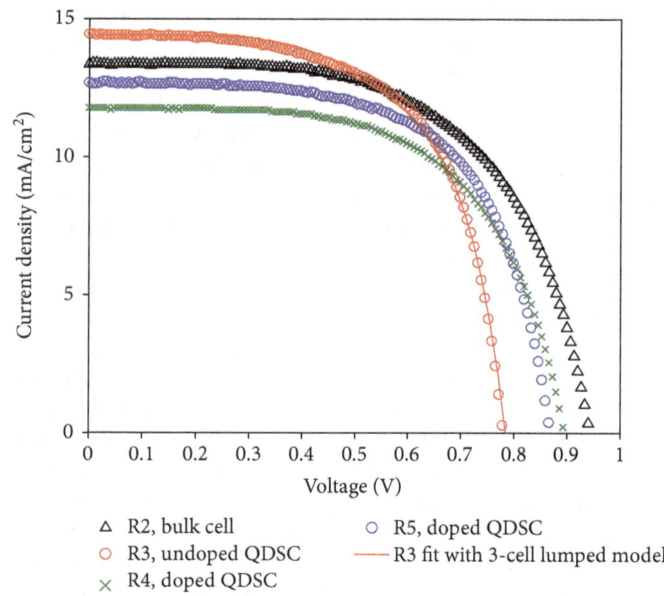

FIGURE 3: Current density versus voltage characteristics measured under one-sun illumination (AM 1.5 G) for the reference GaAs bulk cell (R2), undoped QDSC (R3), and two-doped (14 e/dot) QDSC with interdot spacing of about 20 nm (R4) and 25 nm (R5).

TABLE 1: Photovoltaic parameters extracted from the measured $J-V$ characteristics in Figure 3.

Device	J_{sc}, mA/cm^2	V_{oc}, V	FF, %
R2, REF GaAs cell	13.37	0.942	59.7
R3, undoped QD	14.42	0.778	62.7
R4, doped QD	11.74	0.894	61.1
R5, doped QD	12.70	0.868	63.1

simple lumped model with series and shunt resistances: a nonuniform three-section model of the cell [36] provides a good fitting of the measured characteristics, as shown in Figure 3, with extracted series resistances of about 50, 1, and 1 Ωcm^2 for each subcell. The undoped QDSC shows a marked penalty of V_{oc} with respect to the REF cell, which is partially recovered in the samples with directly doped QDs. On the other side, the doped QDSCs have lower J_{sc} that can be in part attributed to the shrinking of the space charge region and subsequent reduction of carrier collection and in part to lower minority carrier lifetime in the base and emitter regions as demonstrated later (see Figure 4) based on measured EQE and simulations. In may be worth noticing that under the assumption of similar material quality, the slightly larger thickness of the undoped region of the R5 sample (520 nm) with respect to the R4 one (420 nm) would produce an absolutely marginal difference in terms of device level behavior. Finally, the higher J_{sc} of the undoped QDSC with respect to the REF cell is not related to QD photogeneration but rather to a worse carrier collection efficiency in the emitter causing a decreased EQE in the GaAs absorption range.

The electronic structure of QDs is estimated based on room temperature steady-state PL. The measured PL spectra at low and room temperature are shown in Figures 5(a) and

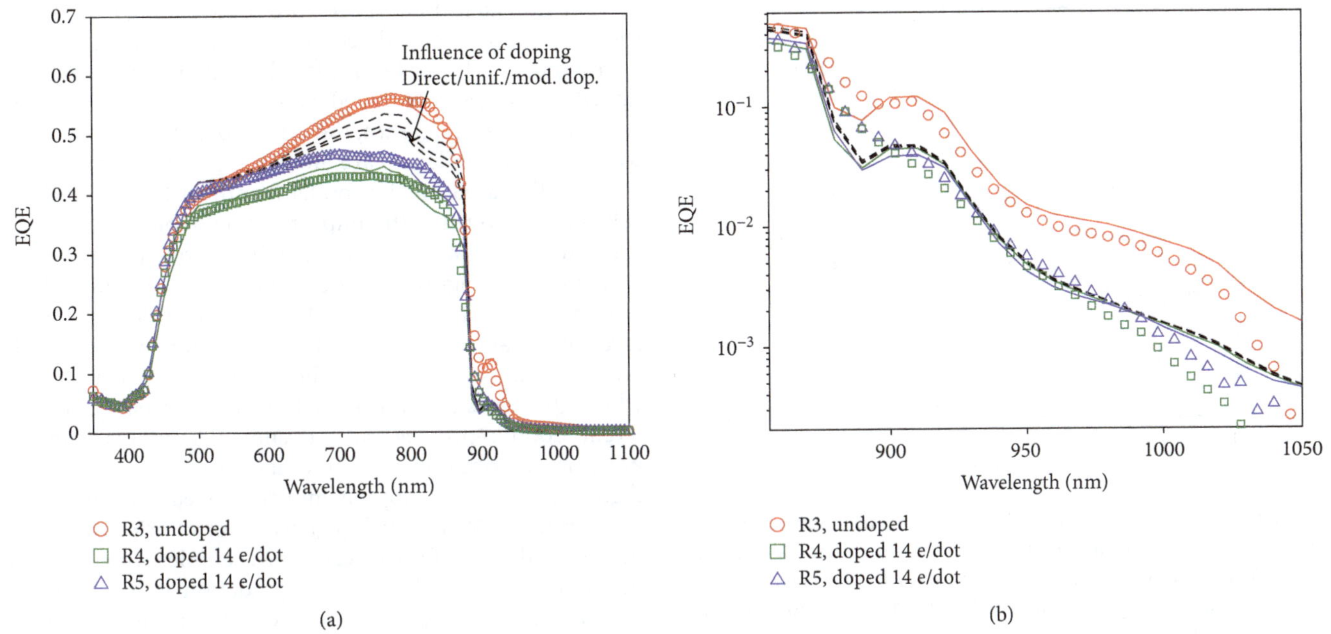

FIGURE 4: (a) Comparison between measured and simulated EQE. (b) Zoom in the QD wavelength region. The black dashed lines show the predicted decrease of EQE according in the doped (14 e/dot) QD solar cell assuming the same minority carrier lifetime as in the undoped sample and under different hypotheses of dopant impurity distribution: well confined in the QD region (direct), uniformly distributed across the QD stack, and located at the center of the GaAs interdot layers. Symbols show experimental data. Solid lines inidicate simulation and best fit.

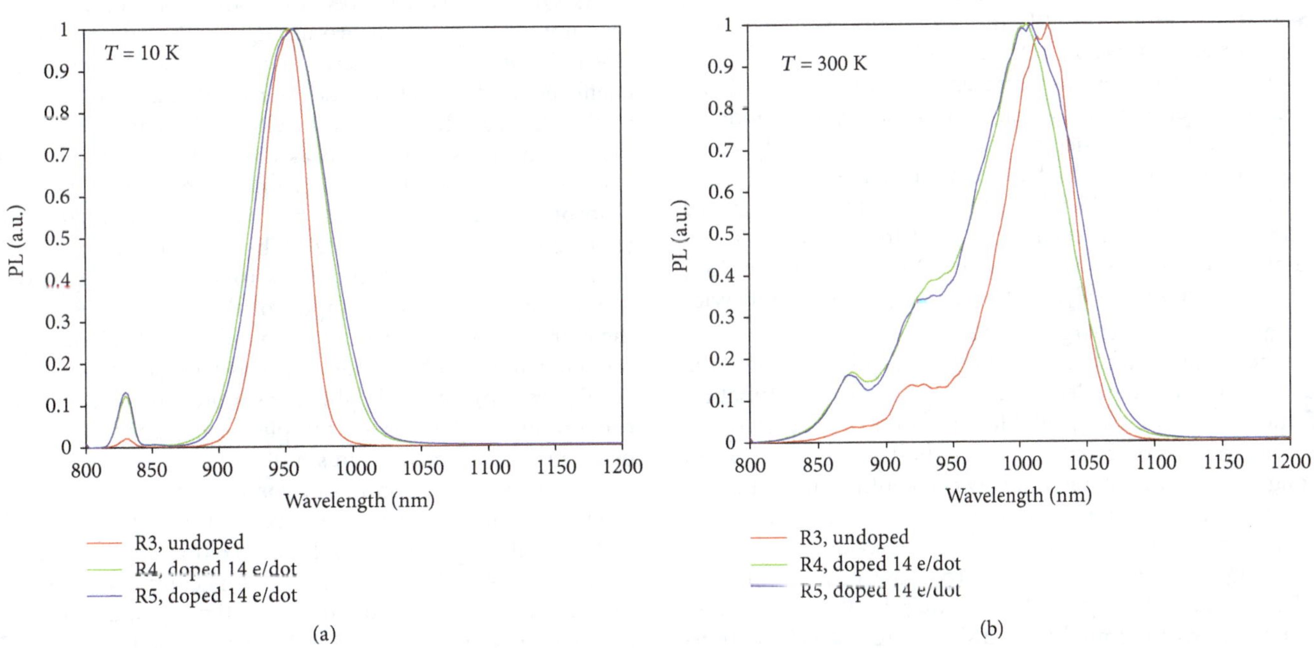

FIGURE 5: Normalized PL spectra measured at $T = 10$ K (a) and $T = 300$ K (b).

5(b), respectively. At low temperature, only the GS state is visible, with a slight redshift of the PL peak and an increase of the full width at half maximum (FWHM) for the doped samples with respect to the undoped one especially in the longer wavelength region. The observed behavior suggests that in the doped samples the distribution of the dot size is less homogeneous, with an increased density of larger (i.e., more confined) QDs which emit at lower energies [37]. At 300 K, the slight blue shift of the GS emission of the doped samples may be indicative of a higher interband transition energy and of a weaker quantum confinement of electrons in the GS. More importantly, the PL spectrum makes well visible the increase of the WL emission with respect to the GS one with doping. This indicates less efficient ES and GS

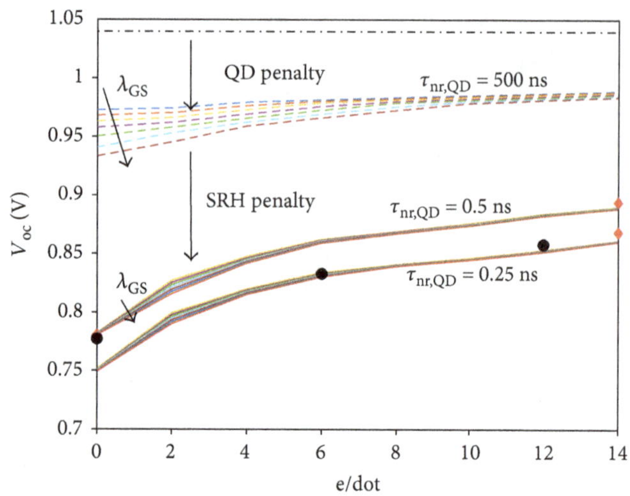

FIGURE 6: Analysis of V_{oc} penalty in QDSC as a function of the per dot doping density, for different GS energy confinement (λ_{GS} ranges from 990 nm to 1050 nm) and different minority carrier lifetime in the interdot layers ($\tau_{nr,QD}$). The black dash-dot line indicates the V_{oc} of the reference bulk GaAs cell. The red bullet indicates experimental data in this work. The black bullet indicates experimental data from Lam et al. [43].

relaxation pathways for the WL state of doped QDSCs that can be understood as a consequence of blocking of electron relaxation in the ES and GS due to state filling effects [18]. From the PL measurements, we identified energy gaps of 1.24 eV, 1.28 eV, and 1.37 eV for GS, ES, and WL, respectively, as representative values for all the samples under study. The impact of slight variations in the QD electronic structure and GS energy is analyzed in detail later (see Figure 6), based on the analysis of V_{oc} data. Based on literature data [38], the confinement energy for electron and holes in each state is set to 80% and 20% of the difference between barrier and QD level energy gap [38] (e.g., $\Delta E^{e}_{GS} = 0.8(E_g - E_{g,GS})$, E_g and $E_{g,GS}$ being the barrier energy gap and GS energy gap, resp.).

The analysis of the experimental results through the device level model in Section 2 provides also an estimation of the Shockley-Read-Hall lifetimes in (3) characterizing nonradiative recombination across the barrier states. Even though the strong interaction between continuum and localized states makes not obvious to single out the impact of such parameters on the cell photovoltaic behavior, some basic knowledge of device operation (as routinely observed in bulk cells) allows identifying a suitable strategy for their extraction. In fact, it is worth reminding that short-circuit SRH recombination is more detrimental in the doped regions, whereas it marginally affects carrier collection efficiency in the undoped one. In contrast, at open-circuit condition, the most detrimental loss arises from SRH recombination in the undoped region. Thus, once a reasonably accurate optical model of the cell is established, the analysis of EQE spectra—in particular in the GaAs wavelength range—allows identifying lower bounds for the SRH lifetimes in the base and emitter, whereas the analysis of the open-circuit voltage provides an estimation of SRH lifetimes in the undoped region.

The measured EQE spectra are shown in Figure 4. Fitting the EQE of the undoped cell in the GaAs wavelength region, we estimated a lower bound for the SRH lifetimes in the p-doped emitter and n-doped base of about 8 ns and 10 ns, respectively. For reference, the expected lifetimes, accounting for doping dependence, are about 20 ns and 300 ns, respectively [39]. Based on the model identified for the undoped cell, we have verified the impact of QD doping on J_{sc}. Since the actual distribution of dopant impurities is not known, we compare in Figure 4 different hypotheses of dopant distribution: dopant atoms at the QD site (direct doping), a thin layer of dopant atoms located in the middle of the interdot layer (mod. dop), and a uniform distribution across the whole QD stack. A detailed analysis of the impact of the different doping profiles on the energy band diagram and potential distribution may be found in [18]. Simulations in Figure 4 show that doping induces a clear penalty in the GaAs-range EQE, but such penalty is not sufficient to explain the reduced EQE of the direct doped samples with respect to the undoped ones. The EQE degradation is in fact reproduced by accounting for a reduction of minority carrier lifetime in the doped regions, with fitted SRH lifetimes of 500 ps in the emitter of the R5 sample and 50 ps (emitter)/200 ps (base) for the R4 sample.

Finally, the optical properties of the QD layer are estimated from the analysis of the subband gap EQE of the undoped cell reported in Figure 4(b). To this aim, the InAs/GaAs QD layer is described as a homogeneous equivalent medium with optical absorption $\alpha_{GaAs} + \Delta\alpha$ and real refractive index $n_{GaAs} + \Delta n$, where $\Delta\alpha$ and Δn are the absorption coefficient and refractive index of the QD material, weighted by the QD areal density. For the GS and ES states, a Gaussian distribution is used, with a peak absorption of 500 cm^{-1} and 1000 cm^{-1} and FWHM of 50 nm, respectively. The WL absorption profile is modeled by convolution of a Heaviside step function with a Lorentzian broadening function, with an absorption peak of about 2×10^4 cm^{-1} and FWHM of 40 nm. Representative examples of the spectral behavior of the extracted $\Delta\alpha$ and Δn can be found in [40]. The complex refractive index of bulk materials was taken from [41].

Figure 4(b) also highlights the suppression of QD photogeneration in the doped samples, with good agreement between measurements and simulations.

At open-circuit condition, based on the QD electronic model, we first analyze the predicted V_{oc} penalty of the QDSCs with respect to the bulk cell. To this aim, we take into account possible fluctuations of the QD confinement as seen from the PL spectra. Figure 6 reports the V_{oc} of the QD solar cell as a function of doping for different QD families with GS transition energy between 1.18 and 1.25 eV. SRH lifetimes in the doped regions are those extracted from the short-circuit analysis; in the undoped region, a SRH lifetime of 500 ns is assumed. For the sake of comparison, the V_{oc} (design value) of the REF cell is also reported. The analysis shows that the *inherent* (i.e., for negligible defectivity of the QD material) penalty due to the inclusion of the QDs is about 120 mV for the largest GS confinement, in agreement with previous theoretical works [20] and experimental data of QDSC with record V_{oc} [42]. Such penalty can be fully attributed

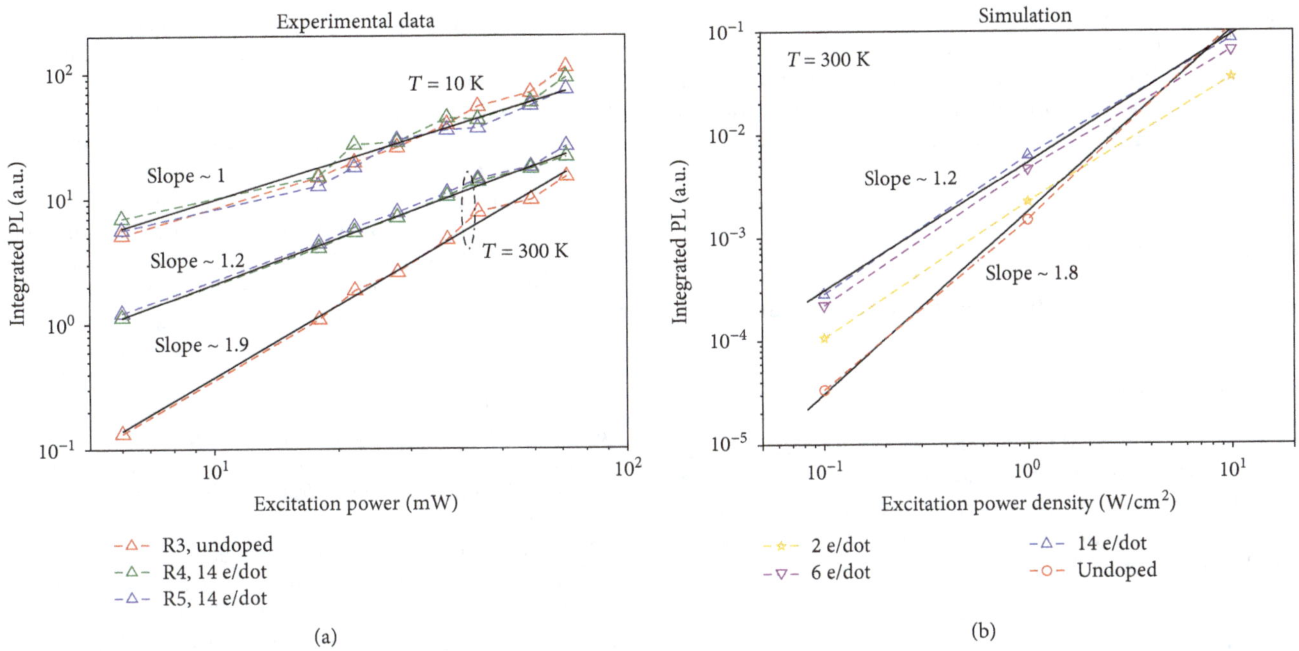

FIGURE 7: (a) Measured IPL intensity versus excitation power. (b) Simulated IPL intensity versus excitation power density. In the experiments, the equivalent excitation power density was about 1 W/cm^2 at power of 16 mW/cm^2.

to radiative recombination through the QD states [15, 18]. On the other hand, the comparison with the experimental data suggests that some other loss mechanism is in play in the devices under study, causing an added penalty of about 180 mV for the undoped case. Besides the measured data of the samples described in Section 3, we included also experimental data of very similar QDSCs, reported by the same research group in [43]. A very good agreement between simulated and measured data is obtained by assuming a significant defectivity of the interdot spacer layers, causing SRH lifetime (labeled as $\tau_{nr,QD}$ in Figure 6) on the order of fractions of ns, that is, lower than the QD radiative lifetime. The calculated trend of V_{oc} as a function of the doping density closely follows the measured one, suggesting that dopant atoms are well incorporated in the QDs and provide electrons per dot in line with the nominal doping value. Overall, the results highlight that doping is always beneficial in terms of V_{oc}, since it mitigates both QD radiative recombination and nonradiative recombination. QD radiative recombination is in fact suppressed by QD state filling; SRH recombination is suppressed because doping shifts the electron quasi-Fermi level far away from the intrinsic Fermi level reducing the capture/emission probability of electrons and holes by the midgap defect. Further analysis of doped-induced SRH suppression and examples of calculated SRH recombination rates in QDSCs may be found in [18, 44].

Finally, further supporting the interpretation that the QDSCs under study are limited by SRH recombination, Figure 7 compares the measured and simulated behavior of the integrated PL (IPL) versus the excitation power (P_{exc}): at $T = 10$ K, all the cells present a linear dependence of IPL with P_{exc} (slope ~ 1 in the log-log plot) regardless of doping, since they are dominated by the radiative recombination through the QDs. On the other hand, at $T = 300$ K, the undoped cell shows a superlinear dependence of IPL with P_{exc} (slope ~ 1.8 in the log-log plot), while the doped ones have again almost linear behavior. Our comprehension of the mechanism is that at $T = 300$ K nonradiative recombination through the barrier is an effective loss mechanism for photogenerated carriers that tends to be suppressed (or saturated) as the excitation power increases: thus, at low excitation power, IPL scales almost quadratically with P_{exc} [18, 45, 46]. The introduction of doping changes the electrostatic potential and electrical field profiles of the cell in such a way that SRH recombination results minimized. As a result, the doped QD cells have again a radiative limited behavior and their IPL scales linearly with P_{exc}.

5. Conclusion

We have applied physics-based numerical simulations to investigate the photovoltaic behavior of quantum dot solar cells which demonstrate a large increase of the open-circuit voltage with doping. The analysis of the experimental results points out that doping has profound effects on the cell behavior, which are difficult to disentangle without the support of device level simulations, since they involve the microscale, in terms of QD carrier dynamics, and the macroscale, since doping affects the electric field distribution across the photoactive region. The analysis allows also assessing the margin of improvement for the open-circuit voltage, taking into account that the cells under study operate under thermally limited regime; that is, two-step photon absorption is negligible. The presented device level simulations are developed within a multiscale modeling framework which combines in a self-consistent fashion transport equations in the bulk, rate equations for carrier dynamics in the QD states, and an accurate electromagnetic model of the optical field across the cell. Further physical mechanisms relevant to QD-based solar cells, such as two-photon absorption, Auger processes, and hot carrier relaxation can be incorporated in this simulation framework allowing us to gain a deeper understanding of the physics of real devices and directions for their development.

Conflicts of Interest

The authors declare that they have no conflicts of interest.

Acknowledgments

This project has received funding from the European Union's Horizon 2020 Research and Innovation Programme under Grant Agreement no. 687253. F. Cappelluti acknowledges useful meetings and discussions supported by COST Action MP1406 "MultiscaleSolar".

References

[1] S. Hubbard and R. Raffaelle, "Boosting solar-cell efficiency with quantum-dot-based nanotechnology," *SPIE Newsroom*, 2010.

[2] A. Luque and A. Martí, "Increasing the efficiency of ideal solar cells by photon induced transitions at intermediate levels," *Physical Review Letters*, vol. 78, no. 26, pp. 5014–5017, 1997.

[3] S. Asahi, H. Teranishi, K. Kusaki, T. Kaizu, and T. Kita, "Two-step photon up-conversion solar cells," *Nature Communications*, vol. 8, p. 14962, 2017.

[4] D. M. Tex, I. Kamiya, and Y. Kanemitsu, "Control of hot-carrier relaxation for realizing ideal quantum-dot intermediate-band solar cells," *Scientific Reports*, vol. 4, p. 4125, 2014.

[5] U. Aeberhard, "Simulation of nanostructure-based high-efficiency solar cells: challenges, existing approaches, and future directions," *IEEE Journal of Selected Topics in Quantum Electronics*, vol. 19, no. 5, pp. 1–11, 2013.

[6] M. Y. Levy and C. Honsberg, "Solar cell with an intermediate band of finite width," *Physical Review B*, vol. 78, no. 16, article 165122, 2008.

[7] L. Cuadra, A. Marti, and A. Luque, "Influence of the overlap between the absorption coefficients on the efficiency of the intermediate band solar cell," *IEEE Transactions on Electron Devices*, vol. 51, no. 6, pp. 1002–1007, 2004.

[8] R. Strandberg and T. Reenaas, "Optimal filling of the intermediate band in idealized intermediate-band solar cells," *IEEE Transactions on Electron Devices*, vol. 58, no. 8, pp. 2559–2565, 2011.

[9] V. Aroutiounian, S. Petrosyan, A. Khachatryan, and K. Touryan, "Quantum dot solar cells," *Journal of Applied Physics*, vol. 89, no. 4, pp. 2268–2271, 2001.

[10] A. S. Lin and J. D. Phillips, "Drift-diffusion modeling for impurity photovoltaic devices," *IEEE Transactions on Electron Devices*, vol. 56, no. 12, pp. 3168–3174, 2009.

[11] K. Yoshida, Y. Okada, and N. Sano, "Self-consistent simulation of intermediate band solar cells: effect of occupation rates on device characteristics," *Applied Physics Letters*, vol. 97, no. 13, p. 133503, 2010.

[12] I. Tobías, A. Luque, and A. Martí, "Numerical modeling of intermediate band solar cells," *Semiconductor Science and Technology*, vol. 26, no. 1, article 014031, 2011.

[13] R. Strandberg and T. Reenaas, "Drift-diffusion model for intermediate band solar cells including photofilling effects," *Progress in Photovoltaics*, vol. 19, no. 1, pp. 21–32, 2012.

[14] V. Aroutiounian, S. Petrosyan, and A. Khachatryan, "Studies of the photocurrent in quantum dot solar cells by the application of a new theoretical model," *Solar Energy Materials and Solar Cells*, vol. 89, no. 2, pp. 165–173, 2005.

[15] M. Gioannini, A. P. Cedola, N. Di Santo, F. Bertazzi, and F. Cappelluti, "Simulation of quantum dot solar cells including carrier intersubband dynamics and transport," *IEEE Journal of Photovoltaics*, vol. 3, no. 4, pp. 1271–1278, 2013.

[16] M. Gioannini, A. P. Cedola, and F. Cappelluti, "Impact of carrier dynamics on the photovoltaic performance of quantum dot solar cells," *IET Optoelectronics*, vol. 9, no. 2, pp. 69–74, 2015.

[17] T. Sogabe, Q. Shen, and K. Yamaguchi, "Recent progress on quantum dot solar cells: a review," *Journal of Photonics for Energy*, vol. 6, no. 4, article 040901, 2016.

[18] F. Cappelluti, M. Gioannini, and A. Khalili, "Impact of doping on InAs/GaAs quantum-dot solar cells: a numerical study on photovoltaic and photoluminescence behavior," *Solar Energy Materials and Solar Cells*, vol. 157, pp. 209–220, 2016.

[19] D. Bimberg, M. Grundmann, and N. N. Ledentsov, *Quantum Dot Heterostructures*, John Wiley & Sons, 1999.

[20] G. Jolley, L. Fu, H. F. Lu, H. H. Tan, and C. Jagadish, "The role of intersubband optical transitions on the electrical properties of InGaAs/GaAs quantum dot solar cells," *Progress in Photovoltaics*, 2012.

[21] G. A. Kosinovsky, "Threshold current and modulation response of semiconductor lasers," PhD Thesis, 1995.

[22] T. R. Nielsen, P. Gartner, and F. Jahnke, "Many-body theory of carrier capture and relaxation in semiconductor quantum-dot lasers," *Physical Review B*, vol. 69, no. 23, article 235314, 2004.

[23] G. A. Narvaez, G. Bester, and A. Zunger, "Carrier relaxation mechanisms in self-assembled (In, Ga) As/Ga As quantum dots: efficient $P \rightarrow S$ Auger relaxation of electrons," *Physical Review B*, vol. 74, no. 7, article 075403, 2006.

[24] K. Schuh, P. Gartner, and F. Jahnke, "Combined influence of carrier-phonon and coulomb scattering on the quantum-dot population dynamics," *Physical Review B*, vol. 87, no. 3, article 035301, 2013.

[25] S. Tomić, "Intermediate-band solar cells: influence of band formation on dynamical processes in InAs/GaAs quantum dot arrays," *Physical Review B*, vol. 82, no. 19, article 195321, 2010.

[26] J. Siegert, S. Marcinkevičius, and Q. X. Zhao, "Carrier dynamics in modulation-doped InAs/GaAs quantum dots," *Physical Review B*, vol. 72, no. 8, 2005.

[27] F. Cappelluti, A. Khalili, and M. Gioannini, "Open circuit voltage recovery in quantum dot solar cells: a numerical study on the impact of wetting layer and doping," *IET Optoelectronics*, vol. 11, no. 2, pp. 44–48, 2017.

[28] P. W. Fry, J. J. Finley, L. R. Wilson et al., "Electric-field-dependent carrier capture and escape in self-assembled InAs/GaAs quantum dots," *Applied Physics Letters*, vol. 77, no. 26, pp. 4344–4346, 2000.

[29] T. Müller, F. Schrey, G. Strasser, and K. Unterrainer, "Ultrafast intraband spectroscopy of electron capture and relaxation in InAs/GaAs quantum dots," *Applied Physics Letters*, vol. 18, no. 17, pp. 3572–3574, 2003.

[30] L. A. Pettersson, L. S. Roman, and O. Inganäs, "Modeling photocurrent action spectra of photovoltaic devices based on organic thin films," *Journal of Applied Physics*, vol. 86, no. 1, pp. 487–496, 1999.

[31] F. Bertazzi, F. Cappelluti, S. D. Guerrieri, F. Bonani, and G. Ghione, "Self-consistent coupled carrier transport full-wave EM analysis of semiconductor traveling-wave devices," *IEEE Transactions on Microwave Theory and Techniques*, vol. 54, no. 4, pp. 1611–1618, 2006.

[32] D. Kim, M. Tang, J. Wu et al., "Si-doped InAs/GaAs quantum-dot solar cell with AlAs cap layers," *IEEE Journal of Photovoltaics*, vol. 6, no. 4, pp. 906–911, 2016.

[33] H. Y. Liu, I. R. Sellers, T. J. Badcock et al., "Improved performance of 1.3μm multilayer InAs quantum-dot lasers using a high-growth-temperature GaAs spacer layer," *Applied Physics Letters*, vol. 85, no. 5, pp. 704–706, 2004.

[34] H. Y. Liu, S. L. Liew, T. Badcock et al., "p-doped 1.3μm InAs/GaAs quantum-dot laser with a low threshold current density and high differential efficiency," *Applied Physics Letters*, vol. 89, no. 7, article 073113, 2006.

[35] F. K. Tutu, I. R. Sellers, M. G. Peinado et al., "Improved performance of multilayer InAs/GaAs quantum-dot solar cells using a high-growth-temperature GaAs spacer layer," *Journal of Applied Physics*, vol. 111, no. 4, article 046101, 2012.

[36] M. Wolf and H. Rauschenbach, "Series resistance effects on solar cell measurements," *Advanced Energy Conversion*, vol. 3, no. 2, pp. 455–479, 1963.

[37] H. L. Wang, F. H. Yang, and S. L. Feng, "Photoluminescence in Si and Be directly doped self-organized InAs/GaAs quantum dots," *Journal of Crystal Growth*, vol. 212, no. 1-2, pp. 35–38, 2000.

[38] A. Markus, M. Rossetti, V. Calligari, J. Chen, and A. Fiore, "Role of thermal hopping and homogeneous broadening on the spectral characteristics of quantum dot lasers," *Journal of Applied Physics*, vol. 98, no. 10, article 104506, 2005.

[39] M. P. Lumb, M. A. Steiner, J. F. Geisz, and R. J. Walters, "Incorporating photon recycling into the analytical drift-diffusion model of high efficiency solar cells," *Journal of Applied Physics*, vol. 116, no. 19, article 194504, 2014.

[40] A. Musu, F. Cappelluti, T. Aho, V. Polojärvi, T. K. Niemi, and M. Guina, "Nanostructures for light management in thin-film GaAs quantum dot solar cells," in *Solid-State Lighting*, no. article JW4A.45, 2016Optical Society of America, 2016.

[41] Sopra database. 1995-2016 Software Spectra, Inc.http://sspectra.com/sopra.html.

[42] C. G. Bailey, D. V. Forbes, R. P. Raffaelle, and S. M. Hubbard, "Near 1 V open circuit voltage InAs/GaAs quantum dot solar cells," *Applied Physics Letters*, vol. 98, no. 16, article 163105, 2011.

[43] P. Lam, S. Hatch, J. Wu et al., "Voltage recovery in charged InAs/GaAs quantum dot solar cells," *Nano Energy*, vol. 6, pp. 159–166, 2014.

[44] S. J. Polly, D. V. Forbes, K. Driscoll, S. Hellstrom, and S. M. Hubbard, "Delta-doping effects on quantum-dot solar cells," *IEEE Journal of Photovoltaics*, vol. 4, no. 4, pp. 1079–1085, 2014.

[45] S. Sanguinetti, D. Colombo, M. Guzzi et al., "Carrier thermodynamics in InAs/$In_xGa_{1-x}As$ quantum dots," *Physical Review B*, vol. 74, no. 20, article 205302, 2006.

[46] T. Kita, R. Hasagawa, and T. Inoue, "Suppression of nonradiative recombination process in directly Si-doped InAs/GaAs quantum dots," *Journal of Applied Physics*, vol. 110, no. 10, article 103511, 2011.

16

A Five-Level H-Bridge STATCOM for an Off-Grid PV Solar Farm under Two Controllers PI and PI$^{\lambda}$-MPC Hybrid

Onur Ozdal Mengi

Department of Energy Systems Engineering, Engineering Faculty, Giresun University, 28100 Giresun, Turkey

Correspondence should be addressed to Onur Ozdal Mengi; onurmengi@yahoo.com

Academic Editor: Mahmoud M. El-Nahass

Investigations were presented in order to eliminate the reactive power on microgrid loads fed by an off-grid and mid-power photovoltaic solar energy system (PVSES) with a static synchronous compensator (STATCOM) device. The electric network is specifically characterized by P-Q loads, ambient temperature, and widely variable solar radiation levels. Two main innovations are developed. Firstly, the STATCOM apparatus is a 5-level H-bridge inverter with capacitances as load and must totally compensates the reactive power in the network load. Secondly, this compensation is controlled by a set of fractional PI (PI$^{\lambda}$) and model predictive control (MPC) hybrid. The efficiencies of these controllers were compared with classical PI controllers. Large simulations, without and with reactive power compensation, in steady and transient states, are carried out to underline the merits of the presented works, by performing in the MATLAB-Simulink environment.

1. Introduction

The sun warms our planet with the resulting energy released from nuclear explosions taken place within itself and makes our planet a place to live for people, animals, and plants. The sun, which is our source of life along with the rapid progress of technology, now also serves as an energy source. Photovoltaic solar cells are systems that convert solar energy from sunlight into electricity and thus produce electricity. Solar cells come together to form panels, and the panels come together to form PVSESs [1]. The solar panels generate DC voltage at the output. The DC voltage and power obtained from the solar panels are used to feed either DC loads or AC loads by converting it to AC voltage via an inverter. Various solar cell production techniques and materials have been developed. There are many solar panels made of numerous materials such as amorphous silicon, copper indium gallium selenide, and cadmium telluride [2]. The costs and efficiencies of these panels are quite different. Nowadays, solar cells yielding 35% of efficiency have been produced [3].

Large scale applications have become more economical with the development of solar panel production techniques, the increase in production, and also the increase in panel power values and yields [4]. As a result of these, large power plants have been begun to build from large scale photovoltaic solar panels [5]. These power plants are on-grid or off-grid [6]. In off-grid PVSESs, compensation must be applied to the system for reducing harmonics caused by the loads having reactive power and preventing injecting of the reactive power from the source [7, 8]. There are many different compensation techniques. Among them, voltage source converter- (VSC-) based compensators are the Static VAR Compensator (SVC), STATCOM, Thyristor Controlled Series Compensator (TCSC), Static Synchronous Series Compensator (SSSC), and Unified Power Flow Controller (UPFC) [9]. STATCOMs are one of the dynamic systems capable of doing this [10, 11]. STATCOMs are able to do compensation by appropriately switching the reactive power generated by capacitors. In general, IGBT, thyristor, and transistor are used for switching. These components are switched by techniques such as Pulse Width Modulation (PWM) and Sinusoidal Pulse Width Modulation (SPWM) [12]. There are different topologies in the inner structure of the STATCOM. These systems, which are usually multilevel inverters, are switched by appropriate control techniques. Current source,

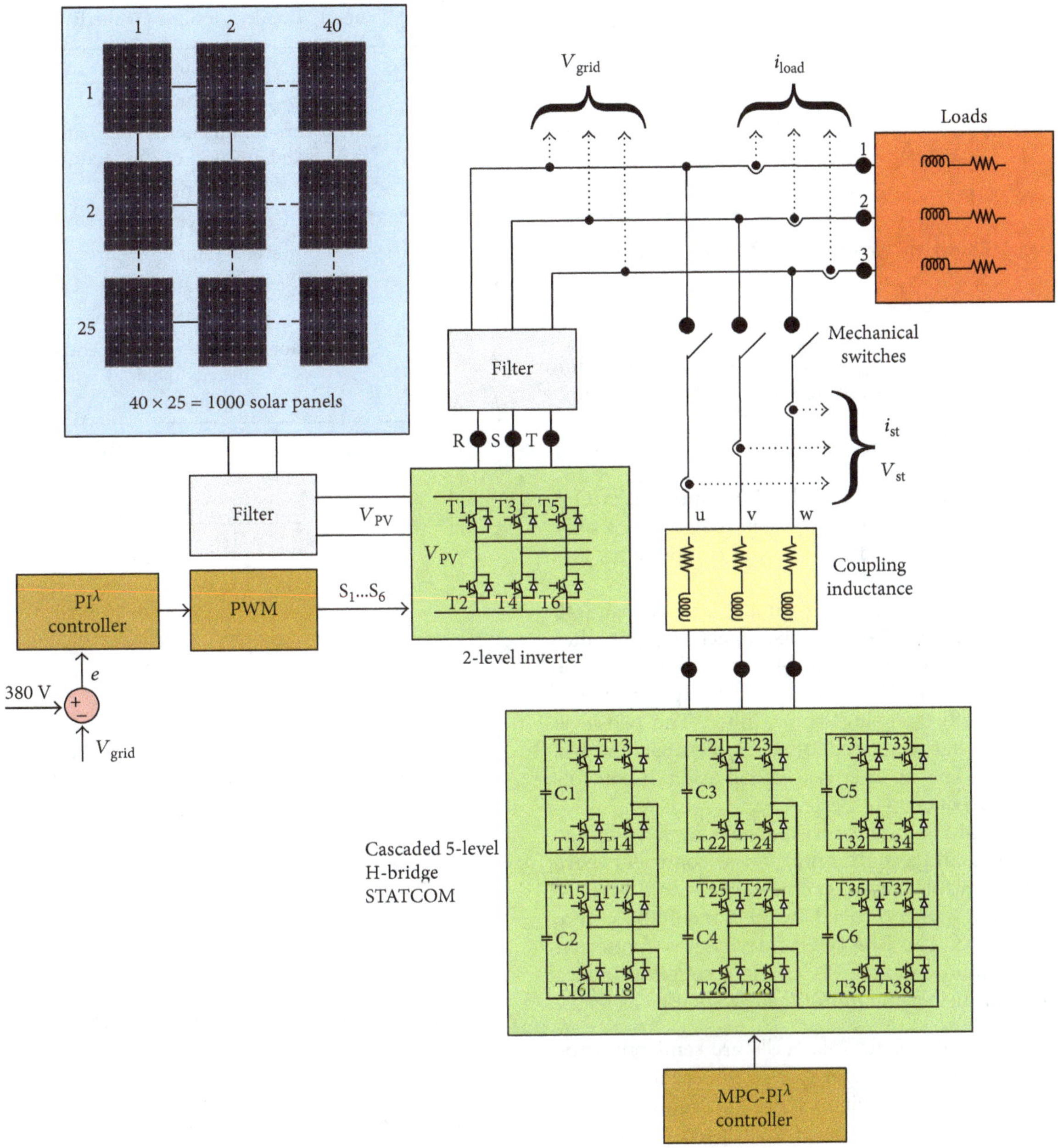

FIGURE 1: PVSES with a STATCOM device.

voltage source, impedance source, quasi impedance source inverters, and multilevel inverters are commonly used [13]. These inverters are controlled by using classical and advanced control methods such as a PID controller [14], fuzzy logic controllers [15], fractional PID [16], artificial neural networks [17], vector control methods [18], model predictive control [19], and genetic algorithms [20].

The innovation carried out by this study is the implementation of a five-level H inverter for the STATCOM converter (including 24 IGBT-s transistors). In addition, this device allows eliminating the load reactive powers, to regulate the network voltage by increasing its efficiency and reducing harmonics contained in the electric network. The plant is fed by an off-grid photovoltaic solar power generation system, and the STATCOM converter is controlled by PI^{λ} and MPC. The efficiency of the obtained advanced control structure was compared using classical PI controllers. The voltages on loads were tried to keep constant at a value of 380 V/50 Hz with a classical 2-level three-phase voltage. It has been tried to create challenging conditions by continuously changing the loads, ambient temperature, and the solar irradiation level. In this environment, the performances of the controllers were tested and the contribution of the advanced control techniques was addressed by comparing the results obtained. All of the simulation was performed in the MATLAB/Simulink environment.

This paper is organized in four sections. Section 1 contains the introduction. The STATCOM device and controllers

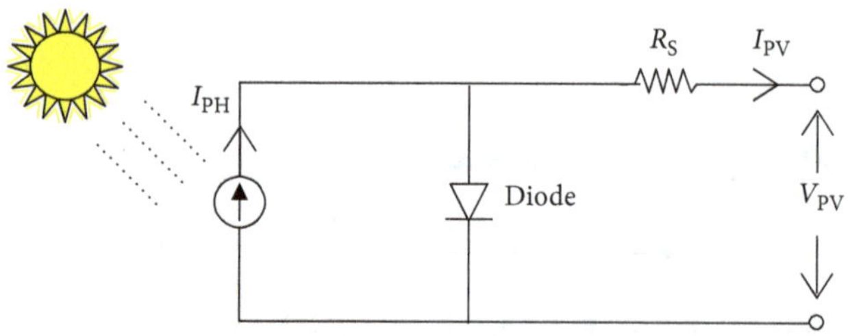

FIGURE 2: Equivalent circuit model of a PV solar cell.

are described in Section 2. Section 3 presents simulation results. The last section gives some remarks and conclusion.

2. Materials and Methods

The overall scheme of the proposed PVSES and STATCOM system was given in Figure 1. The system consisted of solar panels, filters, inverters, loads, measuring components, and a STATCOM device.

The solar panels contained 25 parallel branches and 40 panels in each branch. The PV solar panels' output filter has a 6.5 μF capacitor. The voltage obtained from these panels was converted to 380 V/50 Hz AC voltage through a 2-level inverter before passing from a filter. The two-level inverter output filter consisted of a 22 μF capacitor and 2 mH inductance. This inverter is switched using PI controllers, and carrier frequency is 2 kHz.

The STATCOM device consisted of 5-level H-bridge IGBTs, capacitors, and a control unit. These controllers here were used to control the capacitor voltages and IGBTs. Capacitor voltage was controlled by PI^{λ}, and IGBTs were controlled by MPC. The STATCOM device was connected to point of common coupling (PCC) via a coupling inductance. Different resistive and motor loads were used as loads.

2.1. PV System. Photovoltaic solar cells are semiconductor components that turn solar light to electrical energy. There are different equivalent circuit models. The equivalent circuit model with one diode of a solar panel is seen in Figure 2 [21, 22].

The output voltage of the PV cell is as in (1) [23].

$$V_{PV} = \frac{N}{\lambda} \ln\left(\frac{I_{SC} - I_{PV} + MI_0}{MI_0}\right) - \frac{N}{M} R_s I_{PV}. \quad (1)$$

Descriptions of the variables used here are seen in Table 1.

2.2. STATCOM Device. A STATCOM is a parallel-connected static synchronous compensator. The STATCOM is a FACTS device that continuously supplies reactive power to or injects reactive power from the system that it is connected. STATCOM has important features such as having a fast response time and fewer space requirements and showing very good dynamic characteristic under different operating conditions [24].

TABLE 1: The descriptions for the PV cell.

Expression	Definition	Unit
V_{PV}	Cell output voltage	V
I_{PH}	Photocurrent, function of irradiation level, and junction of temperature	A
I_{PV}	Cell output current	A
I_{SC}	Cell short-circuit current	A
I_0	Reverse saturation current	A
R_S	Series resistance of the cell	Ω
N	Series cells per string	
λ	Constant coefficient and depends upon the cell material	
M	Parallel strings	

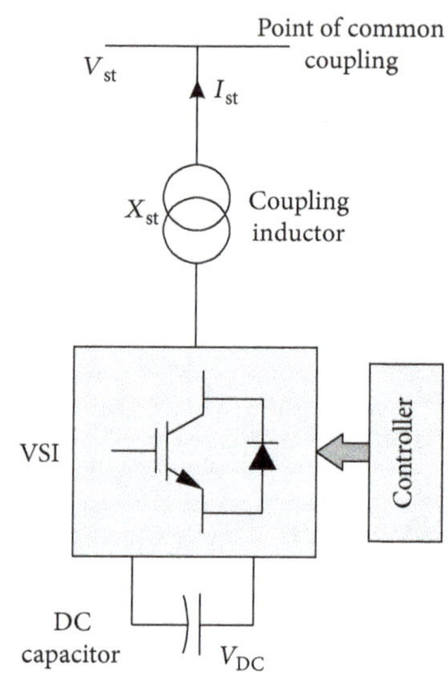

FIGURE 3: Basic structure of the STATCOM device.

The active and reactive power exchanges between the STATCOM and the system are given by (2).

$$P_{st} = \frac{V_{st} V_s}{X_{st}} \sin \delta,$$
$$Q_{st} = \frac{V_s}{X_{st}} (V_s - V_{st} \cos \delta). \quad (2)$$

V_s is system voltage, V_{st} is STATCOM voltage, and δ is the phase angle between V_{st} and V_s. X_{st} is reactance between the STATCOM and the system. L_{st} is an inductive part of the X_{st}.

The basic structure of the STATCOM device is shown in Figure 3.

The values of the passive elements of the STATCOM are calculated in (3) and (4).

$$C_{dc} = \frac{\sqrt{2} \cdot I_{RMS}}{\omega \cdot \Delta V_{pp}} \left[1 - \sin\left(\arccos\left(\frac{M_{max}\pi}{4}\right)\right)\right]. \quad (3)$$

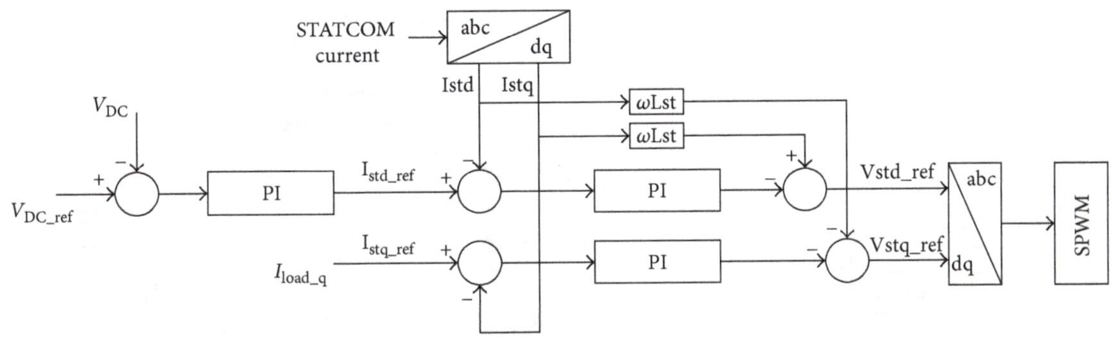

FIGURE 4: General block diagram of an indirect current controller.

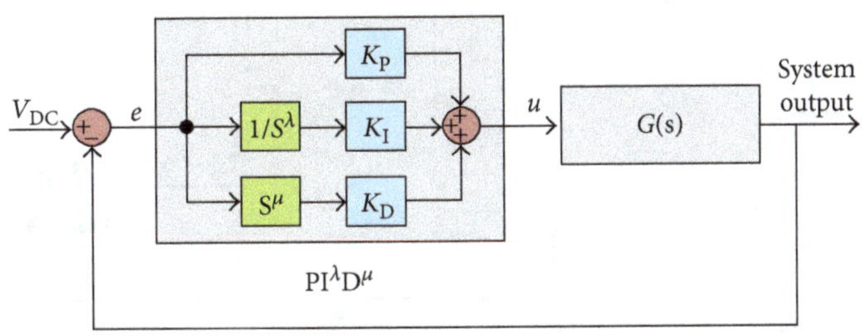

FIGURE 5: General block diagram of a $PI^{\lambda}D^{\mu}$ controller.

I_{RMS} is the RMS value of the system current, ΔV_{pp} is peak to peak variation of the three-phase system voltage, M_{max} is the maximum value of the modulation index.

$$\frac{V_{DC}}{8f_s(n_{Level}-1)\Delta I_r} \leq L \leq \frac{V_{st}-V_s}{I_{load\_f}\omega I_{st}}. \quad (4)$$

V_{DC} is DC voltage of the STATCOM, f_s is PWM frequency, ΔI_r is current ripple, V_{st} is VSI voltage, V_s is system voltage, I_{load_f} is the fundamental harmonic component of the load current, ω is angular velocity, I_{st} is the RMS value of VSI output current, and n_{Level} is the number of the inverter level [25, 26].

In this study, the five-level H-bridge STATCOM device is used. The reason of this, 3-level inverters give lower quality results and 7-level inverters have much more complicated results because of using too much IGBTs.

2.3. STATCOM Device Control Structure with a Classical PI Controller. Various structures for the STATCOM device control are used in literature. One of them is the indirect current control method [26]. In this method, generally, classical PI controllers are used. This control structure is shown in Figure 4.

In Figure 4, I_{load_q} is defined as the reactive component of load current.

2.4. Fractional PID Controller ($PI^{\lambda}D^{\mu}$). $PI^{\lambda}D^{\mu}$ controllers are advanced controllers that can give better results than classical PI controllers. They are widely used because they are able to adjust more precisely than PID controllers. The best knowns are the descriptions of Caputo, Grunwald-Letkinov, and Reimann-Liouville [27, 28]. The $PI^{\lambda}D^{\mu}$ controller general block diagram is shown in Figure 5.

System output, $C(s)$, is as seen in (5).

$$C(s) = \frac{U(s)}{E(s)} = K_P + \frac{K_I}{s^{\lambda}} + K_D s^{\mu}. \quad (5)$$

Here, λ and $\mu \geq 0$. λ is the order of integration, and μ is the order of the differentiator. K_P, K_I, and K_D are the PID controller gains. $U(s)$ and $E(s)$ are the control and error signals, respectively. In this study, $PI^{\lambda}D^{\mu}$ controller software was performed by FOMCON toolbox [29].

2.5. Model Predictive Controller (MPC). MPC is a control technique designed on the basis of the prediction of behavior of the systems in the next step. It is aimed at estimating the next possible switching states and providing the selection of switching that minimizes the error over the existing states. To achieve this, a correct model of the system has to be revealed. At each sampling time, the control of a particular topology defines a cost function that is minimized as an optimization problem. The general application of MPC in the STATCOM device is as shown in Figure 6.

MPC generally consists of two groups: finite control set MPC and continuous control set MPC. They are divided into two groups among themselves. Finite control set MPC is divided into optimal switching vector MPC

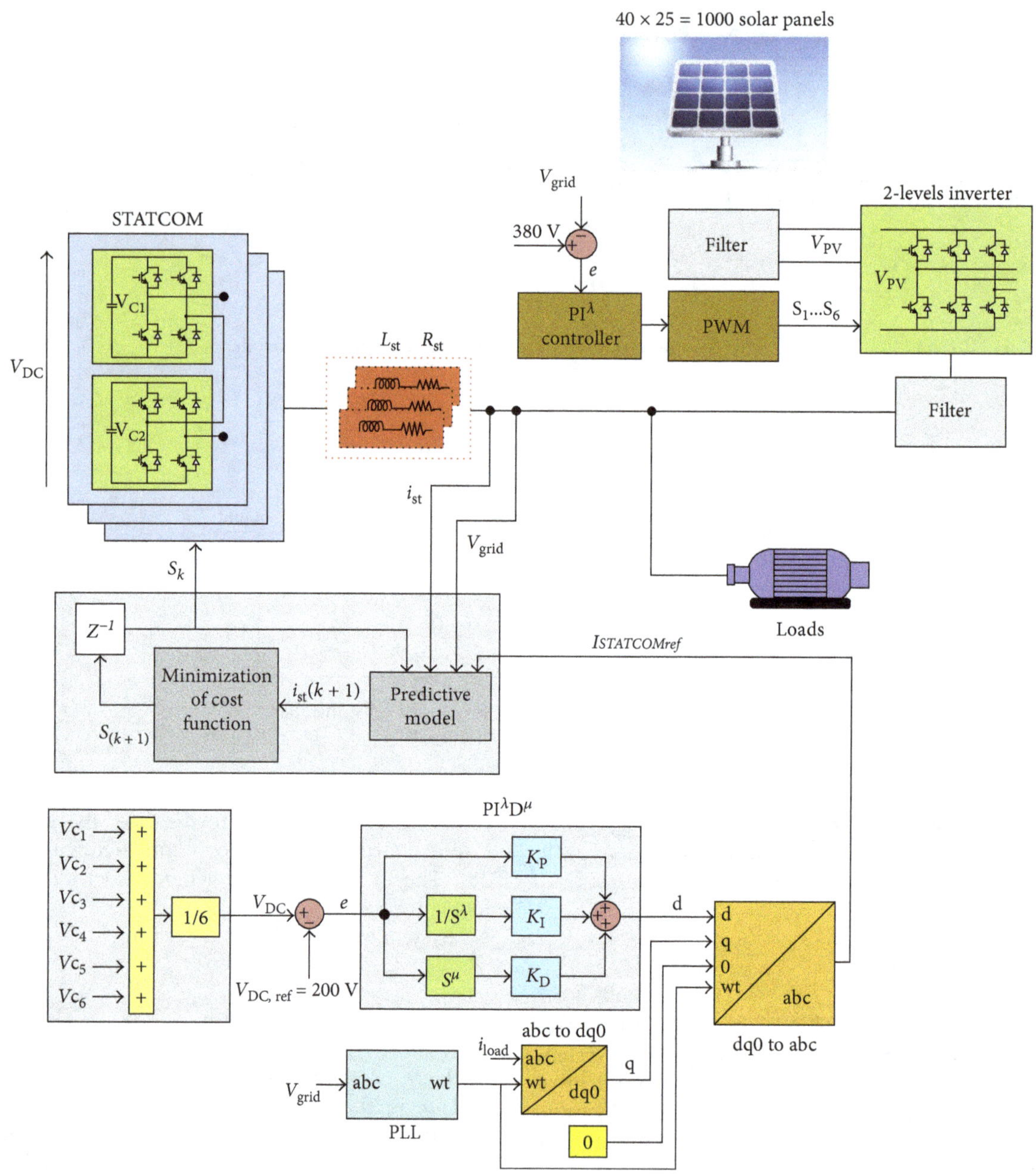

FIGURE 6: Block diagram of MPC for the STATCOM device.

and optimal switching sequence MPC, and continuous control set MPC is divided into generalized predictive control and explicit MPC. In the simulation, an optimal switching vector MPC technique was used because it does not need a modulator and fixed switching frequency, it is intuitive, and it can perform online optimization [30]. The predictive control technique can be used easily in multilevel STATCOM applications.

For multilevel STATCOMs, the AC side equality is as in (6).

$$V_s^{abc} = L_{st}\frac{d}{dt}i_{st}^{abc} + R_{st}i_{st}^{abc} + V_{st}^{abc}. \quad (6)$$

TABLE 2: Photovoltaic solar panel parameters.

Value	Unit
Open-circuit voltage	36.3 V
Short-circuit current	7.84 A
Maximum power	213.15 W
Voltage at maximum power point	29 V
Current at maximum power point	7.35 A
Temperature coefficient of short-circuit current	0.102%/deg.C
Temperature coefficient of open-circuit voltage	−0.36%/deg.C
Serial resistance	0.39 ohm
Diode saturation current	2.92e-10 A
Diode ideality factor	0.98

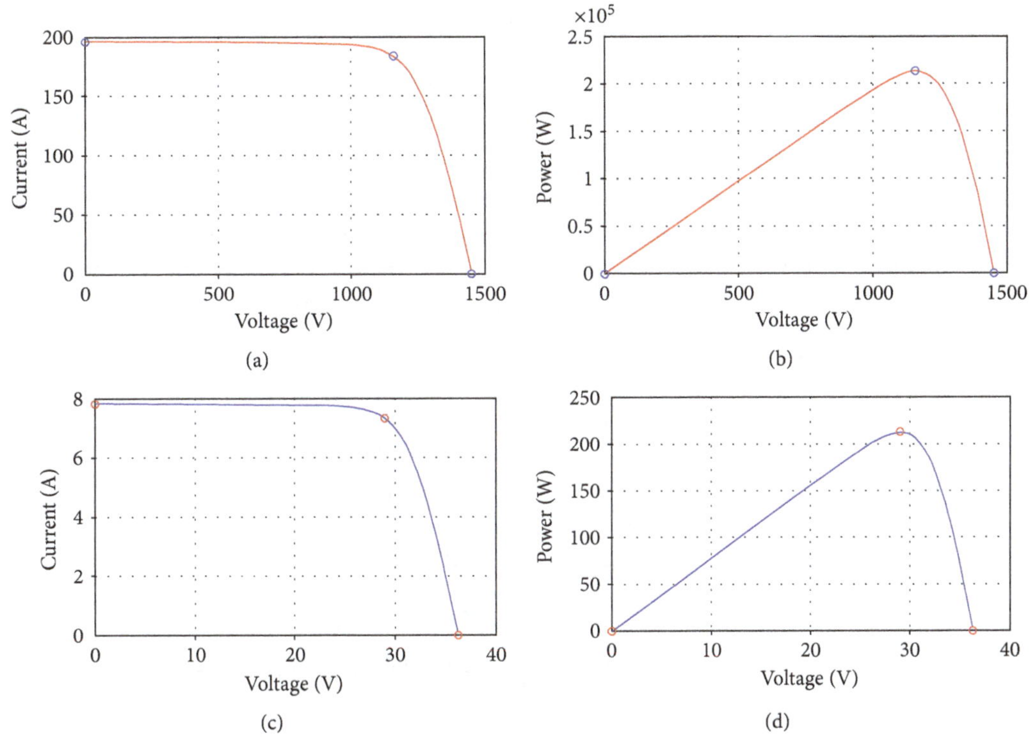

Figure 7: (a) Total PVSES output V-I diagram, (b) total PVSES output V-P diagram, (c) a photovoltaic solar panel V-I diagram, (d) a photovoltaic solar panel V-P diagram.

If i_{st}^{abc} is extracted and discretized from (6), its next value can be computed from (7).

$$i_{st}^{abc}(k+1) = \left(1 - T_s \frac{R_{st}}{L_{st}}\right) i_{st}^{abc}(k) + \frac{T_s}{L_{st}} \left\{ V_s^{abc}(k) - V_{st}^{abc}(k) \right\}. \quad (7)$$

Here, T_s is the sampling time. k expresses the current states of the variables. $k+1$ is used to predict the future behavior of the system. R_{st} is a coupling resistor, L_{st} is a coupling inductance, i_{st}^{abc} is 3-phase STATCOM current, V_s^{abc} is 3-phase grid voltage, and V_{st}^{abc} is 3-phase STATCOM voltage.

A cost function should then be used. This cost function should evaluate the all possible switching combinations, choose the option minimizing cost function, and apply it in the next step. The cost function used to minimize the current error is as in (8) [31].

$$g = \left| i_{st}^{abc,ref}(k+1) - i_{st}^{abc}(k+1) \right|. \quad (8)$$

3. Results and Discussion

Firstly, the installed system was run without the STATCOM device. What happened in this case was investigated. Then, the STATCOM device with a PI controller and finally the PI$^\lambda$-MPC hybrid controller were used, and their results were shown. In all the simulations, the total power of the photovoltaic solar panels and the system used in PVSES were kept same.

3.1. Data of PV Panels. The solar panels are commercially available and have the characteristics as defined in Table 2. The passive elements of the STATCOM device consist of 4700 μF capacitors and of coupling inductance with 56 mΩ and 12 mH [32]. In Figure 7 are shown the exploitation curves of each panel and the set of panels (40 in series and 25 in parallel).

Figure 7 and Table 2 define the data used in commercial PV panels.

Also, in all the simulations, the ambient temperature level and solar irradiation level were changed according to Figure 8.

3.2. Simulation Results and Discussion. Here, the PV filter has a 6.5 μF capacitor. The voltage obtained from these panels was converted to RMS phase-to-phase voltage 380 V under 50 Hz AC voltage through a 2-level inverter before passing from a filter. This filter consists of a 22 μF capacitor and 2 mH inductance. The system sampling time is 5 μs. PV panel output voltage changes between 1200 V and 1350 V, and output current changes up to 150A.

The installed system was tested using variable loads, according to Table 3 for their connection and disconnection times.

In addition, during the time simulation (between 0 and 1 s), ambient solar irradiation and temperature change as shown in Figure 8.

3.2.1. Case 1: Installation without the STATCOM Device (Mechanical Switch Open). The two-level inverter is switched using PI controllers, and carrier frequency is 2 kHz, involving

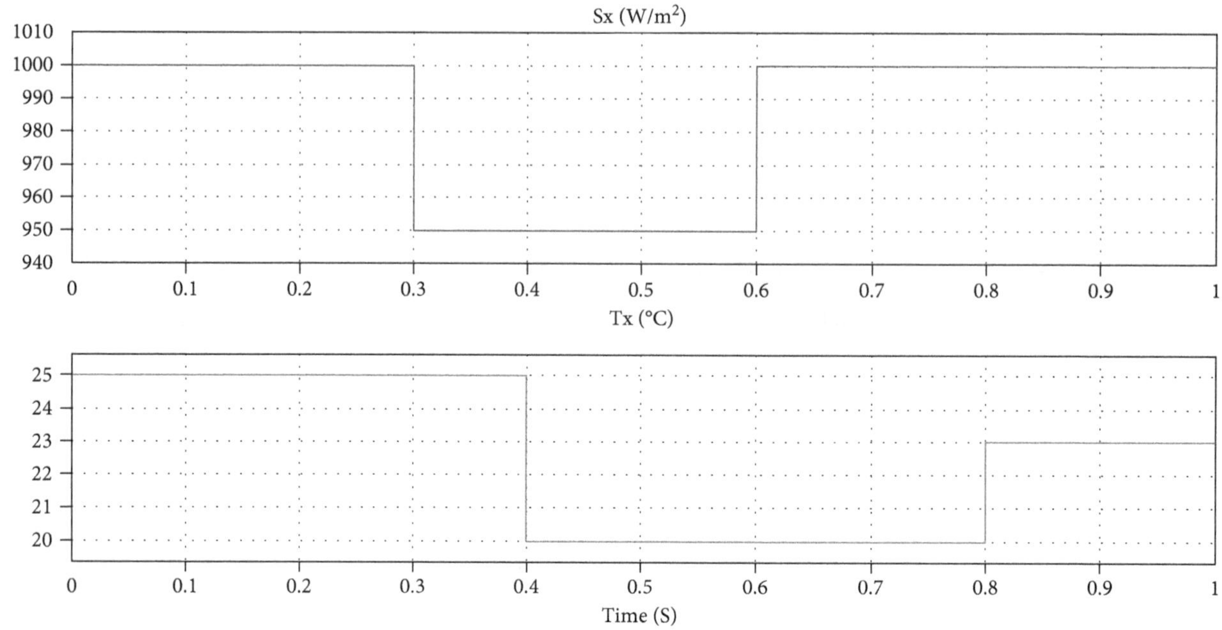

Figure 8: Ambient solar irradiation level and temperature changes.

Table 3: Loads used in the system and times that they entered the circuit.

Load type	Value	Commutation times (s)
R	+50 kW	t1 = 0: connection
RL	+10 kW/+10 kVAr	t2 = 0.3: connection
RL	−10 kW/−10kVAr	t3 = 0.45: disconnection
R	+5 kW	t4 = 0.6: connection
L	+5 kVAr	t5 = 0.8: connection

a pulse width modulation principle. In the absence of the STATCOM device, the compensation of reactive power cannot not be performed. The graphs obtained in this case are shown in Figure 9.

Figure 9 shows that the reactive power taken by the load cannot be compensated when there is no STATCOM device in the system. The same reactive power was present on the load and the grid side.

In Figure 9, change in the voltage amplitude on the loads is shown. PI controllers kept the load voltage constant at 380 V. Here, 2-level IGBTs inverters were used. PI controllers that set the load voltage were also set to $K_P = 0.5$, $K_I = 100$. These values were found by different simulations work on this system. The controller could hold magnificently the RMS phase-to-phase load voltage constant at 380 V. The change was around ±5 V.

As shown in this figure, the dispersion of RMS phase-to-phase voltage is very negligible: around 1 V/380 V and after sudden changes of load, we notice: for $\Delta_2 = -5$ V; for $\Delta_3 = +5.8$ V; $\Delta_5 = -1$ V. And the disturbance duration are lesser 25 ms.

3.2.2. Case 2: STATCOM with a PI Controller. The results obtained when an indirect current control method with PI controllers were used to control both the V_{DC} voltage and the switching signals are shown in Figures 10 and 11. The ambient conditions and also the constant change in loads forced the system. PI controllers used in an indirect current control method were set to $K_P = 5$ and $K_I = 40$. For PI controllers that adjust the load voltage, $K_P = 0.5$ and $K_I = 100$ were also taken. This value is the same as the values used in the case without the STATCOM device.

Figure 10 shows the switching states belonging to a phase when the STATCOM with a PI controller is enabled. The first two graphs show the switchings of IGBTs on the first H-bridge (triggers of T11 and T12) and the other two on the second H-bridge (triggers of T15 and T16). While two IGBTs on each bridge were switched, the opposite way of the switchings was sent to the other two.

Figure 12 shows the efficiency of the STATCOM device. The reactive power compensation was performed successfully. The reactive power take from the grid was zero.

Also, Figure 12 shows change in the tension on loads. Although PI controllers had a high overshoot value, they kept the load voltage at 380 quite well. The PI controller constants were $K_P = 0.5$ and $K_I = 100$. These values are deduced by trial and error.

As shown in this figure, overshoot is too bad. This value reach 465 V. So, ΔV_O is +85 V. Nevertheless, ΔV_2 and ΔV_3 are negligible. Between t_2 and t_3, the reactive power residue is approximately +300 VAr.

In Figure 13, each DC capacitor voltage V_{DC} in the STATCOM device were kept at 200 V and the capacitor voltage on H-bridge converter is perfectly balanced. This underlines that the more the controllers efficiently operate, the compensation action is more performing.

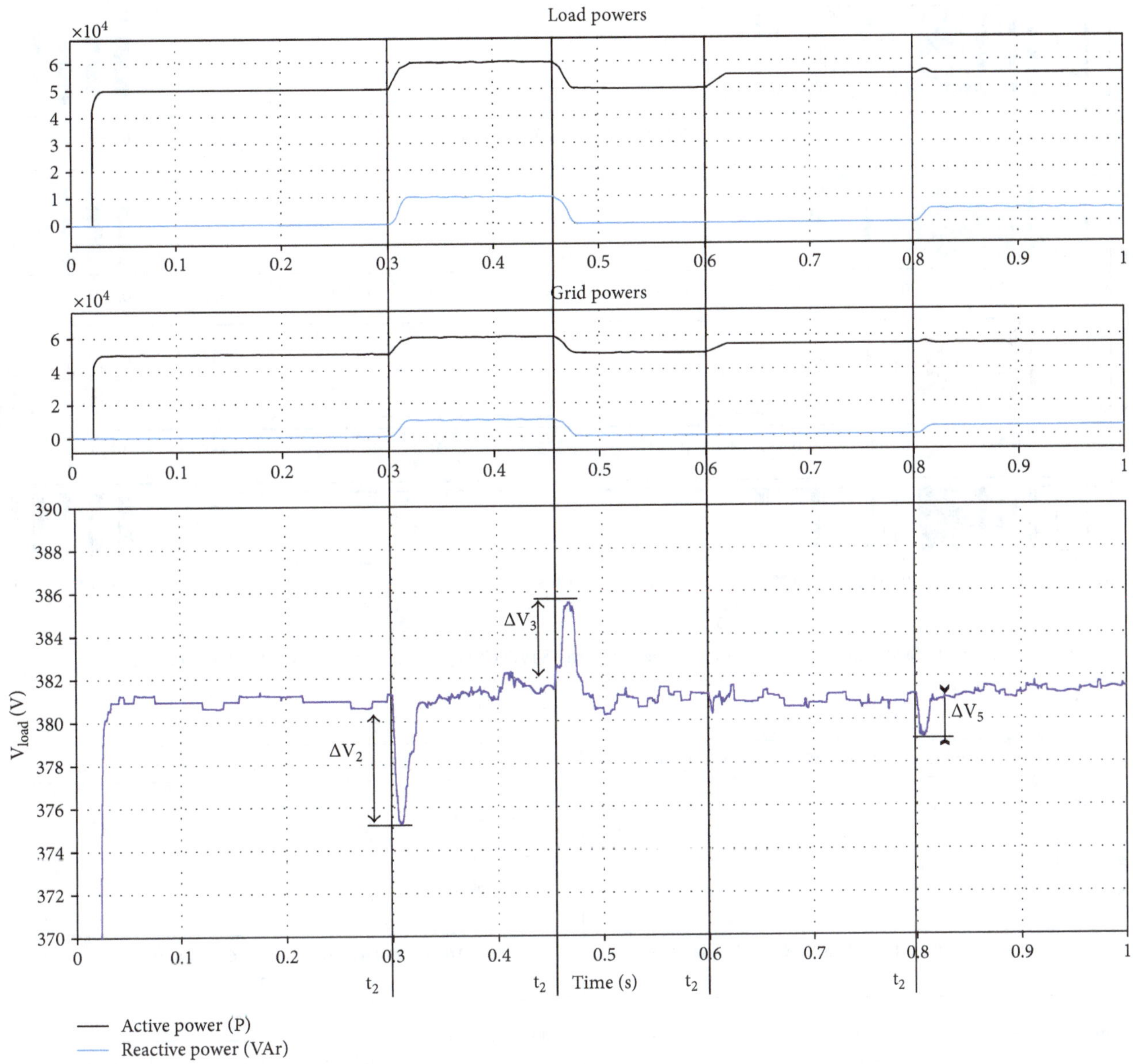

FIGURE 9: Active, reactive power, and load voltage states on the system.

Figure 11 shows FFT analysis of the PVSES system output voltage and current. For the phase-to-neutral voltage, the shape is perfectly sinusoidal and the module gap difference is lesser than 1%.

3.2.3. Case 3: STATCOM with a PI$^\lambda$-MPC Hybrid Controller. The results obtained in using PI$^\lambda$-MPC hybrid controllers to control both V_{DC} voltage and switching signals are shown in Figures 14 and 15. Here, V_{DC} voltage was controlled by PI$^\lambda$ and the switching signals in a 24 IGBTs 5-level STATCOM device were controlled by MPC. Both ambient conditions and loads are constantly changing.

The switching states belonging to a phase when the STATCOM device with a PI$^\lambda$-MPC hybrid controller is enabled are shown in Figure 14 (the first two graphs: triggers of T_{11} and T_{12}, and the other two: triggers of T_{15} and T_{16}). The switching signals are different than in the case using PI controllers. These signals took randomly in the time scale.

Figure 16 shows the efficiency of the STATCOM device with a PI$^\lambda$-MPC hybrid controller. The reactive power compensation was performed successfully. The reactive power taken from the grid was zero. The fluctuation was less than in the case using a PI controller.

This figure shows change in the voltage on loads. Overshoot and settling time values were better than PI controllers. For PI$^\lambda$ controllers that adjust the load voltage, $K_P = 0.5$, $K_I = 100$, and $\lambda = 1.2$ were taken. As shown in this figure, overshoot is very low. This value reach 384 V. So, ΔV_O is +4 V. But, sudden load changes like at ΔV_2, ΔV_3, and ΔV_5 are important. Despite the change in voltage on the load, the reactive power is well compensated. At this time,

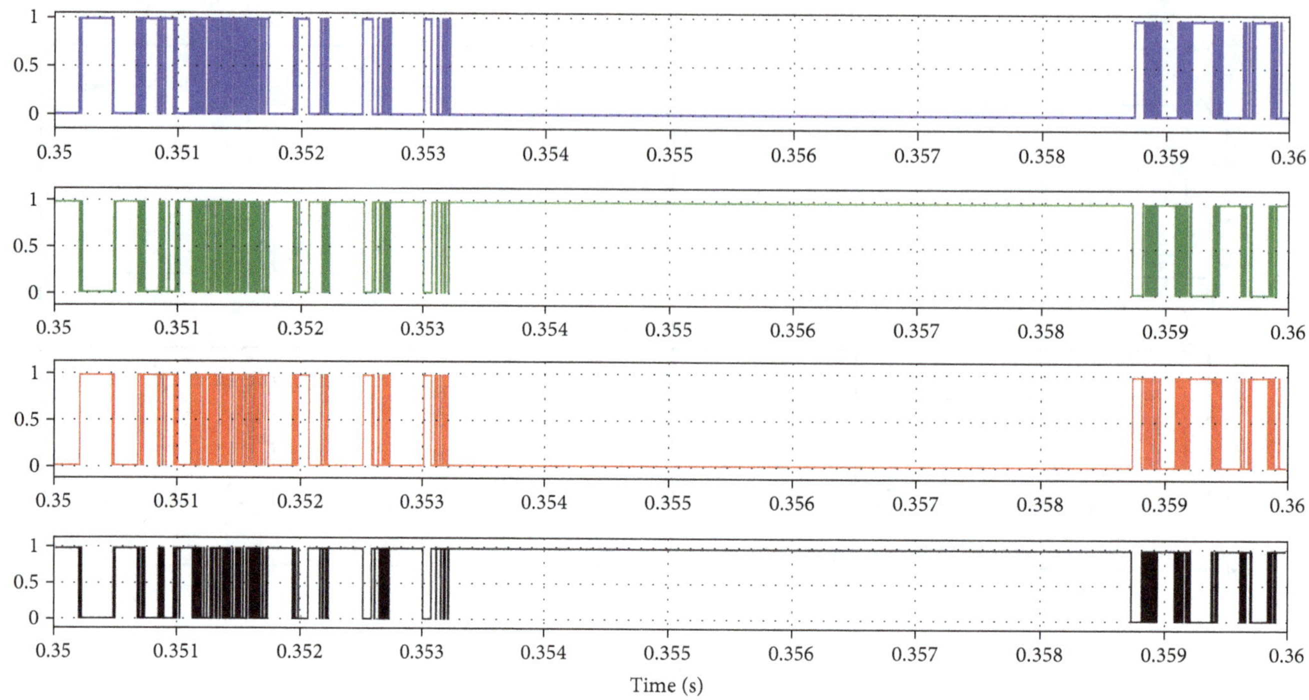

FIGURE 10: Switchings belonging to a phase when the STATCOM device with a PI controller is enabled.

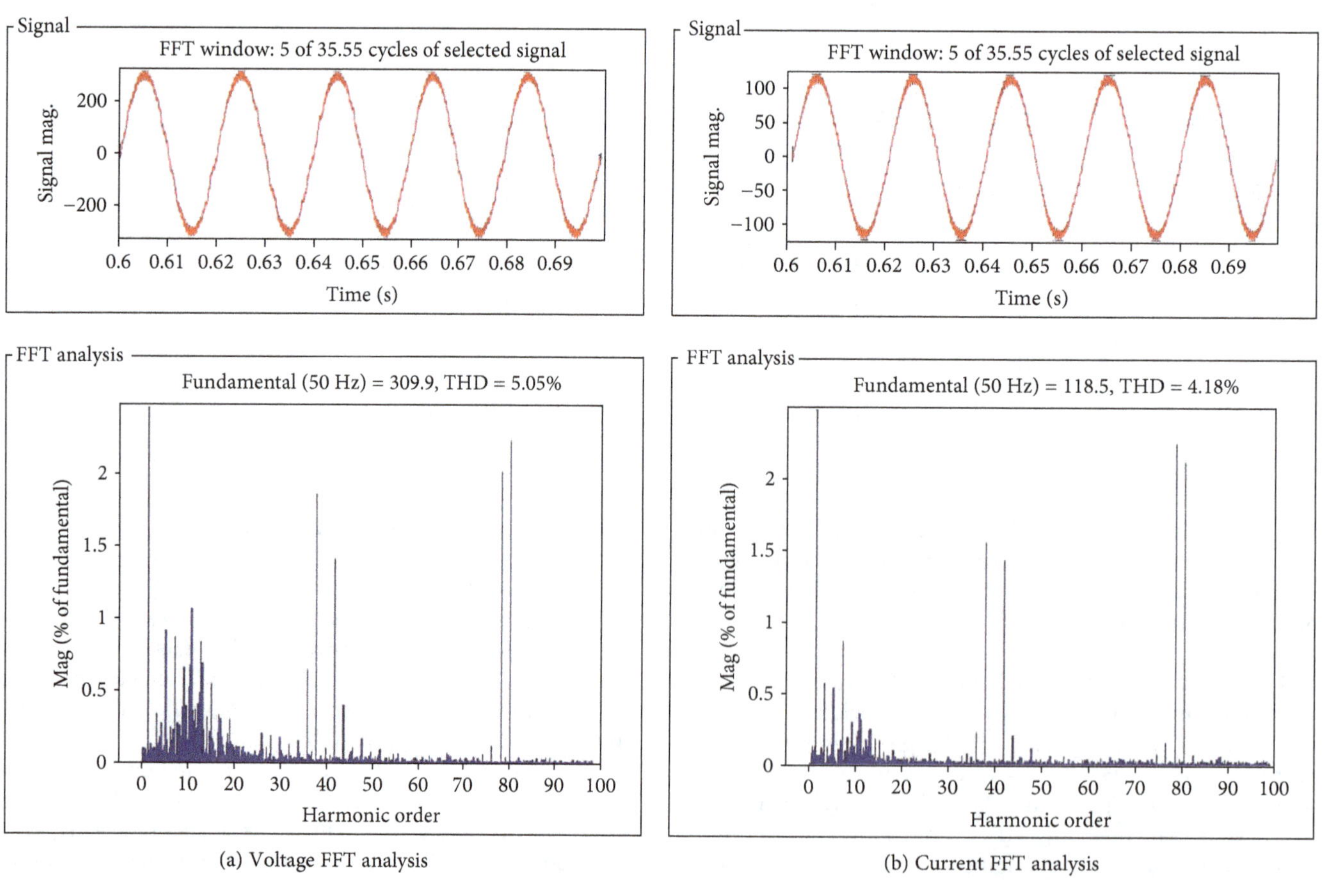

FIGURE 11: FFT analysis of phase to neutral voltage and line current on the electric network of the system.

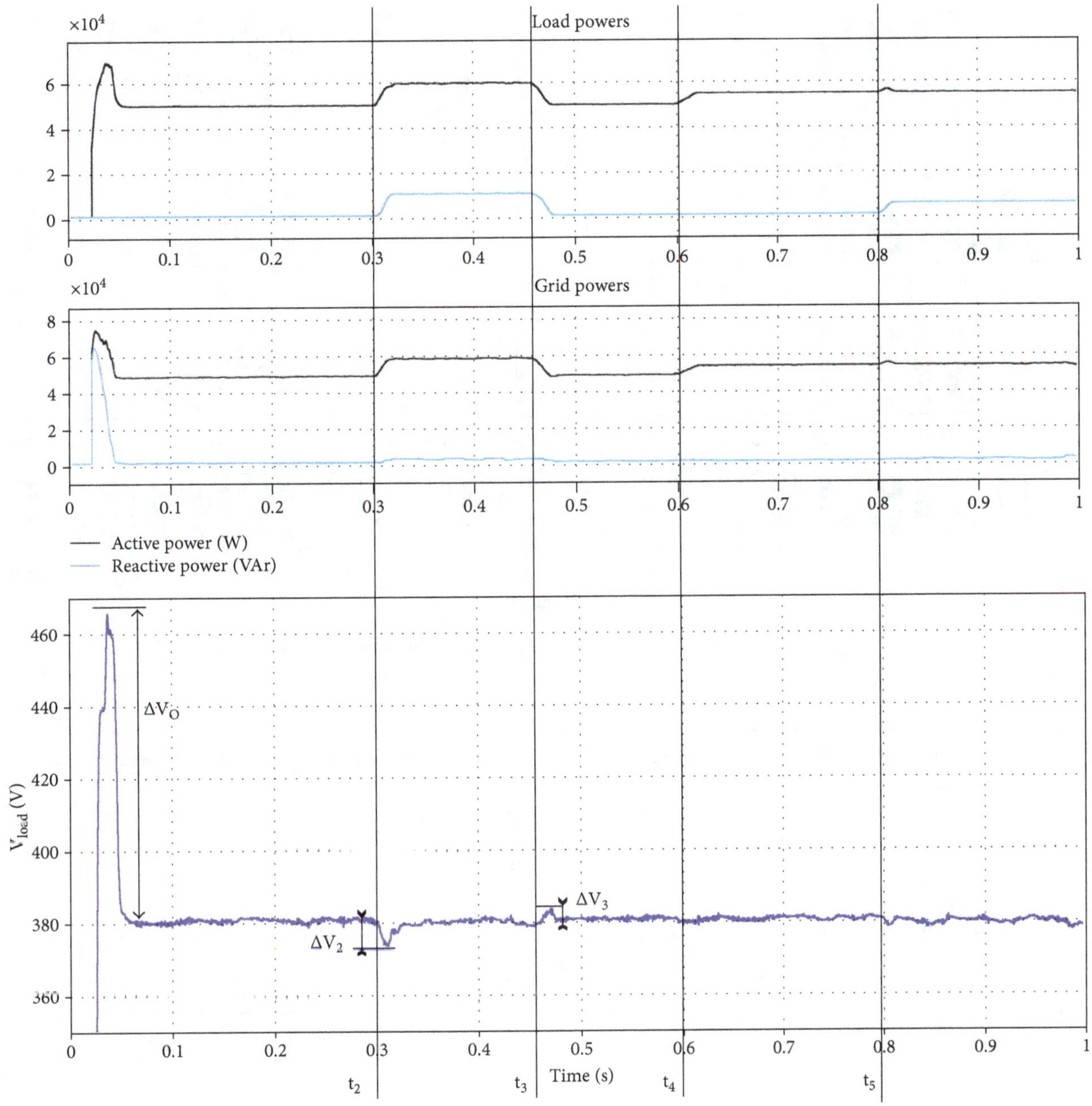

FIGURE 12: Reactive power compensation when the STATCOM device with a PI controller is enabled and voltage change on loads.

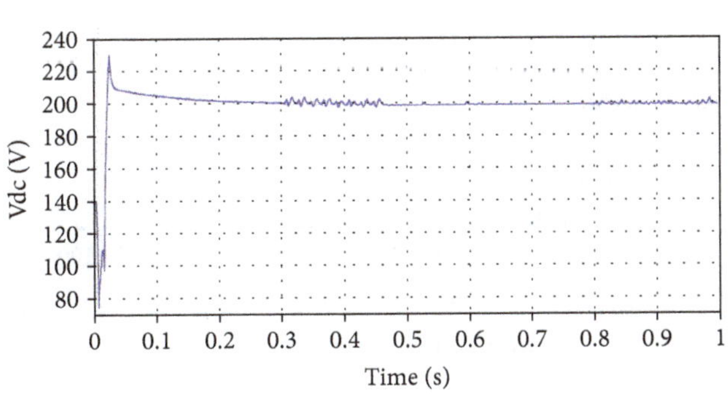

FIGURE 13: Capacitor voltage control in the STATCOM device with a PI-controller.

reactive power on the system is approximately +50 VAr. This is a too low value.

Figure 17 shows that the capacitor voltages in the STATCOM device are kept at 200 V. Both settling time and overshoot values were much better than the case using a PI controller. Also, the fluctuations in the load changes were less. Here, the PI^{λ}-type controller was used. In PI^{λ} controllers used both in the setting of voltage on load and in the controlling of V_{DC} voltage in the STATCOM, controller parameters are $K_P = 5$, $K_I = 40$, ve $\lambda = 1.2$ were taken.

Figure 15 underlines the good results shown in Figure 11 (the same values about the modulus of phase-to-neutral and its THD).

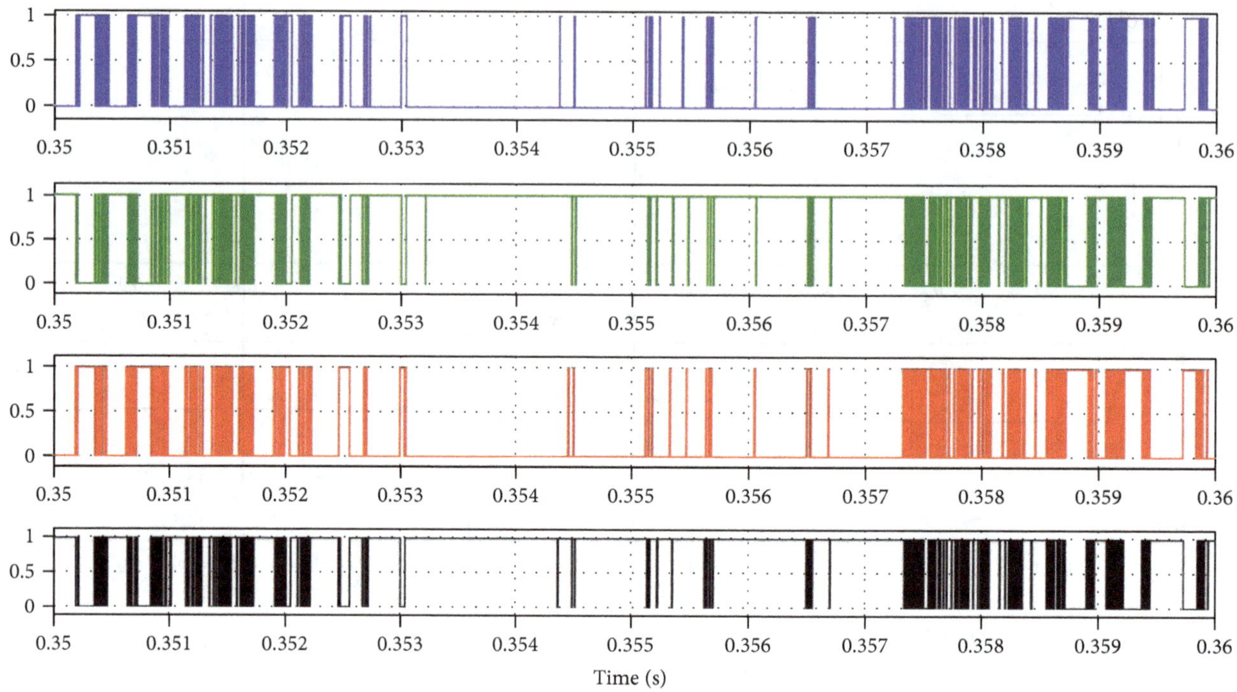

FIGURE 14: Switchings belonging to a phase when the PI$^{\lambda}$-MPC hybrid controller is enabled.

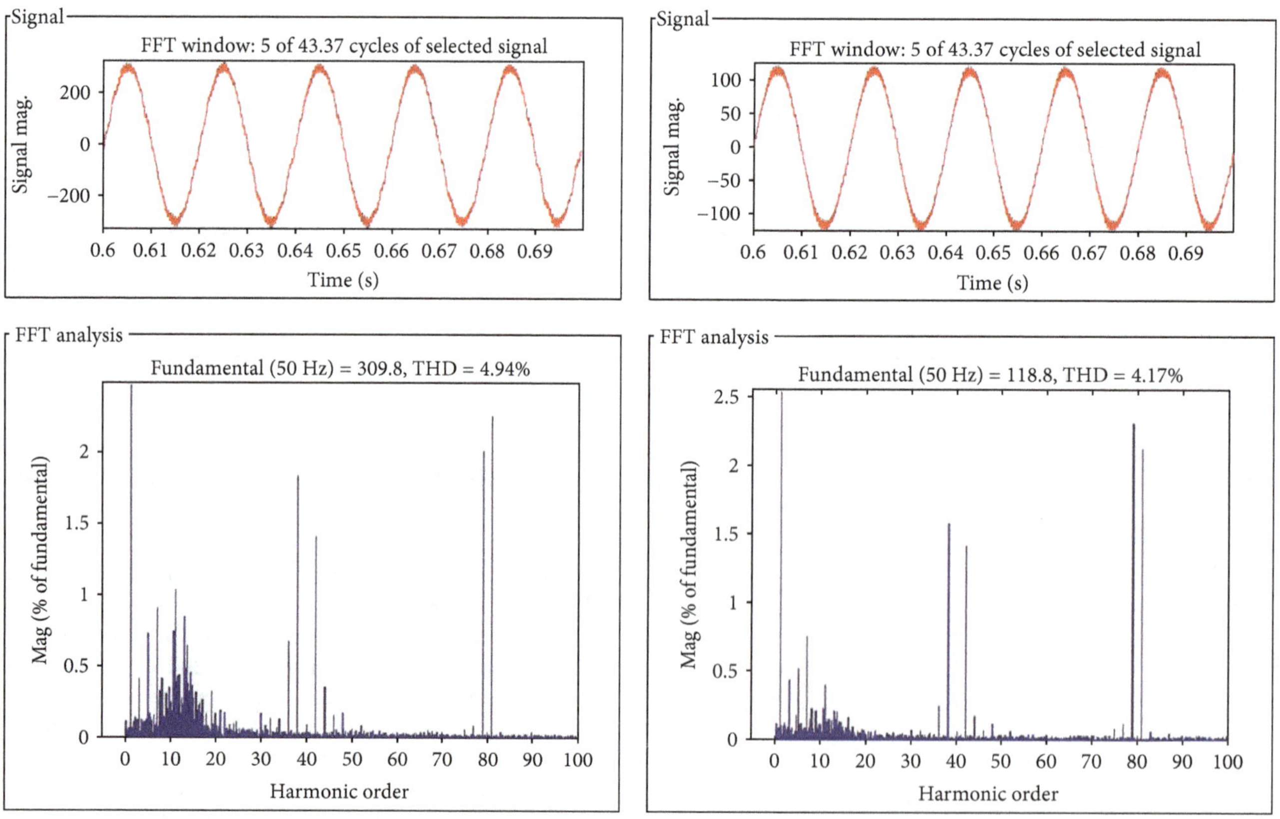

(a) Voltage FFT analysis

(b) Current FFT analysis

FIGURE 15: FFT analysis of a PVSES system.

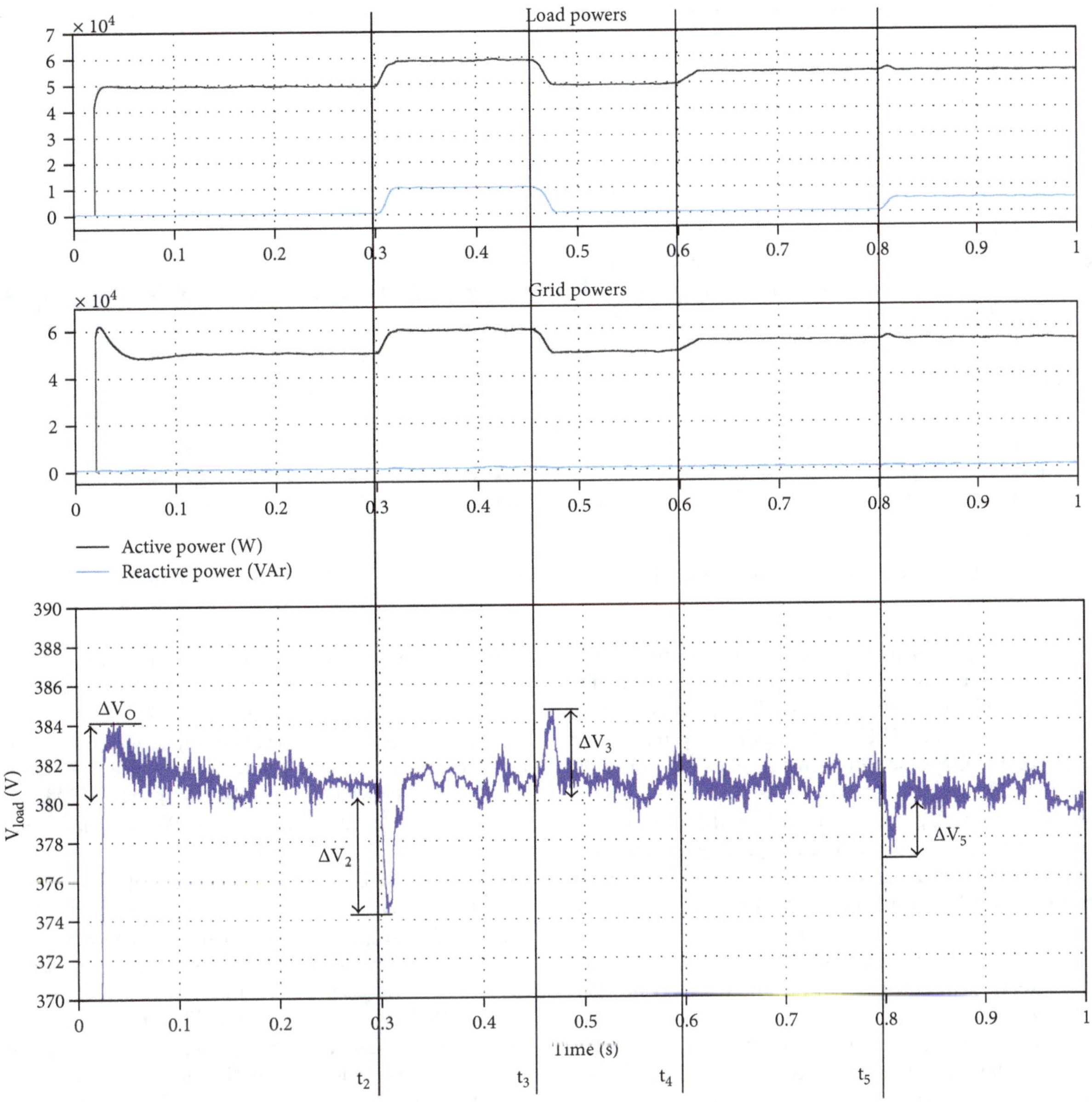

FIGURE 16: Reactive power compensation when the STATCOM device with a PI^{λ}-MPC hybrid controller is enabled and load voltage change.

4. Conclusion

In the study, the performance of the STATCOM device using a PI^{λ}-MPC hybrid controller has been investigated and compared with the version using a PI-type controller to eliminate the reactive power on loads fed by off-grid solar panels. The system was forced by constantly changing the ambient temperature and the solar irradiation level as well as the loads, and the efficiencies of the controllers were tried to be shown meanwhile.

In the system, V_{DC} voltage was controlled by PI^{λ} and the switching signals in a 24 IGBTs 5-level STATCOM device were controlled by MPC. The most important reason for this is that PID-type controllers can be placed everywhere comfortably, while it is necessary to use detailed mathematical equations for designing MPCs. This makes MPCs difficult to design. However, the controllers with different PID types can be placed easily everywhere that errors are in. MPC controls the switching signals in the STATCOM device. It works more efficiently because it chose the right and the best one from the switching possibilities. V_{DC} voltage was controlled by PI^{λ} that is able to adjust more precisely than PI. By combining these two controllers, a hybrid controller was created.

The comparison of STATCOM device results obtained from PI and PI^{λ}-MPC controllers shows that PI^{λ} V_{DC} overshoot parameters are better than PI because PI^{λ} overshoot is 213 V and PI overshoot is 230 V. Similarly, the reactive power overshoot value for a PI controller is 70 kVAr. But MPC has no overshoot. PI^{λ} load voltage overshoot is 384 V; PI controller load voltage overshoot is 465 V.

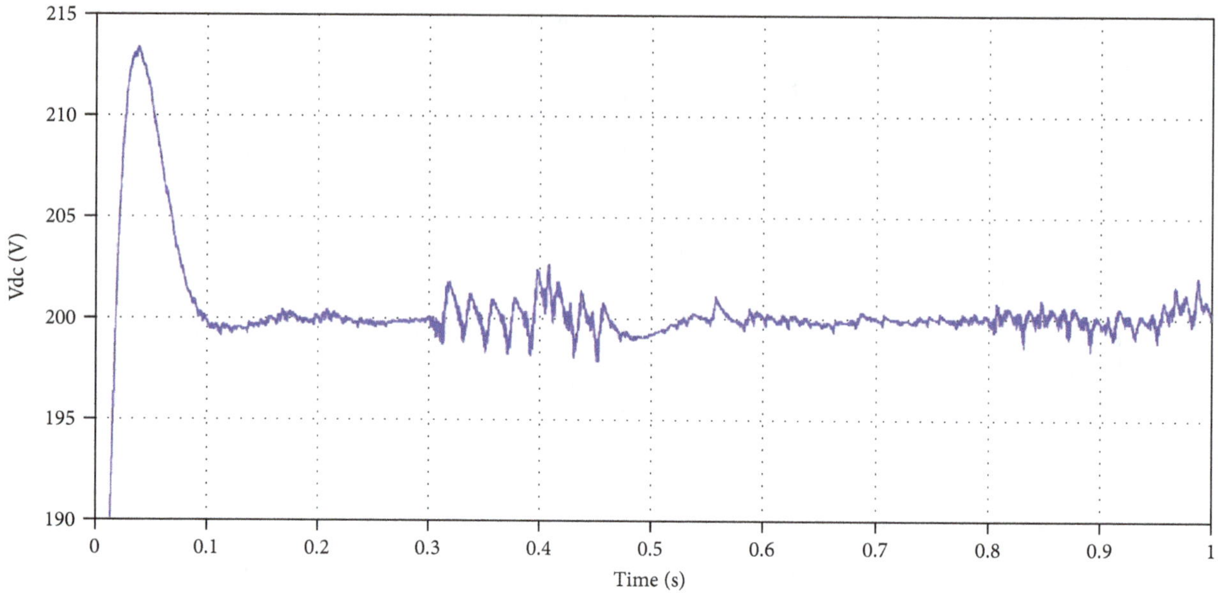

FIGURE 17: Capacitor voltage control in the STATCOM device with a PI^λ-MPC hybrid controller.

PI^λ V_{DC} settling times are better than those of PI. The PI^λ settling time is 0.1 ms, and the PI settling time is 0.3 ms for the steady state. The reactive power settling time is 0.02 ms for a PI controller and MPC is approximately 0 ms, and finally, the load voltage settling time is 0.05 ms for a PI controller and 0.02 ms for a PI^λ controller.

If we compare steady state errors, PI^λ and PI controllers are equal for the V_{DC} steady state error parameter and load voltage steady state error parameter. V_{DC} steady state error is 0 in both PI and PI^λ controllers. The load voltage steady state error change range is ±6V in both PI and PI^λ controllers. But reactive power steady state error is better than MPC. MPC steady state error is 50 VAr, and the PI controller value is 300 VAr.

The results of the FFT analysis show that the PI^λ-MPC controller is better than the PI controller. Total harmonic distortion of the PI^λ-MPC controller is 4.94% for voltage analysis and 4.17% for current analysis, and then PI controller FFT analysis for voltage is 5.05% and current is 4.18%.

As mentioned in the previous paragraph, the reactive power compensation performed very well and the error rates were very weak. The overshoot values of PI controllers were higher than those of PI^λ-MPC hybrid controllers. Also, the settling time values were shorter on the PI^λ-MPC hybrid controller. In general, the performance of the PI^λ-MPC hybrid controller is much better than those of PI controllers.

Conflicts of Interest

The author declares that they have no conflicts of interest.

References

[1] I. H. Altas and A. M. Sharaf, "A novel on-line MPP search algorithm for PV arrays," *IEEE Transactions on Energy Conversion*, vol. 11, no. 4, pp. 748–754, 1996.

[2] T. D. Lee and A. U. Ebong, "A review of thin film solar cell technologies and challenges," *Renewable and Sustainable Energy Reviews*, vol. 70, pp. 1286–1297, 2017.

[3] J. Praveen and V. VijayaRamaraju, "Materials for optimizing efficiencies of solar photovoltaic panels," *Materials Today: Proceedings*, vol. 4, no. 4, Part D, pp. 5233–5238, 2017.

[4] S. C. W. Krauter, *Solar Electric Power Generation*, Springer, New York, NY, USA, 2006.

[5] S. Sumathi, L. A. Kumar, and P. Surekha, *Solar PV and Wind Energy Conversion Systems*, Springer, Switzerland, 2015.

[6] V. Chamola, B. Krishnamachari, and B. Sikdar, "Green energy and delay aware downlink power control and user association for off-grid solar-powered base stations," *IEEE Systems and Journal*, vol. PP, no. 99, pp. 1–12, 2017.

[7] L. Luo, W. Gu, X.-P. Zhang et al., "Optimal siting and sizing of distributed generation in distribution systems with PV solar farm utilized as STATCOM (PV-STATCOM)," *Applied Energy*, vol. 210, pp. 1092–1100, 2018.

[8] P. R. Yelpale, H. T. J. Hod, and C. Charmole, "New control strategy for grid integrated PV solar farm to operate as STATCOM," in *2017 International Conference on Circuit, Power and Computing Technologies (ICCPCT)*, Kollam, India, April 2017.

[9] R. Sirjani and A. R. Jordegi, "Optimal Placement and Sizing of Distribution Static Compensator (D-STATCOM) in Electric Distribution Networks: A review," *Renewable and Sustainable Energy Reviews*, vol. 77, no. 4, pp. 688–694, 2017.

[10] H. Yonezawa, M. Tsukada, and J. J. Paserba, "Study of a STATCOM application for voltage stability evaluated by dynamic PV curves and time simulations," in *2000 IEEE Power Engineering Society Winter Meeting*, pp. 1471–1476, Singapore, Singapore, January 2000.

[11] S. Eshtehardiha, M. B. Poodeh, and M. R. Zare, "Improvement of STATCOM performance with multi variable control methods," in *IET-UK International Conference on Information and Communication Technology in Electrical Sciences (ICTES 2007)*, pp. 105–111, Tamil Nadu, India, December 2007.

[12] K. Rao and N. Patel, "Effectuation in SPWM technique for 3-Φ VSI using STM32F4 discovery board interfaced with MATLAB," in *2016 IEEE 7th Power India International Conference (PIICON)*, Bikaner, India, November 2016.

[13] F. L. Luo and H. Ye, *Advenced DC/AC Inverters*, CRC Press, USA, 2013.

[14] Z. Wu, C. Xu, and Y. Yang, "Adjustable PID control based on adaptive internal model and application in current shared control of multi inverters," *Journal of the Franklin Institute*, vol. 354, no. 7, pp. 2699–2711, 2017.

[15] A. Saygın and A. Kerem, "Fuzzy logic based control of a loaded asynchronous motor using a 6-switched 3-level inverter," in *2017 18th International Conference on Computational Problems of Electrical Engineering (CPEE)*, pp. 1–4, Kutna Hora, Czech Republic, September 2017.

[16] I. Rasoanarivo, K. Arab-Tehrani, and F. M. Sargos, "Fractional order PID and modulated hysteresis for high performance current control in multilevel inverters," in *2011 IEEE Industry Applications Society Annual Meeting (IAS)*, pp. 1–7, Orlando, FL, USA, October 2011.

[17] H. Soliman, I. Abdelsalam, H. Wang, and F. Blaabjerg, "Artificial neural network based DC-link capacitance estimation in a diode-bridge front-end inverter system," in *2017 IEEE 3rd International Future Energy Electronics Conference and ECCE Asia (IFEEC 2017 - ECCE Asia)*, pp. 196–201, Kaohsiung, Taiwan, June 2017.

[18] M. R. B. Fathima, P. M. J. Princy, and S. R. Prasath, "Mathematical modeling of SVPWM ınverter fed 3 phase induction motor vector control in MATLAB/Simulink environment," in *2017 International Conference on Circuit, Power and Computing Technologies (ICCPCT)*, pp. 1–8, Kollam, India, April 2017.

[19] J. Su and D. Sun, "Model predictive torque-vector control for four-switch three-phase inverter-fed PMSM with capacitor voltage offset suppression," in *Electrical Machines and Systems (ICEMS)*, pp. 1–5, Sydney, New South Wales, Australia, August 2017.

[20] S. Qian, Y. Ye, H. Wu, and Z. Zhuang, "Optimizing PWM switching sequence of inverters using an immune genetic algorithm," in *2016 8th International Conference on Intelligent Human-Machine Systems and Cybernetics (IHMSC)*, pp. 7–10, Hangzhou, China, August 2016.

[21] I. H. Altas and A. M. Sharaf, "A photovoltaic array simulation model for MATLAB-Simulink GUI environment," in *2007 International Conference on Clean Electrical Power*, pp. 341–345, Capri, Italy, May 2007.

[22] I. H. Altas and A. M. Sharaf, "A novel maximum power fuzzy logic controller for photovoltaic solar energy systems," *Renewable Energy*, vol. 33, no. 3, pp. 388–399, 2008.

[23] M. A. S. Masoum, H. Dehbonei, and E. F. Fuchs, "Theoretical and experimental analyses of photovoltaic systems with voltage and current-based maximum power-point tracking," *IEEE Transactions on Energy Conversion*, vol. 17, no. 4, pp. 514–522, 2002.

[24] B. Singh, K. Al-Haddad, R. Saha, and A. Chandra, "Static synchronous compensators (STATCOM): a review," *IET Power Electronics*, vol. 2, no. 4, pp. 297–324, 2009.

[25] N. G. Hingorani and L. Gyugti, *Understanding FACTS, Consepts and Technology of Flexible AC Transmission Systems*, IEEE Press, USA, 2000.

[26] K. Yanmaz, I. H. Altas, and O. O. Mengi, "Five level cascaded H-bridge D-STATCOM using a new fuzzy and PI controllers model for wind energy systems," *Advances in Electrical and Computer Engineering*, vol. 17, no. 4, pp. 49–58, 2017.

[27] P. Shah and S. Agashe, "Review of fractional PID controller," *Mechatronics*, vol. 38, pp. 29–41, 2016.

[28] I. Rasoanarivo and F. M. Sargos, "Multi-objective analysis for designing and controlling microgrids under multi-control with PID, MHCC and FOPID controllers," in *2013 IEEE Industry Applications Society Annual Meeting*, pp. 7–10, Lake Buena Vista, FL, USA, October 2013.

[29] A. Tepljakov, E. Petlenkov, and J. Belikov, "FOMCON: fractional-order modeling and control toolbox for MATLAB," in *2011 Proceedings of the 18th International Conference Mixed Design of Integrated Circuits and Systems (MIXDES)*, pp. 684–689, Gliwice, Poland, June 2011.

[30] S. Vazquez, J. Rodriguez, M. Rivera, L. G. Franquelo, and M. Norambuena, "Model predictive control for power converters and drives: advances and trends," *IEEE Transactions on Industrial Electronics*, vol. 64, no. 2, pp. 935–947, 2017.

[31] J. Rodriguez and P. Cortes, *Predictive Control of Power Converters and Electrical Drives*, Wiley-IEEE Press, United Kingdom, 2012.

[32] I. Rasoanarivo, P. Boos, and B. Davat, "Improvements of parallel active filter performances," in *EPE'97*, pp. 898–902, Trondheim, Norway, September 1997.

17

Design, Electrical, and Optical Modelling of Bulk Heterojunction Polymer Solar Cell

Muhammad Ali,[1] Ahmed Shuja,[1] Ahsan Baig,[1,2] Erum Jamil,[1,2] and Muhammad Amjad[3]

[1]*Center for Advanced Electronics and Photovoltaic Engineering (CAEPE), International Islamic University, Islamabad 44000, Pakistan*
[2]*Department of Electrical Engineering, International Islamic University, Islamabad 44000, Pakistan*
[3]*Department of Electronics Engineering, Islamia University of Bahawalpur, Bahawalpur 63100, Pakistan*

Correspondence should be addressed to Muhammad Ali; muhammadali@iiu.edu.pk

Academic Editor: Kei Ohkubo

The energy scenario today is focused on the development and usage of solar cells, especially in the paradigm of clean energy. To readily create electron and hole pairs, solar cells utilize either photoactive or photosensitive components. A bulk heterojunction (BHJ) is a nanolayer consisting of donor and acceptor components with a large interpenetrated acceptor and donor contact area. In this context, a mix of P3HT and PCBM offers novelty for its use as an acceptor as well as a donor. In the work presented here, we address the mechanism of modelling and characterization of a BHJ-based polymer solar cell. Here, a new design of BHJ polymer solar cells have been designed, modelled, using Silvaco TCAD in the Organic Solar module, and matched with an already assembled device having similar features. Using this model, we have been able to estimate key parameters for the modelled devices, such as the short-circuit current density, open-circuit voltage, and fill factor with less than 0.25 error index compared to the fabricated counterpart, paving the way for fabless polymer solar cell design and optimization.

1. Introduction

Recently, there is a keen interest in developing electronic devices consisting of organic materials [1]; the examples of which include thin film transistors, memory devices, and LEDs as well as solar cells. The reason for this interest is their greater flexibility as well as easier fabrication, especially over large areas, along with the lower production cost [2, 3]. Of special consideration are polymer photovoltaic solar cells (PSCs) [4], which traditionally contain 3 tiers: (a) polymer zone (for photon absorption), (b) region for the production of electron and hole pairs, and (c) a layer for the charges to be carried through to the contacts.

The BHJ-PSC is a developing technology because it offers improved power conversion efficiency [5, 6]; amongst the BHJ-PSC, the P3HT:PCBM devices have been researched in greater depth [7–9]. It is, therefore, vital to analyse the connection between the organic blend layer and the electrode to estimate the overall performance of the BHJ-PSCs [10]. One problem they face is that of charge buildup, which leads to recombination loss and is dominant with nonohmic contacts [11]. The traditional method of addressing it is by using ohmic contacts. Even so, to develop devices with better performance efficiency, dedicated charge extraction layers (CELs), sandwiched between organic blend layer and electrode, are employed, which facilitate the selective extraction of photogenerated carriers to respective electrodes [12, 13].

We keep in mind that organic polymer components are allowed to be donor- or acceptor-based and also have the flexibility to be doped by p-type or n-type impurities. Therefore, they are the perfect alternatives for semiconductor material [14]. Meanwhile, CELs can also be generalized into two different types: electron extraction layers (EELs) and hole extraction layers (HELs), which have to be chosen carefully for optimal device performance. As an example, for donor-acceptor systems, zinc oxide (ZnO) and titanium oxide (TiO_2) are the favoured EEL materials due to their low work functions, which help the carriage of

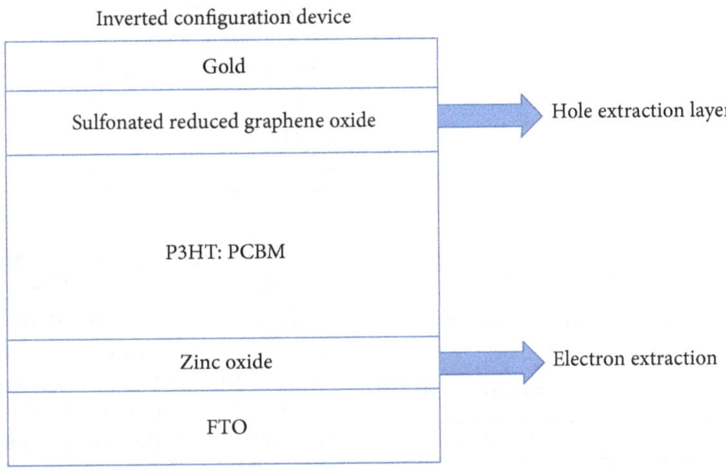

FIGURE 1: Schematic of the P3HT:PCBM device.

electrons in and out of different electro-optical devices [15]. For HELs, however, the search for an efficient, cost-effective, and easy-to-fabricate HEL is still going on. While one novel material, used widely in organic photovoltaics, for HEL is poly (3,4-ethylenedioxythiophene)-poly(styrenesulfonate) (PEDOT:PSS), its inclination to absorb water and resistance to bar electrons create an issue, while also showing inhomogeneous electro-optical behaviour leading to an ongoing search for new HEL materials [16–18]. In this instance, while the use of graphene-based materials is proving to be a viable option for creating HEL in many solar cell applications [19–21], the use of graphene as a stand-alone material leads to issues due to its zero bandgap. In its stead, the introduction of sulfonated functionalized graphene materials, e.g., graphene oxide and sulfonated graphene oxide [22–24], offers greater adaptability of bandwidth.

Recently, Ali et al. [25] have demonstrated groundbreaking work which demonstrates different P3HT:PSC devices which use different reduced sulfonated graphene oxide (rSGO) as the HEL. Their work shows that with different recipes of rSGO, the electro-optical properties can be customized, by modifying the bandgap. By using the different variants of the rSGO, they were able to demonstrate that the variation of the sulphur and oxygen content allows for the efficient hole extraction and transport.

While the work of [25] provides us the experimental methods of using the sulfonated graphene-based solar cells, there is a need to streamline this process for the purpose of optimization and time-saving capabilities. In this work, therefore, we will demonstrate the development of the fabless model for P3HT:PSC, using Organic Solar, a modelling tool developed by Silvaco TCAD. Further along, we will discuss the starting conditions, as well as the optimization of the parameters. Finally, the results of our modelling will be discussed in comparison with the experimental devices tested in [25].

2. Structural Design

The structure of the BHJ polymer solar cell we consider here comprises five layers, as shown in Figure 1. These are the anode made of fluorine-doped tin oxide, FTO, electron extraction layer using ZnO, photoactive organic layer utilizing P3HT:PCBM, and hole extraction layer making use of sulfonated reduced graphene oxide as well as a gold cathode. The use of FTO as a substrate is justified by its optical transparency along with wide bandgap and low electrical resistivity characteristics of ZnO [26, 27]. Moreover, it also has the property of higher carrier density [28, 29] and various fabrication and implementation recipes, routines, and applications of FTO have been developed over the years [30–37]. In our design, the anode consists of FTO and therefore provides a connection for the flow of anions out of the device. The electron transport layer is a semiconductor layer employed as an electron absorber and charge carrier using TiO_2 and ZnO. Traditionally, a dense n-type compact layer of TiO_2 is deposited on FTO/ITO used as a hole blocking layer, where a mesoporous layer of TiO_2 is used. The efficiency of the polymer solar cell can be improved by blocking the direct recombination that may occur at the interface between perovskite and the FTO. This is usually done by introducing a hole blocking layer between the two materials.

The photons, after having acquired enough energy, travel to the P3HT:PCBM active area where the electron hole pair is then created. For the BHJ, the acceptor part of the junction is the P3HT, while the donor part is the PCBM. Upon the incidence of solar radiations, electron-hole pairs are generated from the P3HT:PCBM blend; whereas the electrons move towards ZnO (i.e., electron extraction layer); the holes move towards reduced graphene oxide (r-SGO) layer (i.e., hole extraction layer). This process is depicted in Figure 1. Further along, they pass through the charge selective layers and are then extracted by the anode and cathode. To understand the optical and electrical properties of such a device, we make use of the simple diode circuit [38], depicted in Figure 2. Here, an important adjustment needs to be made, which is when photons of a particular wavelength and intensity impinge on the diode surface, it responds as a current source dependent on the incoming parameters. Using this model, we can obtain key parameters for the characterization of the solar

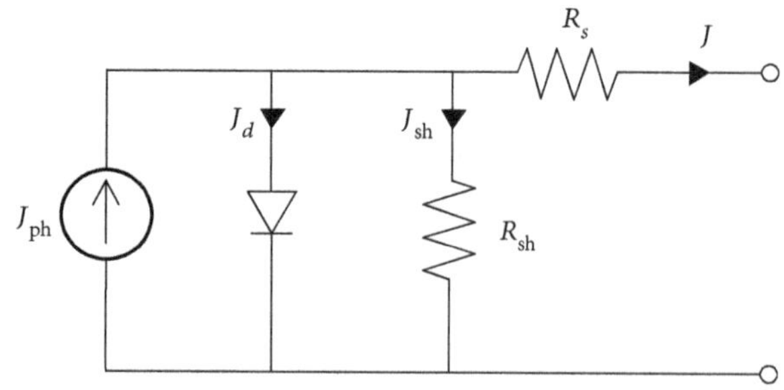

FIGURE 2: The equivalent circuit for a photovoltaic cell [38].

TABLE 1: Materials under test as the hole extraction layer [25].

Material	Bandgap (eV)
Reduced sulfonated graphene oxide (rSGO-1)	1.12
Reduced sulfonated graphene oxide (rSGO-2)	1.38
Reduced sulfonated graphene oxide (rSGO-3)	1.44

cells such as short-circuit currents, open-circuit voltage, and fill factor as well as the solar cell efficiency [38].

3. Modelling

The objective of our work is to develop the modelling methodology for the emulation of the physical device. For this purpose, we analysed three devices with varying bandgaps for the HEL and consisting of reduced sulfonated graphene oxides such as rSGO-1, rSGO-2, and rSGO-3. The aim here is to recreate the physical layers of the device as demonstrated in [25] and then use the transport equations to obtain the current density as a function of the solar cell voltage for these devices. The modelling analysis was performed using Organic Solar module along with the Atlas Simulator (both part of the Silvaco TCAD suite), for the materials under test listed in Table 1. The results from the modelling, i.e., the J-V analysis, are then compared with the J-V curves obtained from the characterization of the physically implemented solar cell [25].

The modelling of the P3HT:PSC device requires us to choose a reasonable starting point for the numerical iteration. While the EEL and HEL can be changed in Athena and Silvaco, the cross-sectional area of the solar cell was kept at 1 cm^2, the HEL = 20 nm, EEL = 50 nm, and photoactive la yer = 200 nm thick, by keeping in mind the physical device tested in [25]. Before we model our device, we consider the mesh of the design under consideration. This is an important parameter as it requires grid intersection points (called nodes), which is where the mathematical models are used to obtain vital parameters of the designed device. These nodes are important to optimize because they directly impact the time required for the model to run as well as its precision. The general rule of thumb is that the mathematical operation needed for achieving a solution can be described as

$$\text{Number of mathematical operations} = (\text{Np})^{\alpha}, \quad (1)$$

where Np is defined as the number of nodes, and α is the iteration parameter (usually of the order of 1.5–2.0) [39]. It is also important to keep in mind that we ensure adequate mesh density in areas with high field as well as to avoid discontinuities in the mesh density. The mobility model that we utilized to estimate the device characteristics is the Poole-Frenkel model [40, 41] in combination with the Langevin recombination model [42] for our devices. This is because these are the models traditionally used for BHJ devices. As part of the verification process, the method was first tested for a solar cell suggested by Koster et al. [43, 44], with our modelling technique following the procedures set in place by Organic Solar. After verifying that the response of the modelling is in line with the experimental techniques, we proceeded to the development of the P3HT:PSC device. The main challenge during the modelling was the repetitive iteration required to obtain a numerical solution for the figures of merit. This required the knowledge of work functions of the materials used to fabricate the device, which for some materials, such as ZnO and TiO_2, was a constant. Thus, we ran different iterations to minimize the error function of the HOMO and LOMO energy levels, binding energies, generation rate, bandgaps, affinity, permittivity, and electron and hole mobility as well as the work function, amongst others. By doing this extensively, we obtained the estimated values for the parameters that will be presented shortly.

TABLE 2: The parameters of the experimental device [25].

HEL materials	V_{oc} (V)	J_{sc} (mA/cm^2)	FF	η (%)
rSGO-1	0.58	5.77	0.46	1.54
rSGO-2	0.60	7.54	0.62	2.80
rSGO-3	0.59	8.29	0.57	2.79

4. Results and Discussion

The figures of merit that we used to analyse the performance of the photovoltaic organic solar cell are the current density, the cell voltage, and the fill factor as well as the solar cell efficiency. These have been listed in Table 2 for the physical devices under test.

The parameters found for the modelled devices are presented in Table 3. Here, we present not only the open-circuit voltage and the short-circuit current density but also other key parameters such as the maximum power output for such devices.

We also present here the relative error in calculation for the modelled and the experimental devices. This is done for all the devices under test and is presented in Table 4. As can be seen, the error in the estimation, compared to the experimental results, is less than 0.15 for almost all of the parameters tested, apart from a few exceptions, showing that as a first approximation, our modelling technique works very well and can be used to design organic solar cells with both electron and hole extraction layers. Here we note that the error index for the solar efficiency, between the modelled and fabricated devices, is greater than all other parameters because it encompasses all the deviations in J-V curves used to calculate the maximum power delivered.

TABLE 3: The parameters of modelled devices via Silvaco TCAD.

HEL materials	J_{max} (mA/cm^2)	V_{max} (V)	P_{max} (mW)	V_{oc} (V)	J_{sc} (mA/cm^2)	Pout (mW)	η (%)	FF
rSGO-1	3.80	0.26	0.93	0.44	6.50	2.92	0.75	0.39
rSGO-2	5.43	0.40	2.17	0.64	8.30	5.32	1.74	0.71
rSGO-3	5.19	0.42	2.18	0.62	8.32	5.17	1.75	0.61

TABLE 4: Relative error between the experimental and modelled results.

HEL materials	Absolute error index			
	V_{oc} (V)	J_{sc} (mA/cm^2)	η (%)	FF
rSGO-1	0.240	0.120	0.51	0.152
rSGO-2	0.068	0.101	0.38	0.1451
rSGO-3	0.052	0.003	0.37	0.070

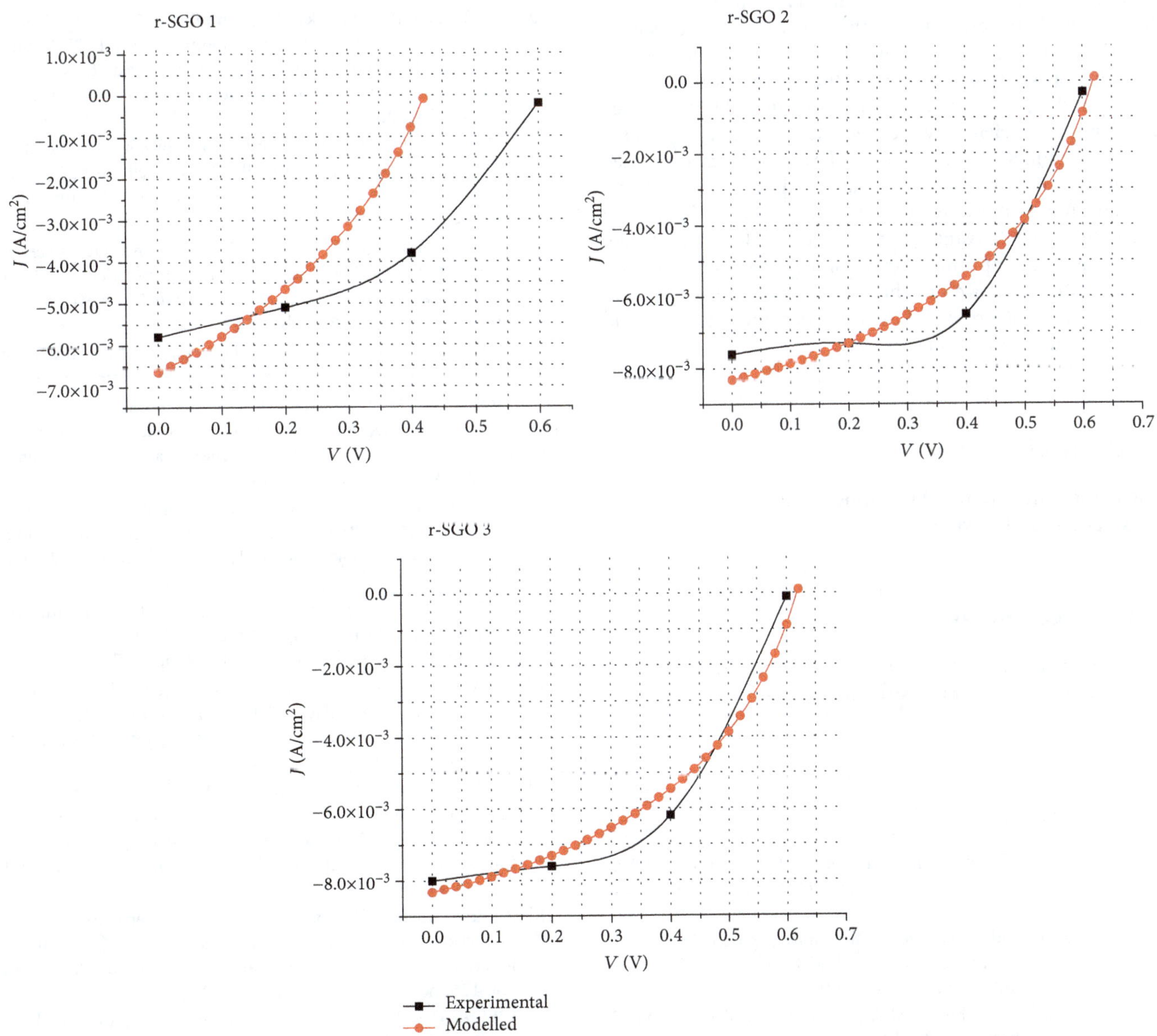

FIGURE 3: The experimental and modelled current density as a function of solar voltage for the devices under test.

The experimental and modelled current densities as a function of the cell voltage have been depicted in Figure 3. We note here that, although Silvaco allows for variation of the HEL and EEL work functions, we kept them constants for simplicity. The fabrication of the device structure (Figure 1) includes physical defects, environmental factors, and grain boundaries associated with the material in its pristine form as well as restructured during the subsequent processing. This is likely to alter some of the conditions which are usually taken fixed while modelling the structures/device within the theoretical framework/governing design laws.

5. Conclusions

Three different BHJ-based polymer solar cells have been designed and tested, against similar fabricated device, for the output characteristics, and the data extracted, to subsequently analyse them. We also characterized the electro-optical properties of the device which includes the current density and cell voltages, power analysis, and fill factor as well as the photo efficiency. Moreover, the relative error indexes for the prominent figures of merit have also been presented which show that the model estimates almost all of the figures of merit in the experimental data with less than 20% error in calculations. The advantage of doing this is that the postdesign analysis and optimization become much easier and convenient. Moreover, we now have the mechanism in place to test variations of presented device much more quickly and in a cost-effective way than earlier.

Conflicts of Interest

The authors declare that no competing interests exist in the presentation of this work.

Acknowledgments

Mr. Asghar Ali [25] is acknowledged for fabricating the polymer solar cell with similar structural design as of this work for comparison.

References

[1] Y. Shirota, "Organic materials for electronic and optoelectronic devices," *Journal of Materials Chemistry*, vol. 10, no. 1, pp. 1–25, 2000.

[2] B. O'Regan and M. Grätzel, "A low-cost, high-efficiency solar cell based on dye-sensitized colloidal TiO_2 films," *Nature*, vol. 353, no. 6346, pp. 737–740, 1991.

[3] M. M. Lee, J. Teuscher, T. Miyasaka, T. N. Murakami, and H. J. Snaith, "Efficient hybrid solar cells based on meso-superstructured organometal halide perovskites," *Science*, vol. 338, no. 6107, pp. 643–647, 2012.

[4] C. Kamble, N. Chide, S. Mhatre, and S. Sukhdeve, "Thin film organic solar cell as an emerging PV technique," in *2013 International Conference on Green Computing, Communication and Conservation of Energy (ICGCE)*, pp. 649–653, Chennai, India, 2013.

[5] J. You, L. Dou, K. Yoshimura et al., "A polymer tandem solar cell with 10.6% power conversion efficiency," *Nature Communications*, vol. 4, no. 1, p. 1446, 2013.

[6] Y. Liu, J. Zhao, Z. Li et al., "Aggregation and morphology control enables multiple cases of high-efficiency polymer solar cells," *Nature Communications*, vol. 5, no. 1, p. 5293, 2014.

[7] M. T. Dang, L. Hirsch, and G. Wantz, "P3HT:PCBM, best seller in polymer photovoltaic research," *Advanced Materials*, vol. 23, no. 31, pp. 3597–3602, 2011.

[8] D. Chen, A. Nakahara, D. Wei, D. Nordlund, and T. P. Russell, "P3HT/PCBM bulk heterojunction organic photovoltaics: correlating efficiency and morphology," *Nano Letters*, vol. 11, no. 2, pp. 561–567, 2011.

[9] J. A. Hauch, P. Schilinsky, S. A. Choulis, R. Childers, M. Biele, and C. J. Brabec, "Flexible organic P3HT:PCBM bulk-heterojunction modules with more than 1 year outdoor lifetime," *Solar Energy Materials and Solar Cells*, vol. 92, no. 7, pp. 727–731, 2008.

[10] I. Litzov and C. Brabec, "Development of efficient and stable inverted bulk heterojunction (BHJ) solar cells using different metal oxide interfaces," *Materials*, vol. 6, no. 12, pp. 5796–5820, 2013.

[11] L. M. Chen, Z. Xu, Z. Hong, and Y. Yang, "Interface investigation and engineering – achieving high performance polymer photovoltaic devices," *Journal of Materials Chemistry*, vol. 20, no. 13, pp. 2575–2598, 2010.

[12] T.-H. Lai, S.-W. Tsang, J. R. Manders, S. Chen, and F. So, "Properties of interlayer for organic photovoltaics," *Materials Today*, vol. 16, no. 11, pp. 424–432, 2013.

[13] S. Lattante, "Electron and hole transport layers: their use in inverted bulk heterojunction polymer solar cells," *Electronics*, vol. 3, no. 1, pp. 132–164, 2014.

[14] M. Reyes-Reyes, D. L. Carroll, W. Blau, and R. López-Sandoval, "Materials and devices for organic electronics," *Journal of Nanotechnology*, vol. 2011, Article ID 589241, 2 pages, 2011.

[15] Y. Zhou, C. Fuentes-Hernandez, J. Shim et al., "A universal method to produce low-work function electrodes for organic electronics," *Science*, vol. 336, no. 6079, pp. 327–332, 2012.

[16] M. Kemerink, S. Timpanaro, M. M. de Kok, E. A. Meulenkamp, and F. J. Touwslager, "Three-dimensional inhomogeneities in PEDOT:PSS films," *The Journal of Physical Chemistry B*, vol. 108, no. 49, pp. 18820–18825, 2004.

[17] Y.-H. Kim, S.-H. Lee, J. Noh, and S.-H. Han, "Performance and stability of electroluminescent device with self-assembled layers of poly(3,4-ethylenedioxythiophene)–poly(styrenesulfonate) and polyelectrolytes," *Thin Solid Films*, vol. 510, no. 1-2, pp. 305–310, 2006.

[18] Z. Zhu, Y. Bai, T. Zhang et al., "High-performance hole-extraction layer of sol–gel-processed NiO nanocrystals for inverted planar perovskite solar cells," *Angewandte Chemie*, vol. 126, no. 46, pp. 12779–12783, 2014.

[19] J. M. Kim, S. W. Seo, D. H. Shin et al., "Ag-nanowires-doped graphene/Si Schottky-junction solar cells encapsulated with another graphene layer," *Current Applied Physics*, vol. 17, no. 8, pp. 1136–1141, 2017.

[20] N. Kim, G. Xin, S. M. Cho, C. Pang, and H. Chae, "Microwave-reduced graphene oxide for efficient and stable hole extraction layers of polymer solar cells," *Current Applied Physics*, vol. 15, no. 9, pp. 953–957, 2015.

[21] G. Jang, W. Yim, Y. H. Ahn, S. Lee, and J.-Y. Park, "Control of device characteristics by passivation of graphene field effect transistors with polymers," *Current Applied Physics*, vol. 16, no. 11, pp. 1506–1510, 2016.

[22] D. He, Z. Kou, Y. Xiong et al., "Simultaneous sulfonation and reduction of graphene oxide as highly efficient supports for metal nanocatalysts," *Carbon*, vol. 66, pp. 312–319, 2014.

[23] J. Liu, Y. Xue, and L. Dai, "Sulfated graphene oxide as a hole-extraction layer in high-performance polymer solar cells," *The Journal of Physical Chemistry Letters*, vol. 3, no. 14, pp. 1928–1933, 2012.

[24] J. Liu, Y. Xue, Y. Gao, D. Yu, M. Durstock, and L. Dai, "Hole and electron extraction layers based on graphene oxide derivatives for high-performance bulk heterojunction solar cells," *Advanced Materials*, vol. 24, no. 17, pp. 2228–2233, 2012.

[25] A. Ali, Z. S. Khan, M. Jamil, Y. Khan, N. Ahmad, and S. Ahmed, "Simultaneous reduction and sulfonation of graphene oxide for efficient hole selectivity in polymer solar cells," *Current Applied Physics*, vol. 18, no. 5, pp. 599–610, 2018.

[26] S. Tongay, K. Berke, M. Lemaitre et al., "Stable hole doping of graphene for low electrical resistance and high optical transparency," *Nanotechnology*, vol. 22, no. 42, article 425701, 2011.

[27] X. Wang, L. Zhi, and K. Müllen, "Transparent, conductive graphene electrodes for dye-sensitized solar cells," *Nano Letters*, vol. 8, no. 1, pp. 323–327, 2008.

[28] K. S. Ramaiah and V. S. Raja, "Structural and electrical properties of fluorine doped tin oxide films prepared by spray-pyrolysis technique," *Applied Surface Science*, vol. 253, no. 3, pp. 1451–1458, 2006.

[29] T. Tesfamichael, G. Will, M. Colella, and J. Bell, "Optical and electrical properties of nitrogen ion implanted fluorine doped tin oxide films," *Nuclear Instruments and Methods in Physics Research Section B: Beam Interactions with Materials and Atoms*, vol. 201, no. 4, pp. 581–588, 2003.

[30] E. Elangovan and K. Ramamurthi, "Studies on micro-structural and electrical properties of spray-deposited fluorine-doped tin oxide thin films from low-cost precursor," *Thin Solid Films*, vol. 476, no. 2, pp. 231–236, 2005.

[31] P. F. Gerhardinger and R. J. Mccurdy, "Float line deposited transparent conductors-implications for the PV industry," *MRS Proceedings*, vol. 426, 1996.

[32] P. M. Gorley, V. V. Khomyak, S. V. Bilichuk, I. G. Orletsky, P. P. Horley, and V. O. Grechko, "SnO_2 films: formation, electrical and optical properties," *Materials Science and Engineering: B*, vol. 118, no. 1–3, pp. 160–163, 2005.

[33] C.-Y. Kim and D.-H. Riu, "Texture control of fluorine-doped tin oxide thin film," *Thin Solid Films*, vol. 519, no. 10, pp. 3081–3085, 2011.

[34] H. Kim, R. C. Y. Auyeung, and A. Piqué, "Transparent conducting F-doped SnO_2 thin films grown by pulsed laser deposition," *Thin Solid Films*, vol. 516, no. 15, pp. 5052–5056, 2008.

[35] T. Maruyama and H. Akagi, "Fluorine-doped tin dioxide thin films prepared by radio-frequency magnetron sputtering," *Journal of the Electrochemical Society*, vol. 143, no. 1, pp. 283–287, 1996.

[36] R. Mientus and K. Ellmer, "Structural, electrical and optical properties of SnO_{2-x}:F-layers deposited by DC-reactive magnetron-sputtering from a metallic target in $Ar–O_2/CF_4$ mixtures," *Surface and Coatings Technology*, vol. 98, no. 1–3, pp. 1267–1271, 1998.

[37] N. Sankara Subramanian, B. Santhi, S. Sundareswaran, and K. S. Venkatakrishnan, "Studies on spray deposited SnO_2, Pd:SnO_2 and F:SnO_2 thin films for gas sensor applications," *Synthesis and Reactivity in Inorganic, Metal-Organic, and Nano-Metal Chemistry*, vol. 36, no. 1, pp. 131–135, 2006.

[38] C. M. Proctor and T.-Q. Nguyen, "Effect of leakage current and shunt resistance on the light intensity dependence of organic solar cells," *Applied Physics Letters*, vol. 106, no. 8, article 083301, 2015.

[39] "ATLAS," http://www.eng.buffalo.edu/~wie/silvaco/atlas_user_manual.pdf.

[40] F. Torricelli and L. Colalongo, "Unified mobility model for disordered organic semiconductors," *IEEE Electron Device Letters*, vol. 30, no. 10, pp. 1048–1050, 2009.

[41] C.-Y. Nam, D. Su, and C. T. Black, "High-performance air-processed polymer–fullerene bulk heterojunction solar cells," *Advanced Functional Materials*, vol. 19, no. 22, pp. 3552–3559, 2009.

[42] I. Hwang, C. R. McNeill, and N. C. Greenham, "Drift-diffusion modeling of photocurrent transients in bulk heterojunction solar cells," *Journal of Applied Physics*, vol. 106, no. 9, article 094506, 2009.

[43] L. J. A. Koster, E. C. P. Smits, V. D. Mihailetchi, and P. W. M. Blom, "Device model for the operation of polymer/fullerene bulk heterojunction solar cells," *Physical Review B*, vol. 72, no. 8, 2005.

[44] https://www.silvaco.com/examples/tcad/section44/example7/index.html.

18

A New Multilevel Inverter Topology for Grid-Connected Photovoltaic Systems

Muhammad Bilal Satti,[1] **Ammar Hasan**,[1] **and Mian Ilyas Ahmad**[2]

[1]*School of Electrical Engineering and Computer Science (SEECS), National University of Science and Technology, Islamabad, Pakistan*
[2]*Research Center for Modeling and Simulation (RCMS), National University of Sciences and Technology (NUST), Islamabad, Pakistan*

Correspondence should be addressed to Muhammad Bilal Satti; msatti.msee15seecs@seecs.edu.pk

Academic Editor: Pierluigi Guerriero

The demand for clean and sustainable energy has spurred research in all forms of renewable energy sources, including solar energy from photovoltaic systems. Grid-connected photovoltaic systems (GCPS) provide an effective solution to integrate solar energy into the existing grid. A key component of the GCPS is the inverter. The inverter can have a significant impact on the overall performance of the GCPS, including maximum power point (MPP) tracking, total harmonic distortion (THD), and efficiency. Multilevel inverters are one of the most promising classes of converters that offer a low THD. In this paper, we propose a new multilevel inverter topology with the motivation to improve all the three aforementioned aspects of performance. The proposed topology is controlled through direct model predictive control (DMPC), which is state of the art in control techniques. We compare the performance of the proposed topology with the topologies reported in literature. The proposed topology offers one of the best efficiency, MPP tracking, and voltage THD.

1. Introduction

At present, one of the key research areas in renewable energy are grid-connected photovoltaic systems (GCPS). The cost reduction of photovoltaic (PV) panels, environmental benefits, increased demand of energy, integration with existing grid, and advances in power electronics are some of the reasons for the focus on GCPS [1]. In a GCPS, the key performance criteria are conversion efficiency, maximum power point (MPP) tracking, and total harmonic distortion (THD) of the power injected into the grid. All these performance criteria are highly dependent on the choice of the inverter topology and the control technique utilized for the inverter.

In the literature, several topologies of inverters have been reported that can be used in a GCPS [2]. Multilevel inverters (MLIs) are one of the most promising classes of converters. As the name suggests, MLIs have a higher number of output voltage levels as compared to the two/three levels in traditional inverters. Due to higher output voltage levels, a main advantage of MLIs is the staircase-like output waveform, which resembles a sinusoid more closely than the traditional inverters. Therefore, MLIs offer a significantly lower THD [3]. Another advantage of MLIs pertinent to GCPS is that they can include multiple PV panels as an input. There are three main subclasses of multilevel inverters: (1) cascaded H-bridge (CHB), (2) diode clamped, and (3) flying capacitor [4]. The most common subclass is the CHB due to its modular structure, absence of energy storing elements, and the ability to isolate faults [5–8]. Practical multilevel inverters have typically five or more output voltage levels. Seven-level converters are most widely used as they offer a reasonable trade-off between performance and complexity. Therefore, in this paper, we have focused on seven-level inverters.

There are different variants of the CHB topology that are being employed in GCPS. The basic CHB topology [7, 8] provides excellent THD and MPP tracking due to the reason that each PV panel could be connected independently. However, it has a large number of semiconductor switches that

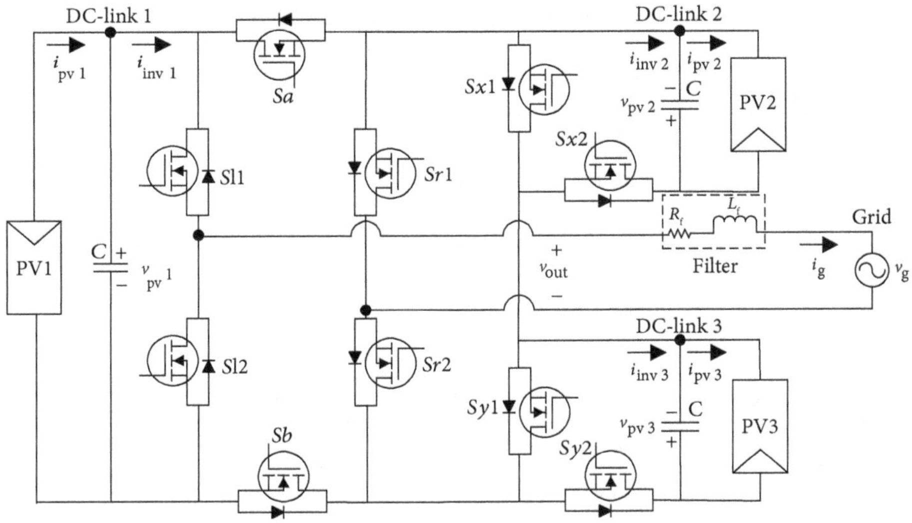

FIGURE 1: Proposed MLI-based grid-connected photovoltaic system.

increase the overall cost of the converter. The topology presented in [9] for a GCPS has decreased the number of switches as a trade-off to lower performance in terms of THD and MPPT tracking [10]. Another topology, called the novel H-bridge, was presented in [3, 11]. The novel H-bridge further reduced the number of switches. However, GCPS based on the topology in [9] and novel H-bridge has a drastically poor performance in MPP tracking when compared to GCPS based on the basic CHB topology. In this paper, we propose a new topology that is inspired from the novel H-bridge [3, 11] and an existing MLI structure in the literature [12]. The new topology has a lower number of switches than the basic CHB. Moreover, like basic CHB, the proposed topology also allows independent MPP tracking of each PV module, thereby giving a comparable performance to basic CHB in terms of MPP tracking.

As mentioned earlier, the employed control technique is the second key design component that affects the performance of a GCPS. Classical control schemes, such as PID, with a variety of modulation methods and compensations, have been used in CHB MLI-based GCPS [8]. A main drawback of classical control schemes is the large transient time and slow response [13]. Direct model predictive control (DMPC) is an emerging control scheme that is being used in power electronic converters. The distinguishing feature of DMPC is that the switches are controlled directly without a demodulator, as in PWM control. Advantages of DMPC include handling of plant nonlinearities, multiple control goals, explicit incorporation of constraints on inputs/states, and optimal control action. For most of the converters, DMPC has been shown to exhibit fast transient response, improved efficiency, and better overall control performance as compared to classical control schemes [14, 15].

In this paper, we have used DMPC for the control of the proposed topology and the GCPS. The key idea is to benefit from both the advantages of proposed MLI and DMPC to have an overall improved performance of the GCPS. To compare like with like, we collate our work with DMPC of GCPS based on three existing topologies mentioned earlier.

In Section 2, we propose a new multilevel inverter topology, describe its application in GCPS, and derive its mathematical model. In Section 3 of the paper, the proposed control scheme is explained in detail. The simulation results and comparison with existing work are given in Section 4 to show that the proposed MLI exhibits better performance than the three other recent converters. Finally, the conclusion is stated in Section 5.

2. Proposed Multilevel Inverter Topology

The circuit of the proposed MLI topology is shown in Figure 1. The converter has ten current bidirectional semiconductor switches labeled $Sl1$, $Sl2$, $Sr1$, $Sr2$, Sa, Sb, $Sx1$, $Sx2$, $Sy1$, and $Sy2$. The converter can have three input sources. For a GCPS, the sources are PV panels. Therefore, the input voltages are labeled v_{pv1}, v_{pv2}, and v_{pv3}. The part of the converter comprising of switches $Sx1$, $Sx2$, $Sy1$, and $Sy2$ is motivated from the structure presented in [12]. The remaining part of the converter is inspired from the novel H-bridge [3, 11].

The switch pair ($Sl1$, $Sl2$) is complementary, that is, if one of the switches in the pair is off, and then, the other must be on. Similarly, the switch pairs ($Sr1$, $Sr2$), (Sa, Sb), ($Sx1$, $Sx2$), and ($Sy1$, $Sy2$) are complementary. The four switches $Sl1$, $Sl2$, $Sr1$, and $Sr2$ constitute an H-bridge that allows inversion. The switches Sa and Sb allow selection of voltages on the right leg of the H-bridge, which enables multiple voltage levels at the output. The switches ($Sx1$, $Sx2$) and ($Sy1$, $Sy2$) allow inclusion/exclusion of sources v_{pv2} and v_{pv3}, respectively.

In the proposed topology, there are a total of 32 possible switching combinations. However, excluding the redundant combinations, only 15 combinations are used in operation. The 15 switching combinations and corresponding output voltages are shown in Table 1. In the table, 0 and 1 represent the off and on position of the switches, respectively. If the magnitude of all the voltage sources is kept equal, the proposed topology is able to generate seven-voltage levels. In

TABLE 1: Switching table for the proposed MLI topology.

Ind. j	Switching positions $S(j)$ $Sl1, Sl2$	$Sr1, Sr2$	Sa, Sb	$Sx1, Sx2$	$Sy1, Sy2$	v_{out}	v_{amp}
1	1, 0	0, 1	0, 1	0, 1	0, 1	v_{pv1}	
2	0, 1	1, 0	0, 1	0, 1	1, 0	v_{pv2}	$1v_{pv}$
3	0, 1	1, 0	0, 1	1, 0	0, 1	v_{pv3}	
4	1, 0	1, 0	0, 1	0, 1	1, 0	$(v_{pv1} + v_{pv2})$	
5	1, 0	1, 0	0, 1	1, 0	0, 1	$(v_{pv1} + v_{pv3})$	$2v_{pv}$
6	0, 1	1, 0	0, 1	0, 1	0, 1	$(v_{pv2} + v_{pv3})$	
7	1, 0	1, 0	0, 1	0, 1	0, 1	$(v_{pv1} + v_{pv2} + v_{pv3})$	$3v_{pv}$
8	0, 1	0, 1	0, 1	0, 1	0, 1	0	$0v_{pv}$
9	0, 1	1, 0	1, 0	0, 1	0, 1	$-v_{pv1}$	
10	1, 0	0, 1	1, 0	0, 1	1, 0	$-v_{pv2}$	$-1v_{pv}$
11	1, 0	0, 1	1, 0	1, 0	0, 1	$-v_{pv3}$	
12	0, 1	0, 1	1, 0	0, 1	1, 0	$-(v_{pv1} + v_{pv2})$	
13	0, 1	0, 1	1, 0	1, 0	0, 1	$-(v_{pv1} + v_{pv3})$	$-2v_{pv}$
14	1, 0	0, 1	1, 0	0, 1	0, 1	$-(v_{pv2} + v_{pv3})$	
15	0, 1	0, 1	1, 0	0, 1	0, 1	$-(v_{pv1} + v_{pv2} + v_{pv3})$	$-3v_{pv}$

the table, v_{out} is the output voltage of the inverter in terms of the voltage of each PV panel module. The voltage v_{amp} denotes the output voltage of the inverter for the case that the voltage of all PV panel modules is equal, that is, $v_{pv1} = v_{pv2} = v_{pv3} = v_{pv}$.

A main aspect of the proposed topology is that all the three input sources can be independently included/excluded in the output voltage. This feature is similar to the basic CHB topology, which is beneficial for independent MPP tracking of the PV panels.

Our main focus in this paper is on the use of the proposed topology in a grid-connected photovoltaic system (GCPS). The GCPS based on the proposed topology is shown in Figure 1. The GCPS consists of the proposed converter, PV panel modules, DC-link capacitors, low-pass filter, and the grid. A simple passive low-pass inductor filter L_f is used at the grid side to suppress the high-frequency harmonics. The resistance R_f is the equivalent series resistance of the inductor. For safety purposes, a transformer may be inserted at the grid side for isolation.

2.1. Modeling of the Proposed GCPS. The states in the proposed system are the DC-link capacitor voltages and grid/inductor current. The equation for the current can be found by applying Kirchhoff's voltage law (KVL) at the output port of the proposed system in Figure 1. The equation is as follows:

$$\frac{L\,\mathrm{d}i_g}{\mathrm{d}t} = v_{out} - i_g * R_f - v_g, \tag{1}$$

where v_{out} is the output voltage of the MLI, v_g is the grid voltage, and i_g is the grid/inductor current. Discretization of the above equation by Euler's forward method yields the prediction equation for the grid current, which is

$$i_g^P = i_g(k+1) = i_g(k)\left(1 - \frac{R_f * T_s}{L}\right) + \frac{T_s}{L}\left(v_{out}(k) - v_g(k)\right), \tag{2}$$

where $i_g(k)$ and $i_g(k+1)$ are the values of grid current at time instant k and $k+1$, and T_s is the sampling time.

The equations for the DC-link voltages v_{pv1}, v_{pv2}, and v_{pv3} can be derived by applying Kirchhoff's current law (KCL) at the nodes of DC-links. The equations are given below:

$$\begin{aligned} \frac{C\,\mathrm{d}v_{pv1}}{\mathrm{d}t} &= i_{pv1}(k) - i_{inv1}(k), \\ \frac{C\,\mathrm{d}v_{pv2}}{\mathrm{d}t} &= i_{pv2}(k) - i_{inv2}(k), \\ \frac{C\,\mathrm{d}v_{pv3}}{\mathrm{d}t} &= i_{pv3}(k) - i_{inv3}(k). \end{aligned} \tag{3}$$

Discretization of the above equations yields

$$\begin{aligned} v_{pv1}^P &= v_{pv1}(k+1) = v_{pv1}(k) + \frac{T_s}{C}\left(i_{pv1}(k) - i_{inv1}(k)\right), \\ v_{pv2}^P &= v_{pv2}(k+1) = v_{pv2}(k) + \frac{T_s}{C}\left(i_{pv2}(k) - i_{inv2}(k)\right), \\ v_{pv3}^P &= v_{pv3}(k+1) = v_{pv3}(k) + \frac{T_s}{C}\left(i_{pv3}(k) - i_{inv3}(k)\right), \end{aligned} \tag{4}$$

where $v_{pv1}(k)$ is the DC-link 1 voltage, $v_{pv2}(k)$ is the DC-link 2 voltage, $v_{pv3}(k)$ is the DC-link 3 voltage, $i_{pv1}(k)$ is the

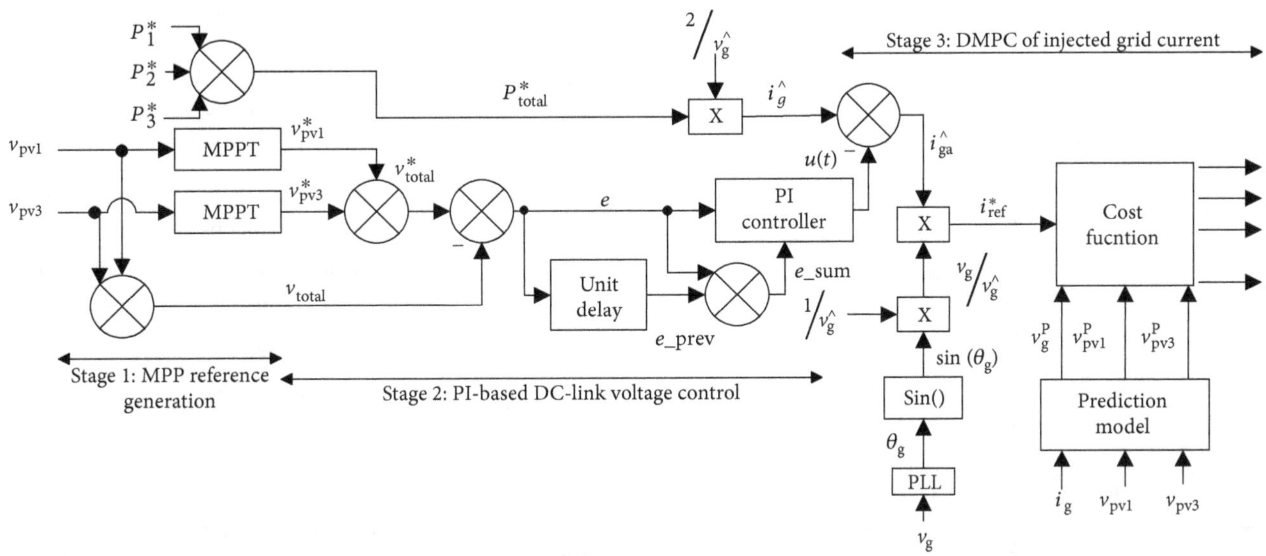

FIGURE 2: The basic controller diagram.

output current of PV1, $i_{pv2}(k)$ is the output current of PV2, $i_{pv3}(k)$ is the output current of PV3, C is the value of DC-link capacitors, and $i_{inv1}(k)$, $i_{inv2}(k)$, and $i_{inv3}(k)$ are the input currents of MLI.

3. Proposed Control Scheme

In this section, we describe the proposed control scheme for our GCPS. The control scheme is depicted in Figure 2. It consists of three main stages: (1) MPP reference generation, (2) PI-based DC-link voltage control, and (3) direct model predictive control of injected grid current. The initial two stages of the proposed scheme are standard in PV systems [7, 10]. However, the third stage of the control is based on the state-of-the-art direct model predictive control [14]. We will briefly describe the first two stages before a detailed explanation of the third stage.

The first stage generates the set point/reference to be used in stages 2 and 3. The first stage calculates the MPP reference values for the DC-link voltages v_{pv1}^*, v_{pv2}^*, and v_{pv3}^* and power P_1^*, P_2^*, and P_3^* for all the three PV panel modules. The reference values for MPP can change with the atmospheric conditions such as irradiance and temperature [16]. Therefore, the reference values need to be updated regularly.

The purpose of the second stage is to generate a reference grid current for the third stage. The reference grid current mainly depends on the total reference power calculated in stage 1 [7]. However, if the DC-link voltages are not at MPP, then the reference grid current has to be adjusted accordingly. The second stage uses a linear PI controller for this purpose that adjusts the reference peak value of the grid current $\hat{i}_g$ in accordance with the difference between reference and actual values of DC-link voltage. The peak value of the reference grid current is calculated as $\hat{i}_g = (P_{total}^*/\hat{v}_g) * 2$. The adjusted peak value of the reference grid current, denoted as $\hat{i}_{ga}$ in Figure 2, is calculated by adding $\hat{i}_g$ to the output of PI controller. Finally, the instantaneous reference i_{ref}^* of the grid current is calculated based on the adjusted peak reference $\hat{i}_{ga}$ and PLL.

The third stage is the DMPC of injected grid current which manipulates the semiconductor switches of the proposed multilevel inverter. The reference for the DMPC is i_{ref}^*. DMPC makes use of the discrete time model of the converter to calculate the optimal switching position among all the possible switching positions such that the reference tracking of grid current is ensured.

In DMPC, a cost function is minimized at each sampling instant k to select the optimal switching combination. We have used the following cost function

$$g = \left| i_{ref}^* - i_g(k+1) \right| + \lambda_v \left[\sum_{l=1}^{n} \left| v_{pv,l}^* - v_{pv,l}(k+1) \right| \right], \tag{5}$$

where i_{ref}^* is the reference for grid current, $i_g(k+1)$ is the actual grid current, $v_{pv,l}^*$ for $l = 1, 2, 3$ is the reference MPP voltage for each PV panel, $v_{pv,l}(k+1)$ for $l = 1, 2, 3$ is the actual voltage of each PV panel, n is the total number of input sources, and λ_v is the weighting factor. The cost function includes two terms. The first term penalizes the deviation of the actual current from the reference, which is the primary control objective of the third stage. The second term helps to regulate the DC-link voltages to their references. The weighting factor λ_v is used to specify the weightage of the second term in relevance to the first term.

Figure 3 depicts the flowchart of the DMPC algorithm. First of all, the controlled variables are measured, which are the grid current and DC-link voltages. The reference values of controlled variables are also taken from the first and second stages. It is also assumed that the best known cost g_{opt} is a very large number. The algorithm has to find the switching combination that gives the lowest cost among all the fifteen switching combinations that are listed in Table 1.

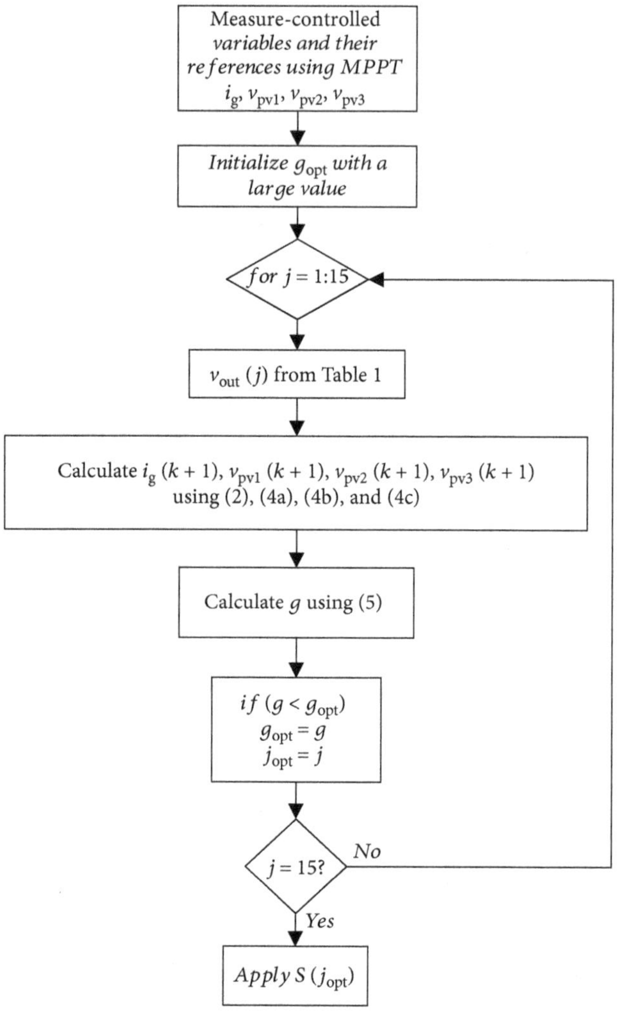

Figure 3: Flow diagram of the proposed control scheme.

For each switching combination, the output voltage of the MLI (v_{out}) is found by using Table 1. The grid current and DC-link voltages are predicted using the converter model (2) and (4). The cost (5) is calculated for the switching combination. If the cost is less than the optimal cost known so far, then the optimal cost g_{opt} and optimal switching sequence index j_{opt} are updated. At the end of the algorithm, the optimal switching combination is applied to the switches of the proposed multilevel inverter.

4. Simulation Results

A GCPS based on the proposed multilevel inverter and the proposed control scheme has been simulated in MATLAB Simulink. The PV panel module "Sharp ND-208U2" has been used as the source at all the three input ports of the converter. The rated maximum power of "Sharp ND-208U2" PV panel module is 208.05 Watts. The value of DC-link capacitors is kept at 15 mF. An inductor of 3 mH is used as filter. The ESR of the inductor is 0.2 Ω. Other parameters used in the simulation are specified in Table 2.

Table 2: System parameters.

Variable	Definition	Value
P_{total}	Nominal input power	624.1 W
$v^*_{pv1}, v^*_{pv2}, v^*_{pv3}$	Reference DC-link voltages	28.5 v
f	Frequency of grid	50 Hz
f_s	Switching/sampling frequency	20 kHz
L_f	Filtering inductor	3 mH
R_f	ESR	0.2 Ω
C	DC-link capacitor	15 mF
$\hat{v}_g$	Peak grid voltage	70 v

To evaluate the performance of the proposed converter, we have compared it with the simulation of three other converters. We have compared our proposed system to GCPS with DMPC based on (1) CHB [7], (2) topology in [9], and (3) novel H-bridge [11]. For each system, we obtained the simulation results for four different scenarios. In scenario 1, all the PV modules are working at 1000 W/m^2 irradiance and 25°C temperature. Scenario 1 is simulated for the first 0.4 seconds of the simulation. In scenario 2, which is from 0.4 s to 0.8 s of simulation time, solar irradiance of the module PV1 is changed from 1000 to 600 W/m^2, whereas in scenario 3, which is from 0.8 s to 1.2 s of simulation time, the solar irradiance of the module PV2 is changed from 1000 to 600 W/m^2. For the 4th scenario, the temperature of the module PV2 is changed from 25 to 35°C. Scenario 4 is from 1.2 s to 1.6 s of the simulation time.

The results of the comparison are given in Table 3. The comparison enlists the THD of grid current, THD of MLI output voltage, efficiency of the converter, and the number of semiconductor switches. Moreover, for each of the aforementioned four scenarios, the comparison table also shows the amount of power harvested from the panels.

The power extraction of the proposed converter is better than both the topology in [9] and the novel H-bridge. The power extraction is comparable to CHB. It is due to the reason that both the proposed converter and CHB can independently track the MPP of each PV panel module. In terms of efficiency, the proposed system fares much better than CHB and the topology in [9]. The efficiency of the proposed converter is comparable to novel H-bridge. The proposed converter offers the lowest THD in output voltage. The only performance criterion in which the proposed converter is outperformed by the others is the current THD. However, it is well within the acceptable limit of 5% in the standards [17]. Considering all the results in the table, the proposed converter arguably offers the best overall performance.

To provide a detailed analysis of the converter, Figure 4 illustrates the reference MPP DC-link voltage, actual DC-link voltages, reference MPP power, and the actual power extracted for each PV module. It can be seen that the proposed converter along with the proposed DMPC is able

Table 3: Comparison of systems.

GCPS Feature	DMPC and basic CHB-based GCPS [7]	DMPC and topology in [9] based on GCPS [10]	Novel H-bridge and DMPC-based GCPS	Proposed topology and DMPC-based GCPS
Grid current harmonics (%)	1.5–2	2.7–3.1	2.5–2.7	3.2
Output voltage harmonics (%)	28–30	29–36	24–32	22
Total power extracted during $0 < t < 0.4$ s (W)	623.50	613	618	623.25
Total power extracted during $0.4 < t < 0.8$ s (W)	543	542	542	542.7
Total power extracted during $0.8 < t < 1.2$ s (W)	462	385	395	462.65
Total power extracted during $1.2 < t \leq 1.6$ s (W)	456.9	383	391.5	456.95
Number of switches for 7-level MLI	12	9	6	10
η (%)	83	89	96	95

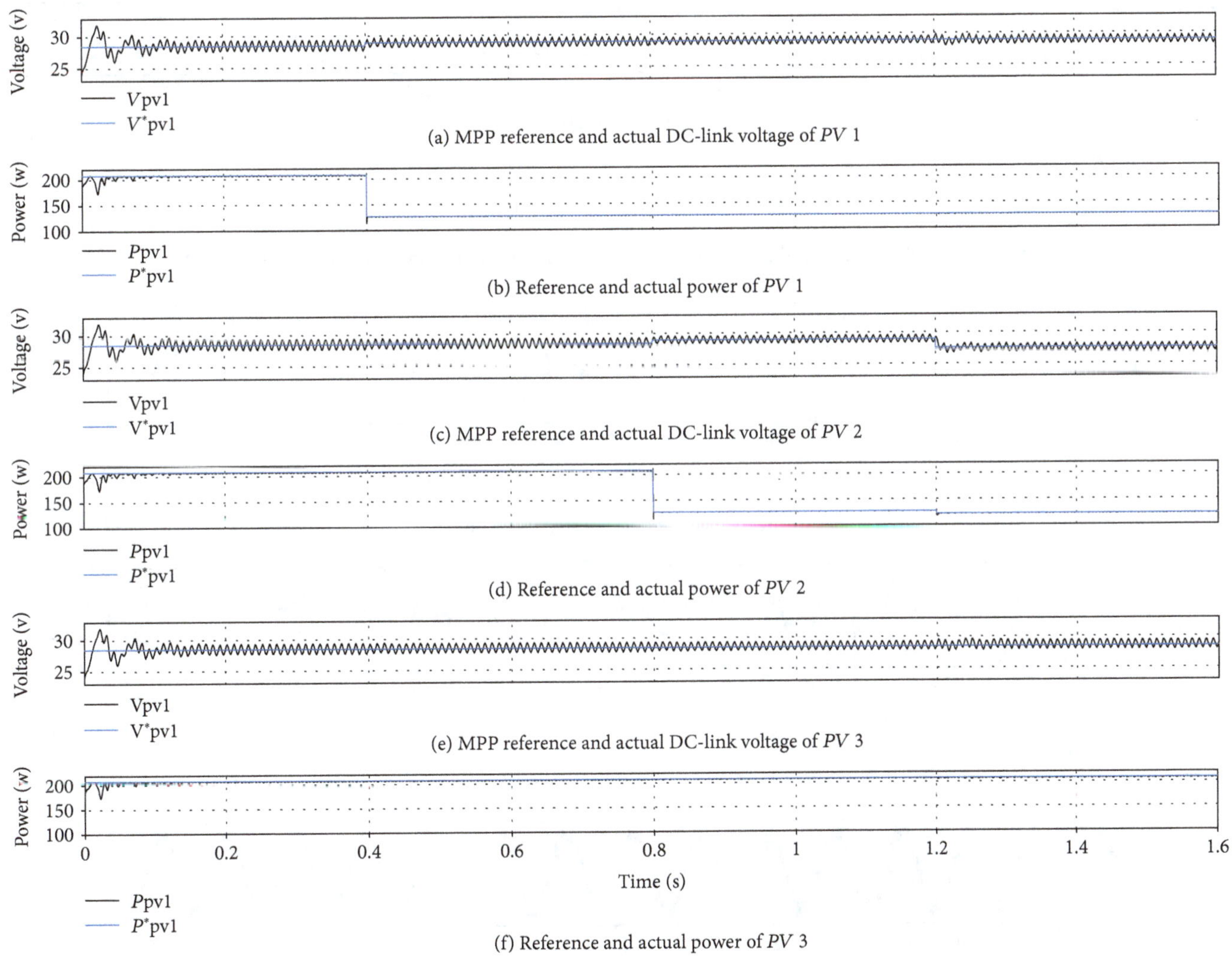

(a) MPP reference and actual DC-link voltage of *PV* 1

(b) Reference and actual power of *PV* 1

(c) MPP reference and actual DC-link voltage of *PV* 2

(d) Reference and actual power of *PV* 2

(e) MPP reference and actual DC-link voltage of *PV* 3

(f) Reference and actual power of *PV* 3

Figure 4: MPPT reference and actual DC-link voltage and power of the proposed MLI topology-based GCPS.

to track both the voltage and power references in all the four scenarios. For the purpose of comparison, we have shown the plots for novel H-bridge in Figure 5. The novel H-bridge has two input ports. In the figure, we see that the reference tracking of novel H-bridge-based GCPS maintains a constant error. Moreover, the effect of changes in atmospheric

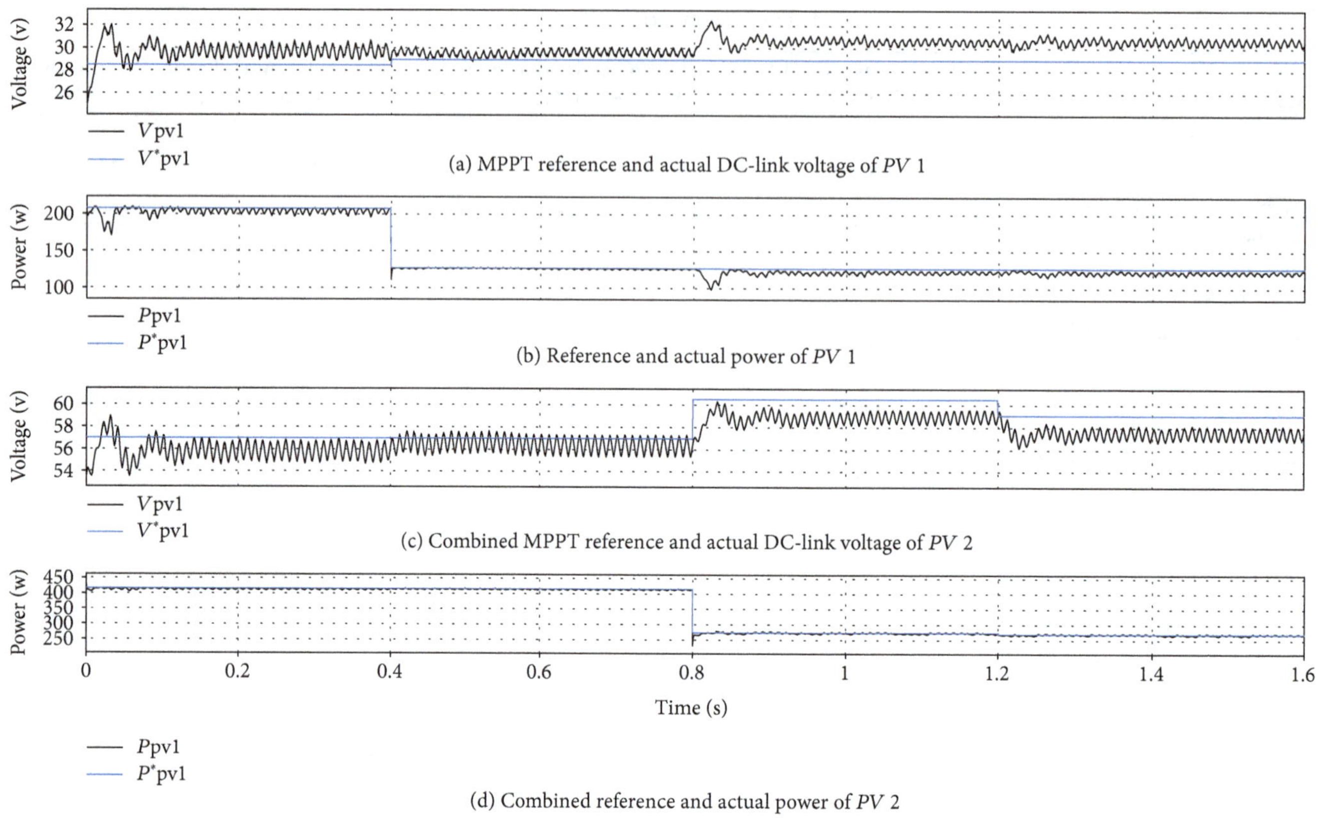

FIGURE 5: MPPT reference and actual DC-link voltage and power of novel H-bridge-based GCPS.

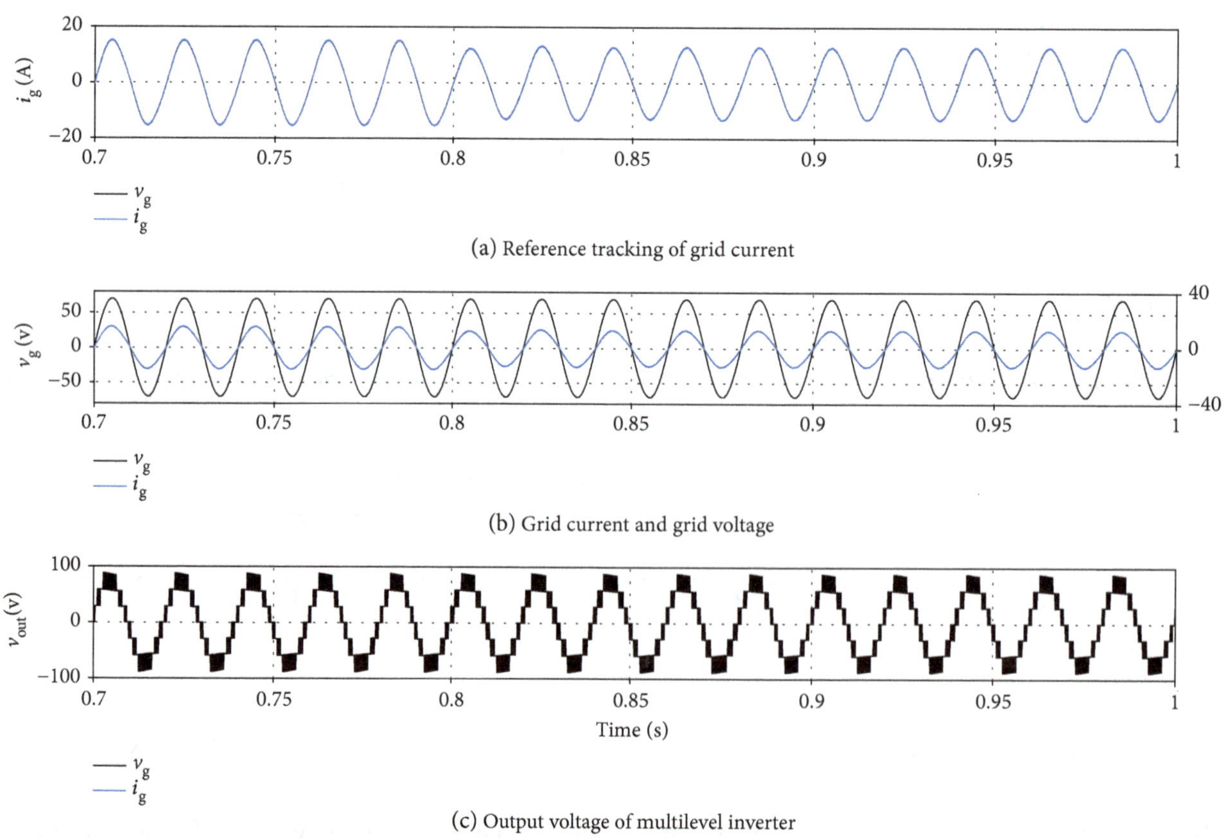

FIGURE 6: Grid current, grid voltage, and output voltage of multilevel inverter.

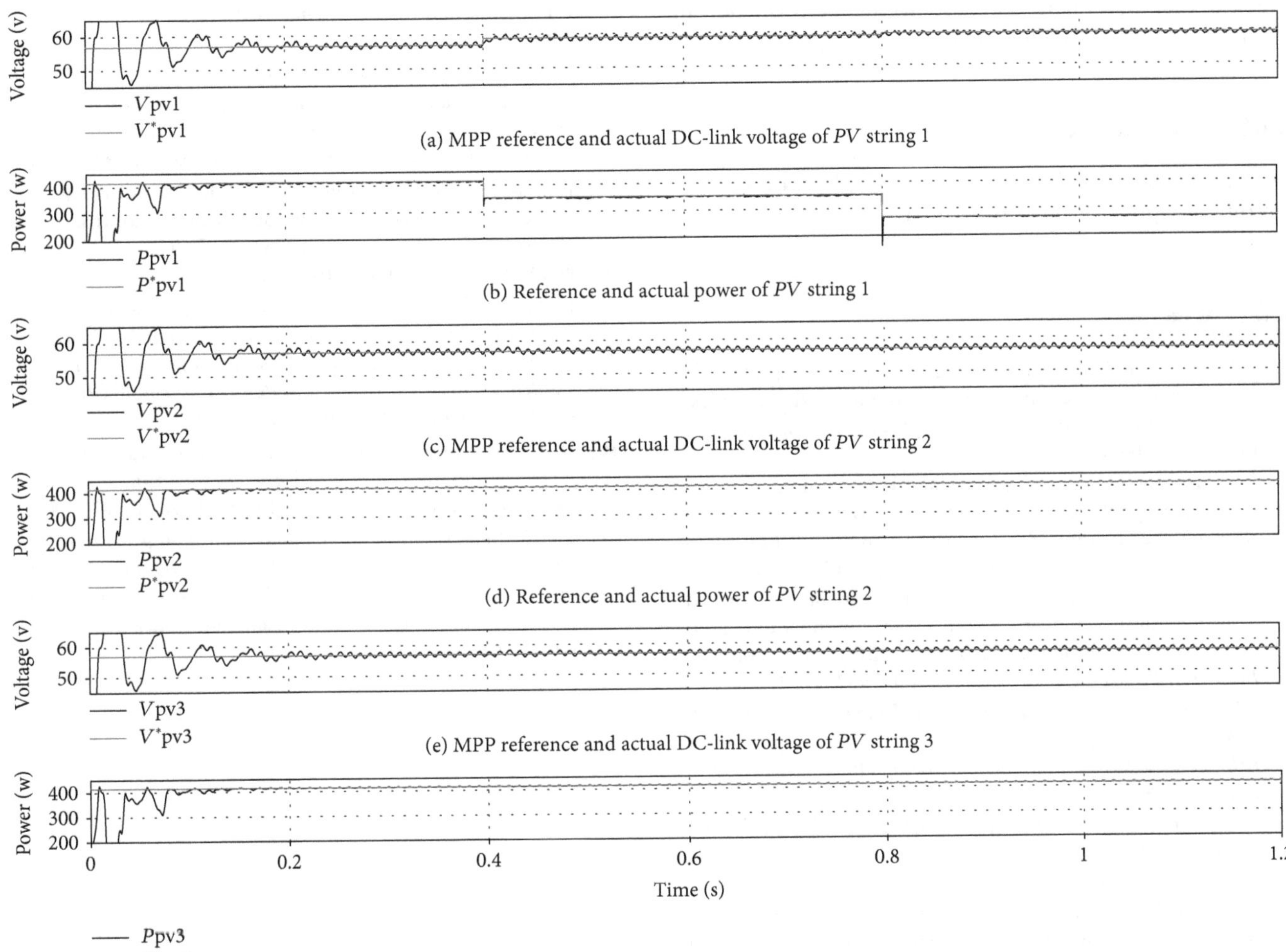

(a) MPP reference and actual DC-link voltage of *PV* string 1

(b) Reference and actual power of *PV* string 1

(c) MPP reference and actual DC-link voltage of *PV* string 2

(d) Reference and actual power of *PV* string 2

(e) MPP reference and actual DC-link voltage of *PV* string 3

(f) Reference and actual power of *PV* string 3

FIGURE 7: MPPT reference and actual DC-link voltage and power of the proposed MLI topology-based GCPS with PV strings as input sources.

condition of one PV module also influences the performance of other PV modules.

The reference grid current tracking of the proposed converter is shown in Figure 6(a). The figure is zoomed around 0.8 s. At $t = 0.8$ s, the reference grid current changes due to the change in reference power. However, the controller is able to quickly track the reference. The combined waveform of grid voltage and current is shown in Figure 6(b), which has no phase difference between them despite of changes in atmospheric conditions. The seven-level output voltage waveform of proposed MLI is shown in Figure 6(c).

4.1. Proposed System with PV Panels in String Formation. A common configuration of PV panels is the string formation [2]. In this section, we provide the simulation results of the proposed system with two series-connected PV modules (PV string) at each input port. The system is simulated for three different scenarios, and the results are shown in Figure 7. At the converter startup, all the PV strings, and their constituting PV panel modules, are operating at 1000 W/m^2 irradiance and 25°C temperature. The system achieves steady state in around 0.2 s. At $t = 0.4$ s, the irradiance of the first PV panel module of the PV string 1 is changed from 1000 W/m^2 to 800 W/m^2, whereas, at $t = 0.8$ s, the irradiance of the second PV panel module of PV string 1 is changed from 1000 W/m^2 to 600 W/m^2, while keeping the irradiance of first PV module at 800 W/m^2. As seen in Figure 7, at both of the above changes in irradiance, the proposed system is able to accurately track the MPP of PV string 1. Moreover, there is no impact on the MPP tracking of PV strings 2 and 3.

5. Conclusion

In this paper, we present a new multilevel inverter topology targeted for the application of grid-connected photovoltaic systems (GCPS). A GCPS based on the proposed converter and controlled via proposed DPMC is simulated. The proposed system is compared with GCPS with DMPC based on three other state-of-the-art topologies reported in the literature. The proposed converter is shown to be one of the best in terms of efficiency, power extraction, and voltage THD. The simulation results allow us to conclude that the proposed topology offers the best overall performance in comparison to other topologies.

Conflicts of Interest

The authors declare that there is no conflict of interest regarding the publication of this paper.

References

[1] S. B. Kjaer, J. K. Pedersen, and F. Blaabjerg, "A review of single-phase grid-connected inverters for photovoltaic modules," *IEEE Transactions on Industry Applications*, vol. 41, no. 5, pp. 1292–1306, 2005.

[2] S. Kouro, J. I. Leon, D. Vinnikov, and L. G. Franquelo, "Grid-connected photovoltaic systems: an overview of recent research and emerging PV converter technology," *IEEE Industrial Electronics Magazine*, vol. 9, no. 1, pp. 47–61, 2015.

[3] E. Babaei, S. Alilu, and S. Laali, "A new general topology for cascaded multilevel inverters with reduced number of components based on developed H-bridge," *IEEE Transactions on Industrial Electronics*, vol. 61, no. 8, pp. 3932–3939, 2014.

[4] J. Rodriguez, Jih-Sheng Lai, and Fang Zheng Peng, "Multilevel inverters: a survey of topologies, controls, and applications," *IEEE Transactions on Industrial Electronics*, vol. 49, no. 4, pp. 724–738, 2002.

[5] L. G. Franquelo, J. Rodriguez, J. Leon, S. Kouro, R. Portillo, and M. Prats, "The age of multilevel converters arrives," *IEEE Industrial Electronics Magazine*, vol. 2, no. 2, pp. 28–39, 2008.

[6] T. M., A. U. Khan, M. Luqman, M. B. Satti, M. Aaqib, and M. F. Khan, "Generation of isolated DC voltage sources for multilevel inverters," in *2015 Power Generation System and Renewable Energy Technologies (PGSRET)*, pp. 1–6, Islamabad, Pakistan, June 2015.

[7] P. Cortes, S. Kouro, F. Barrios, and J. Rodriguez, "Predictive control of a single-phase cascaded h-bridge photovoltaic energy conversion system," in *Proceedings of The 7th International Power Electronics and Motion Control Conference*, vol. 2, pp. 1423–1428, Harbin, China, June 2012.

[8] E. Villanueva, P. Correa, J. Rodriguez, and M. Pacas, "Control of a single-phase cascaded H-bridge multilevel inverter for grid-connected photovoltaic systems," *IEEE Transactions on Industrial Electronics*, vol. 56, no. 11, pp. 4399–4406, 2009.

[9] M. Mousa, M. E. Ahmed, and M. Orabi, "New converter circuitry for PV applications using multilevel converters," in *INTELEC 2009 - 31st International Telecommunications Energy Conference*, pp. 1–6, Incheon, South Korea, October 2009.

[10] M. Mosa, H. Abu-Rub, M. E. Ahmed, A. Kouzou, and J. Rodríguez, "Control of single phase grid connected multilevel inverter using model predictive control," in *4th International Conference on Power Engineering, Energy and Electrical Drives*, pp. 624–628, Istanbul, Turkey, May 2013.

[11] E. Babaei, S. Laali, and S. Alilu, "Cascaded multilevel inverter with series connection of novel H-bridge basic units," *IEEE Transactions on Industrial Electronics*, vol. 61, no. 12, pp. 6664–6671, 2014.

[12] E. Babaei and S. H. Hosseini, "New cascaded multilevel inverter topology with minimum number of switches," *Energy Conversion and Management*, vol. 50, no. 11, pp. 2761–2767, 2009.

[13] S. Vazquez, J. Rodriguez, M. Rivera, L. G. Franquelo, and M. Norambuena, "Model predictive control for power converters and drives: advances and trends," *IEEE Transactions on Industrial Electronics*, vol. 64, no. 2, pp. 935–947, 2017.

[14] J. Rodriguez, M. P. Kazmierkowski, J. R. Espinoza et al., "State of the art of finite control set model predictive control in power electronics," *IEEE Transactions on Industrial Informatics*, vol. 9, no. 2, pp. 1003–1016, 2013.

[15] R. P. Aguilera, Y. Yu, P. Acuna et al., "Predictive control algorithm to achieve power balance of cascaded H-bridge converters," in *2015 IEEE International Symposium on Predictive Control of Electrical Drives and Power Electronics (PRECEDE)*, pp. 49–54, Valparaiso, Chile, October 2015.

[16] D. P. Hohm and M. E. Ropp, "Comparative study of maximum power point tracking algorithms," *Progress in Photovoltaics: Research and Applications*, vol. 11, no. 1, pp. 47–62, 2003.

[17] M. Castilla, J. Miret, A. Camacho, J. Matas, and L. G. de Vicuna, "Reduction of current harmonic distortion in three-phase grid-connected photovoltaic inverters via resonant current control," *IEEE Transactions on Industrial Electronics*, vol. 60, no. 4, pp. 1464–1472, 2013.

19

Life Cycle Cost Comparison Study of PV and Concrete Rooftop in Jakarta

Religiana Hendarti

Architecture Department, Faculty of Engineering, Bina Nusantara University, Jakarta, Indonesia

Correspondence should be addressed to Religiana Hendarti; rhendarti@binus.edu

Academic Editor: Gabriele Battista

This paper presents a comparative study of "life cycle cost" or LCC of a building school rooftop element in Jakarta. The simulation applied two different types of roof: a concrete roof and a PV rooftop. The aim of this study is to investigate the electricity production of the solar panels, the saving to investment ratio or SIR, and the total life cycle cost of each rooftop element. To accommodate those objectives, the calculation utilized a software called "Building Life Cycle Cost (BLCC) version 5" which is a product of the US Department of Energy. The simulation results showed that the LCC can be improved by 27.6%, and the "discounted payback" is reached at year 15. Indeed, this indicates that a roof made of solar panels is promising to replace the existing concrete roof.

1. Introduction

Sustainable development has been proclaimed for decades, and one of the goals is the use of alternative energy as an alternative source of electricity to replace the use of fossil fuel. In correlation to that, the report of the "Indonesian Energy Outlook 2012" states that a future building is expected to use alternative energy as a source of electricity by a minimum of 5% from the whole building energy usage. One of the most promising alternative energies according to the Indonesian Ministry of Energy and Mineral Resources is solar energy (Table 1).

However, the application of solar energy in Indonesia is still in its early stages. Essentially, there are several considerations in applying this technology; one of them is the high initial cost of the installation.

There are little studies on the economic factors in the application of solar energy in Indonesia, particularly the life cycle cost of a PV system. This study, therefore, focused on investigating the improvement of life cycle cost of a rooftop by replacing the current concrete rooftop with solar panels.

Life cycle cost or LCC is a method of an estimated economic cost calculation that focuses on all resources used by a project during its lifetime [1]. Additionally, according to Hin and Zmeureanu [2], the LCC includes operating and maintenance cost, initial investment, replacement, and disposal cost, so that it calculates current and predicted future cost. There are two parameters to estimate the future cost; they are net present value and internal rate of return. The internal rate of return is a discount rate that makes the net present value (NPV) of all cash flows from a particular project equal to zero. And as stated by Spertino et al. [3], the NPV and the internal rate of return are used to compare various alternative investments.

The purpose of this life cycle cost study can be listed as follows: (1) to simulate the electricity production of the PV rooftop, (2) to compare the life cycle cost between the concrete roof and the PV rooftop, (3) to study the saving to investment ratio when using solar panels as the rooftop, and (4) to conduct a study of the implications of the initial cost of using solar cells as compared to concrete roofs.

In order to fulfill the objectives, a calculation and comparative analysis of life cycle cost between a roof using solar panel and a concrete roof were conducted using Building Life Cycle Cost (BLCC) version 5 software. The software is a product of the Department of Energy, USA. The Bina Nusantara campus building was selected as the case study. The building is located in Jakarta, Indonesia.

Table 1: Comparison of energy power potential.

Type of renewable energy	Power source	Expected production
Geothermal	29.614 MW	1.341 MW
Hydro	75.000 MW	6.848 MW
Biomass	49,810 MW	1644 MW
Solar	4.8 kWh/m2/day	22.45 MW
Wind	3–6 m/s	1.87 MW
Ocean	49 GW	0.01 MW
Oil	7408 billion barrels	0.314 billion barrels
Natural gas	150 TCF	2.98 CF
Coal	161 billion tons	0.317 billion tons

Source: Indonesia Energy Outlook 2013, Ministry of Energy and Mineral Resources.

2. The Profile of the Case Study

The Bina Nusantara campus building has a coordinate position of 6°12′6.54″ S for the latitude position and 106°46′54.87″ E for the position of the longitude, and the orientation of the building is about 80 degrees of the north (Figure 1).

The area of the building is around 6650 m^2 and is divided into two main buildings, the lecture building and the parking building. The solar panel was only simulated over the lecture building which has an area of around 4900 m^2.

3. Simulation Method

3.1. Orientation of Solar Panels. In order to know the optimum orientation of the solar panel to obtain maximum sunlight, Ecotect software was used. The simulation results can be seen in Figure 2 and show that the region of Jakarta has maximum sunlight at the point of 77.5 degrees from the north direction.

3.2. Heat Transfer Model. The heat transfer method used in this study, particularly in defining the cell temperature, was the "decoupled" method. The decoupled method essentially is divided into two methods of condition. First is based on the nominal operating temperature condition (NOCT) and second is based on dynamic condition. The NOCT method in estimating the cell temperature at each time step is based on the method from Duffie and Beckman (1991). The method assumes the simulated environment with the wind speed of 1 m/s, the solar radiation of 800 W/m^2, the ambient temperature of 20°C, and no electrical load and a certain specified insolation. In this method, the input data of "module heat capacity" and "module heat loss coefficient" are not considered and the detailed equation is as follows:

$$T_c = T_a + \frac{(1 - (\eta_c/\tau\alpha))}{(G_T(\tau\alpha/U_L))}, \quad (1)$$

where T_c is the cell temperature, T_a is the ambient temperature, η_c is the cell efficiency which varies with the ambient condition, G_T is the total radiation incident of the PV array,

Figure 1: The location of Bina Nusantara University building.

U_L is the array thermal loss coefficient, and $\tau\alpha$ is the user defined constant. Based on 1, therefore, the cell temperature is only influenced by the cell efficiency that varies with the ambient condition. As for the other method, namely, "Decoupled Ulleberg Dynamic", the cell temperature is defined by a function of the previous cell temperature and the thermal capacity of the PV module (C). It can be seen from the equation below that is developed by Ulleberg:

$$T_{\text{cell}}|_t = T_{\text{ambient}} + (T_{\text{cell}}|_{t-1} - T_{\text{ambient}}) * e^{(-UL/\text{cap})\Delta t}. \quad (2)$$

Since the simulation was assumed to be conducted in a more real condition, therefore, the study used the second method.

3.3. Simulation

3.3.1. Electricity Production. The simulation was conducted in EnergyPlus version 8.0. The geometry of the building was built on Google Sketchup where OpenStudio software had been integrated to allow visualization of the selected building. Solar panels with monocrystalline silicone technology was used in the simulation, with the following technical specifications: (1) the watt peak was 240 watts, (2) the efficiency of the module was 21.6%, (3) the temperature coefficient was −0.3%/°K, and (4) the panel dimension was 1580 × 798 × 35 mm which consists of 72 cells. Based on the previous research, to maximize the work of solar panels, the selected roof was flat shaped (flat roof) with 10-degree angle from the horizontal line (see Figure 3) [4].

3.3.2. Building Life Cycle Cost. As explained earlier, this study applies a software called Building Life Cycle Cost (BLCC) version 5. This software is particularly useful for analyzing and predicting the "cost" and benefits of energy conservation-oriented projects. In addition, another advantage of using this software is that it can evaluate two or three alternative systems to see the lowest life cycle cost. The aspects required for the calculation are as follows: (1) the "energy cost" and (2) the "capital component" which includes (a) investment cost, (b) replacement cost, and (c) operating, maintenance, and repairs, including annually and nonannually recurring costs. The calculation also needs cost parameters such as: (1) service life, (2) inflation rate and (3) discount rate. The concept of analysis of this study can be seen in Figure 4.

3.4. Energy Cost. The "energy cost" here is determined by the production of electricity generated by solar panels and then multiplied by the price of electricity per kilowatt hour (kWh) of the local State Electricity Company. The price of electricity inputted into the simulation system was assumed

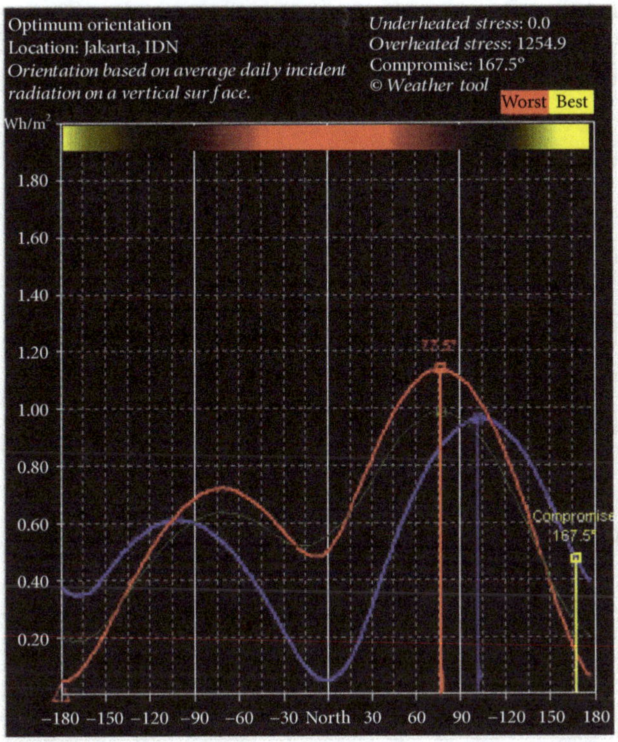

FIGURE 2: The direction of most sunlight throughout the year. Source: Weather Tool

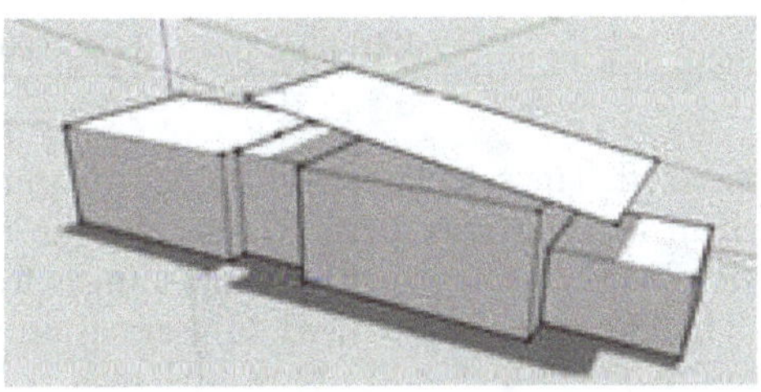

FIGURE 3: Flat roof form made of solar panels. Source: [4].

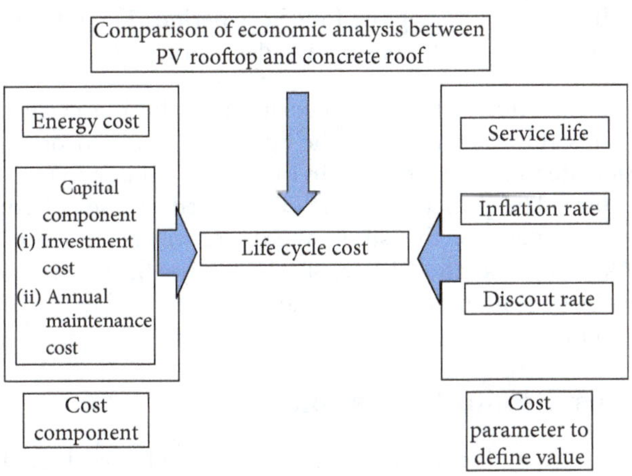

FIGURE 4: Analysis approach.

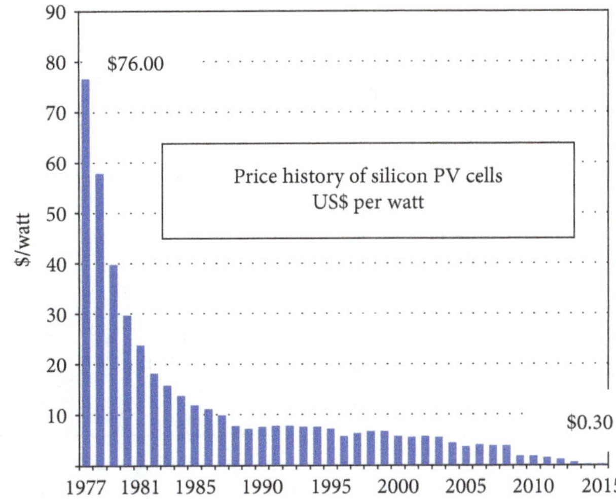

FIGURE 5: Graph of the development of the price of solar panels per watt peak. Source: https://commons.wikimedia.org/wiki/File:Price_history_of_silicon_PV_cells_since_1977.svg.

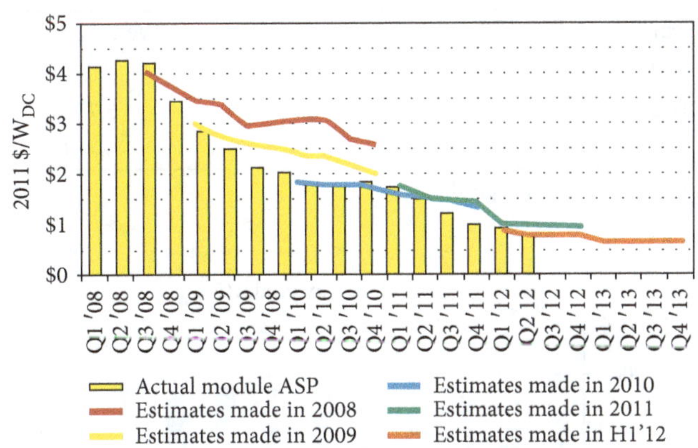

FIGURE 6: Approximate prices according to the average actual prices on the market analysts. Source: Department of Energy, USA, 2012.

to be IDR 1250.59. This price was for a customer that has electrical usage above 200 kVA, and as for the solar panels, the price was based on the specifications of solar panels, in particular was the watt peak. Figure 5 shows the price trend of solar panels per watt peak from 1997 to 2015. It can be seen that the retail price of solar panels declined considerably of about 50% over a period of 9 years, and this is quite promising for further developments.

Furthermore, the declined price of the solar panel was also reported by the Department of Energy, USA. The report can be concluded as follows: in 2012, the indicated price of installed solar panels is $4.39/W for 5.1 kW residential systems, $3.43/W for 221 kW commercial rooftop systems, $2.79/W for 191.5 MW fixed-tilt utility-scale systems, and $3.37/W for 191.5 MW one-axis-tracking utility-scale systems. Indeed, it was projected that prices will continue to decline to around $0.74/W in 2013, as well as the inverter prices (see Figure 6).

3.4.1. Capital Investment. In the simulation, the amount of solar panel system investment (including inverters,

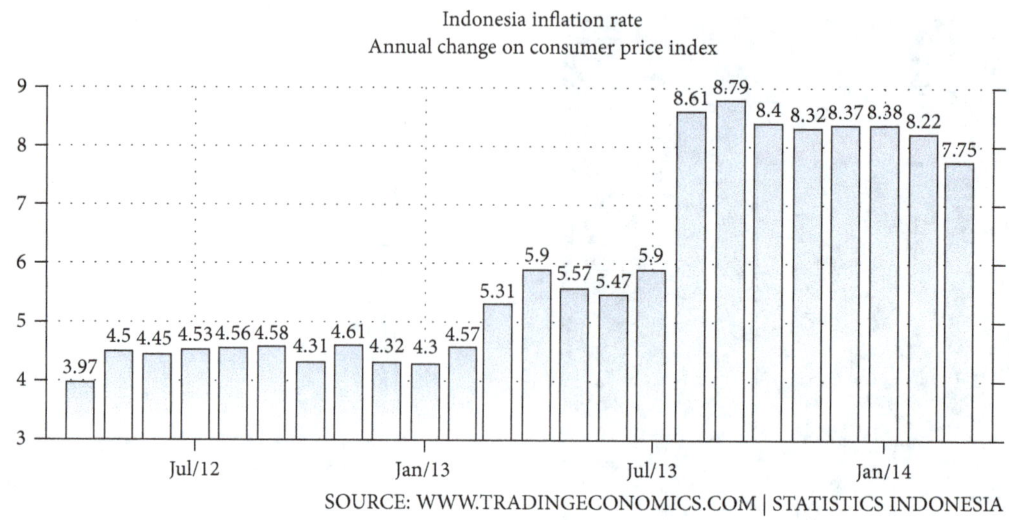

FIGURE 7: Indonesia inflation rate. Source: http://www.tradingeconomics.com.

TABLE 2: History of BI rate in 2011–2014.

Year	Average BI rate
2014 (November)	7.5%
2013	5.75%
2012	5.77%

Source: http://www.bi.go.id.

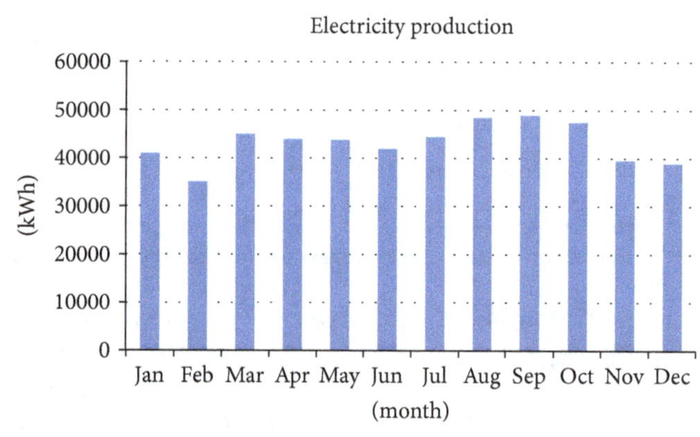

FIGURE 8: Results of simulation of electricity production from PV panels.

structures, and other installation costs) was determined at US $3.00/Watt peak. The watt peak of a solar panel is the maximum amount of power in a predetermined condition which is called "standard test condition (STC)." The STC is a condition in which the solar radiation is 1000 W/m^2, the air mass (AM) is 1.5, and the temperature is 25°C.

3.4.2. Operating and Maintenance Cost. The operation and maintenance costs are calculated in the form of annual fees that also include insurance costs [5]. The annual fee was determined as 1% of the initial investment cost. In general, the cost for maintaining PV systems is quite cheap as compared to the conventional electricity costs.

TABLE 3: Comparison of life cycle cost between the concrete and PV rooftop.

	Concrete roof (IDR)	The total solar panels (1550 panels) (IDR)
Initial cost (IDR)	4,800,000,000	13,400,000,000
Annual energy cost	624,000,000	—
Annual maintenance cost	48,000,000	90,000,000
Nonannual maintenance cost (for 25 years)	2,400,000,000	—
LCC	19,400,000,000	14,100,000,000

3.4.3. Service Life. The calculation for service life in this study was based on the warranty period issued by the manufacturer. Generally, the solar panel warranty ranges from 20 to 25 years. Therefore, this study used 25 years as an input for the service life.

3.4.4. Inflation Rate. The inflation rate in Indonesia can be seen in Figure 7. The average inflation rate in Indonesia was from 3–8% during the four-year period. The rate simulated in this study, therefore, was determined at 7%.

3.4.5. Discount Rate. The calculation of the life cycle cost requires the calculation of the "present value" in order to anticipate the price increases in the service time span (service lifetime). The present value is determined by the "discount rate" number. For this study, the figure was taken from the Bank Indonesia rate (BI rate). Table 2 shows the annual average rate of Bank Indonesia (BI), and the rate used for the simulation is 7.5%.

4. Results and Discussion

4.1. Electricity Production. The results of the EnergyPlus simulation can be analyzed in Figure 8. The total production of electricity from solar cell technology was 519,618.6 kWh per year, and the highest electricity production was in September at 49,278.11 kWh, while the lowest was in February at

Table 4: Comparison of the present value and annual value.

Component	Concrete roof		Solar panel roof	
	Present value (IDR)	Annual value (IDR)	Present value (IDR)	Annual value (IDR)
Initial cost (IDR)	4,800,000,000	566,000,000	13,400,000,000	1,580,000,000
Annual energy cost	13,800,000,000	1,620,000,000	0	0
Annual maintenance cost	364,248,000	42,912,000	683,000,000	80,460,000
Nonannual maintenance cost (for 25 years)	515,868,000	60,648,000	0	0
Total life cycle cost	19,400,000,000	2,290,000,000	14,100,000,000	1,660,000,000

35,426.94 kWh. The result of this simulation is then used as the input data for the energy cost by multiplying with the local electricity price.

4.2. The Life Cycle Cost Comparison of Concrete and PV Rooftop. Table 3 shows that the life cycle cost number can be improved by 27.6% for 25 years or there was a reduction in expenditure of IDR 5,367,072,000 for 25 years.

The reduction of life cycle cost on the solar panel roof was caused by the absence of the annual energy cost and the non-annual maintenance cost. The nonannual maintenance cost of the concrete roof comes from the insulation cost of the concrete roof, such as the recoating work every 10 years in order to protect from the weather effect. The comparison of the present value and "annual value" of each roof can be seen in Table 4.

It can be seen from Table 4 that the present value and annual value had also increased to around IDR 5,367,072,000 and IDR 632,196,000, respectively. Therefore, the saving to investment ratio (SIR) was 1.62. The SIR is the value of comparison between savings and initial investment costs. Based on the results, the solar panel roof was simply "cost-effective" because it had a value above 1. Furthermore, the discounted payback simulation results appeared in the 15th year.

A similar study has been conducted by Nilima et al. [6]. They studied a comparative life cycle cost between an existing building without energy efficient approach and with energy efficient approach. The result showed that a building with energy efficient approach by applying solar panels system as the energy source can reduce the LCC of the existing building effectively. The reduction was around 50% with the average inflation rate as per consumer price index: 7.8%, the average inflation rate as per energy index: 3%, and the average interest rate as per Reserve Bank of India: 8%.

5. Conclusion

This study has simulated a large production of electricity from a roof made of monocrystalline silicone solar panel with an efficiency of 21.6% and the capacity of 240 watt peak. The simulated annual electricity production is around 519,619 kWh. The results have shown that the use of solar panels is more cost-effective since the life cycle cost value was improved at around 27.6%, and the saving to investment ratio was above 1 or 1.62. Furthermore, the investment return value is at year 15 whereas the service lifetime of the solar panel is 25 years.

Conflicts of Interest

The author declares that there are no conflicts of interest regarding the publication of this paper.

References

[1] M. Bagg, "Save cash and energy costs via an LCC model," *World Pumps*, vol. 2013, no. 12, pp. 26-27, 2013.

[2] J. N. Cheng Hin and R. Zmeureanu, "Optimization of a residential solar combisystem for minimum life cycle cost, energy use and exergy destroyed," *Solar Energy*, vol. 100, pp. 102–113, 2014.

[3] F. Spertino, P. Di Leo, and V. Cocina, "Economic analysis of investment in the rooftop photovoltaic systems: a long-term research in the two main markets," *Renewable and Sustainable Energy Reviews*, vol. 28, pp. 531–540, 2013.

[4] H. Religiana and W. Wiyantara, "Analysis on the application of PV technology in terms of tranditional roof house. Case study: Julang Ngapak Sundanese traditional roof form," *Advanced Materials Research*, vol. 1025-1026, no. 4, pp. 1066–1069, 2014.

[5] A. Roy and M. A. Kabir, "Relative life cycle economic analysis of stand-alone solar PV and fossil fuel powered systems in Bangladesh with regard to load demand and market controlling factors," *Renewable and Sustainable Energy Reviews*, vol. 16, no. 7, pp. 4629–4637, 2012.

[6] N. N. Kale, D. Joshi, and R. Menon, "Life cycle cost analysis of commercial buildings with energy efficient approach," *Perspectives in Science*, vol. 8, pp. 452–454, 2016.

PERMISSIONS

All chapters in this book were first published in IJP, by Hindawi Publishing Corporation; hereby published with permission under the Creative Commons Attribution License or equivalent. Every chapter published in this book has been scrutinized by our experts. Their significance has been extensively debated. The topics covered herein carry significant findings which will fuel the growth of the discipline. They may even be implemented as practical applications or may be referred to as a beginning point for another development.

The contributors of this book come from diverse backgrounds, making this book a truly international effort. This book will bring forth new frontiers with its revolutionizing research information and detailed analysis of the nascent developments around the world.

We would like to thank all the contributing authors for lending their expertise to make the book truly unique. They have played a crucial role in the development of this book. Without their invaluable contributions this book wouldn't have been possible. They have made vital efforts to compile up to date information on the varied aspects of this subject to make this book a valuable addition to the collection of many professionals and students.

This book was conceptualized with the vision of imparting up-to-date information and advanced data in this field. To ensure the same, a matchless editorial board was set up. Every individual on the board went through rigorous rounds of assessment to prove their worth. After which they invested a large part of their time researching and compiling the most relevant data for our readers.

The editorial board has been involved in producing this book since its inception. They have spent rigorous hours researching and exploring the diverse topics which have resulted in the successful publishing of this book. They have passed on their knowledge of decades through this book. To expedite this challenging task, the publisher supported the team at every step. A small team of assistant editors was also appointed to further simplify the editing procedure and attain best results for the readers.

Apart from the editorial board, the designing team has also invested a significant amount of their time in understanding the subject and creating the most relevant covers. They scrutinized every image to scout for the most suitable representation of the subject and create an appropriate cover for the book.

The publishing team has been an ardent support to the editorial, designing and production team. Their endless efforts to recruit the best for this project, has resulted in the accomplishment of this book. They are a veteran in the field of academics and their pool of knowledge is as vast as their experience in printing. Their expertise and guidance has proved useful at every step. Their uncompromising quality standards have made this book an exceptional effort. Their encouragement from time to time has been an inspiration for everyone.

The publisher and the editorial board hope that this book will prove to be a valuable piece of knowledge for researchers, students, practitioners and scholars across the globe.

LIST OF CONTRIBUTORS

Qingjun Liu
College of Mechanical and Electrical Engineering, Hohai University, Changzhou 213022, China
College of Water Conservancy and Hydropower Engineering, Hohai University, Nanjing 210098, China

Fei Cao
College of Mechanical and Electrical Engineering, Hohai University, Changzhou 213022, China
Sunshore Solar Energy Company Limited, Nantong 226300, China

Yanhua Liu and Tianyu Zhu
College of Mechanical and Electrical Engineering, Hohai University, Changzhou 213022, China

Deyou Liu
College of Water Conservancy and Hydropower Engineering, Hohai University, Nanjing 210098, China

Takeo Oku, Junya Nomura, Atsushi Suzuki and Hiroki Tanaka
Department of Materials Science, The University of Shiga Prefecture, Hikone, Shiga 522-8533, Japan

Sakiko Fukunishi, Satoshi Minami and Shinichiro Tsukada
Frontier Materials Laboratories, Osaka Gas Chemicals Co., Ltd., Osaka 554-0051, Japan

Kingsley O. Ukoba, Freddie L. Inambao and Andrew C. Eloka-Eboka
Mechanical Engineering, University of KwaZulu-Natal, Durban, South Africa

G. Jia, J. Plentz, A. Gawlik, J. Dellith, A. Dellith and G. Andrä
Leibniz Institute of Photonic Technology (IPHT), Albert-Einstein-Str. 9, 07745 Jena, Germany

A. S. Azar, P. A. Carvalho and A. Ulyashin
SINTEF, Oslo, Norway

G. Stokkan and M. Syvertsen
SINTEF, Trondheim, Norway

Ramsha Iftikhar, Iftikhar Ahmad, Muhammad Arsalan and Neelma Naz
SEECS, National University of Science and Technology, Islamabad, Pakistan

Naghmash Ali and Hammad Armghan
School of Electrical Engineering, The University of Faisalabad, Faisalabad, Pakistan

Yaojie Sun
School of Information Science and Technology, Fudan University, Shanghai 200433, China

Jingsheng Huang
School of Information Science and Technology, Fudan University, Shanghai 200433, China
State Key Laboratory of Operation and Control of Renewable Energy & Storage Systems, China Electric Power Research Institute, Nanjing 210003, China

Hongtao Li
State Key Laboratory of Operation and Control of Renewable Energy & Storage Systems, China Electric Power Research Institute, Nanjing 210003, China

He Wang and Hong Yang
MOE Key Laboratory for Nonequilibrium Synthesis and Modulation of Condensed Matter, School of Science, Xi'an Jiaotong University, Xi'an 710049, China

Gökhan Ömeroğlu
Mechanical Engineering Department, Engineering Faculty, Atatürk University, 25300 Erzurum, Turkey

Qi Wang, Shunxiang Ji and Minqiang Hu
School of Electrical & Automation Engineering, Nanjing Normal University, Nanjing 210042, China
Jiangsu Province Gas-Electricity Integrated Energy Engineering Laboratory, Nanjing 210046, China

Wei Li, Fusuo Liu and Ling Zhu
State Key Laboratory of Smart Grid Protection and Control, Nanjing 211106, China

Yang Hu
State Key Laboratory of Alternate Electrical Power System with Renewable Energy Sources, North China Electric Power University, Changping District, Beijing, China

Honglu Zhu and Jianxi Yao
State Key Laboratory of Alternate Electrical Power System with Renewable Energy Sources, North China Electric Power University, Changping District, Beijing, China
School of Renewable Energy, North China Electric Power University, Changping District, Beijing, China

Cao Yu
School of Renewable Energy, North China Electric Power University, Changping District, Beijing, China
China Three Gorges New Energy Co., Ltd, Beijing, China

Lingxing Lu and Weiwei Lian
School of Renewable Energy, North China Electric Power University, Changping District, Beijing, China

Luigi Costanzo and Massimo Vitelli
Department of Engineering, Università degli Studi della Campania "Luigi Vanvitelli", Aversa, 81031 Caserta, Italy

Joaquín Vaquero and Ivan Soriano
Electronics Technology Area, Rey Juan Carlos University, 28933 Mostoles, Madrid, Spain

Nimrod Vázquez
Electronics Engineering Department, Technological Institute of Celaya, 38010 Celaya, GTO, Mexico

Jeziel Vázquez
Computer Engineering Department, Superior Technological Institute of South of Guanajuato, 38980 Uriangato, GTO, Mexico

Markus Diantoro, Arif Hidayat, Ahmad Taufiq and Abdulloh Fuad
Department of Physics, Faculty of Mathematics and Natural Sciences, Universitas Negeri Malang, Jl. Semarang 5, Malang 65146, Indonesia
Central Laboratory for Minerals and Advanced Materials, Faculty of Mathematics and Natural Sciences, Universitas Negeri Malang, Jl. Semarang 5, Malang 65145, Indonesia

Thathit Suprayogi
Department of Physics, Faculty of Mathematics and Natural Sciences, Universitas Negeri Malang, Jl. Semarang 5, Malang 65146, Indonesia

Risa Suryana
Department of Physics, Faculty of Mathematics and Natural Sciences, Universitas Sebelas Maret, Solo, Jl. Ir. Sutami No. 36 A, Surakarta 57126, Indonesia

Yunfeng Wang, Ming Li, Xun Ma and Guoliang Li
Solar Energy Research Institute, Yunnan Normal University, Kunming, China

Reda Hassanien Emam Hassanien
Solar Energy Research Institute, Yunnan Normal University, Kunming, China
Agricultural Engineering Department, Cairo University, Giza 12613, Egypt

M. Imtiaz Hussain
Green Energy Technology Research Center, Kongju National University, Cheonan, Republic of Korea

Jun-Tae Kim
Department of Architectural Engineering, Kongju National University, Cheonan, Republic of Korea

A. P. Cédola, A. Khalili and F. Cappelluti
Department of Electronics and Telecommunications, Corso Duca degli Abruzzi 24, 10129 Torino, Italy

D. Kim, M. Tang, J. Wu and H. Liu
Department of Electronic and Electrical Engineering, University College London, Torrington Place, London WC1E 7JE, UK

A. Tibaldi
3stituto di Elettronica e di Ingegneria dell'Informazione e delle Telecomunicazioni (IEIIT), Consiglio Nazionale delle Ricerche (CNR), Corso Duca degli Abruzzi 24, 10129 Torino, Italy

Onur Ozdal Mengi
Department of Energy Systems Engineering, Engineering Faculty, Giresun University, 28100 Giresun, Turkey

Muhammad Ali and Ahmed Shuja
Center for Advanced Electronics and Photovoltaic Engineering (CAEPE), International Islamic University, Islamabad 44000, Pakistan

Ahsan Baig and Erum Jamil
Center for Advanced Electronics and Photovoltaic Engineering (CAEPE), International Islamic University, Islamabad 44000, Pakistan
Department of Electrical Engineering, International Islamic University, Islamabad 44000, Pakistan

Muhammad Amjad
Department of Electronics Engineering, Islamia University of Bahawalpur, Bahawalpur 63100, Pakistan

Muhammad Bilal Satti and Ammar Hasan
School of Electrical Engineering and Computer Science (SEECS), National University of Science and Technology, Islamabad, Pakistan

Mian Ilyas Ahmad
Research Center for Modeling and Simulation (RCMS), National University of Sciences and Technology (NUST), Islamabad, Pakistan

Religiana Hendarti
Architecture Department, Faculty of Engineering, Bina Nusantara University, Jakarta, Indonesia

Index

T

U

Z

www.ingramcontent.com/pod-product-compliance
Lightning Source LLC
LaVergne TN
LVHW060610210926
848485LV00005B/1394